THIS LAND

A GUIDE TO EASTERN NATIONAL FORESTS

The publisher gratefully acknowledges the generous contribution to this book provided by the General Endowment Fund of the University of California Press Foundation.

THIS LAND

A GUIDE TO EASTERN NATIONAL FORESTS

Robert H. Mohlenbrock

UNIVERSITY OF CALIFORNIA PRESS

Berkeley   Los Angeles   London

This book is dedicated to Vittorio Maestro, senior editor at Natural History magazine, *who has made all of my "This Land" columns in the magazine readable.*

University of California Press, one of the most distinguished university presses in the United States, enriches lives around the world by advancing scholarship in the humanities, social sciences, and natural sciences. Its activities are supported by the UC Press Foundation and by philanthropic contributions from individuals and institutions. For more information, visit www.ucpress.edu.

University of California Press
Berkeley and Los Angeles, California

University of California Press, Ltd.
London, England

Library of Congress Cataloging-in-Publication Data
Mohlenbrock, Robert H., 1931–
    This land : a guide to eastern national forests / by Robert H. Mohlenbrock.
        p.   cm.
    Includes bibliographical references (p.   ) and index.
    ISBN 0-520-23984-9 (pbk. : alk. paper)
    1. Forest reserves — East (U.S.) — Guidebooks. I. Title: Guide to eastern national forests. II. Title.

    SD428.A2E27 2006
    333.75'0974--dc22
                                                                                              2005034497

Manufactured in Canada
10   09   08   07   06
10   9   8   7   6   5   4   3   2   1

The paper used in this publication meets the minimum requirements of ANSI/NISO Z39.48–1992 (R 1997) (*Permanence of Paper*). ♾

Cover: Second Falls, Graveyard Fields Area, Blue Ridge Parkway, Pisgah National Forest, North Carolina. Photograph by Kevin Adams.

CONTENTS

Plates follow page 208

FOREWORD

As the Civil War came to an end, the United States found itself positioned to become a leader among nations. A country of immigrants with a rich endowment of natural resources, America was already a land of opportunity, but the young nation lacked the cultural marks of achievement that characterized its Old World counterparts. Europe had great temples, cathedrals, and museums filled with artifacts. Asia had great dynasties that embodied its long and glorious past. Though short on history, America did have a powerful national spirit that was expressed especially well through its abundant and bountiful land, much of which was public domain, west of the Mississippi—the great frontier. Historian Frederick Jackson Turner referred to this land as the "greatest free gift ever bestowed on mankind."

Yet around the time of the Civil War, a number of influential Americans were becoming increasingly concerned that some of the country's public lands were being plundered. Although homesteading had served the nation well in bringing territory under effective national control, it had become clear that some of the public domain lands had to be set aside as a legacy for all Americans.

In 1864, Henry David Thoreau called for the establishment of "national preserves" of virgin forests, "not for idle sport or food, but for inspiration and our own true re-creation." That same year, President Abraham Lincoln signed legislation granting Yosemite Valley and the Mariposa Big Tree Grove to the state of California to hold forever "for public use, resort, and recreation." Also in the 1860s, Frederick Edwin Church painted *Twilight in the Wilderness,* which inspired artists to capture on canvas the grandeur of the American landscape.

In 1891, President Benjamin Harrison created the nation's first forest reserve, the 1.2-million-acre Yellowstone Park Timber Land Reservation, just south of Yellowstone National Park. Today this area comprises Shoshone and Teton national forests. Before his term ended, President Harrison proclaimed another 13 million acres of forest reserves in the West, laying the foundation for a National Forest system. However, the reserves were little more than lines drawn on a map, or "paper parks," without managers, regulations, or budgets.

The National Academy of Sciences established a National Forestry Commission in 1896, which issued a report that became the blueprint for forest policy emphasizing that the federal forest reserves belonged to all Americans and should be managed for them and not for any particular class. It went on to say that "steep-sloped lands should not be cleared, the grazing of sheep should be regulated, miners should not be allowed to burn land over willfully, lands better suited for agriculture or mining should be eliminated from the reserves, mature timber should be cut and sold, and settlers and miners should be allowed to cut only such timber as they need."

Just before the close of the 19th century, President Grover Cleveland established another 21 million acres of forest reserves, and in 1897, the Forest Management Act, or Organic Act, was passed, which specified that forest reserves were "to improve and protect the forest, or for the purpose of securing favorable conditions of water flows, and to furnish a continuous supply of timber for the use and necessities of citizens of the U.S."

Although Lincoln was perhaps the first president to see the "people's land" as a legacy to preserve for posterity, President Theodore Roosevelt thrust the nation into its first conservation movement. Roosevelt acted aggressively to expand federal forest reserves and to establish the first national wildlife refuges and national monuments. In 1905, with Gifford Pinchot at the helm, the U.S. Department of Agriculture Forest Service was established. Secretary of Agriculture James Wilson, in 1905, directed Pinchot to manage the national forests "for the greatest good for the greatest number for the long run." Thanks to such visionary leaders in the 19th and early 20th centuries, today we have 192 million acres of national forests that are owned by the people and are the birthright of all American citizens.

The pages that follow in this volume by Robert H. Mohlenbrock, a distinguished botanist, natural historian, and conservationist, are an account of the national forests of the East. Mohlenbrock has spent more than 40 years visiting and working in all of the 155 national forests. State by state, Professor Mohlenbrock describes each of the region's national forests in detail, including their size, location, access routes, basic geology, hydrology, and biota, as well as things to see and do.

He describes the trails that take visitors to wilderness areas and features of special interest and concern. He discusses rare and endangered species, notable historical landmarks and events, and sites of scenic beauty. *This Land: A Guide to Eastern National Forests* contains a wealth of information presented in clear and concise language. It adeptly conveys the sense of awe that characterizes our national forests. In the end, this volume will help us and future generations understand and appreciate the wealth of this land and remind us of the importance of being responsible stewards of the people's land today and for future generations.

Mike Dombeck
Chief Emeritus of the U.S. Forest Service

PREFACE

My family and I began visiting the national forests in 1960, and we have spent all of our vacations and an enormous amount of days in them, eventually visiting each of the 155 national forests. We soon discovered that they contain millions of acres of habitats and scenery that are nearly on a par with that found in national parks and national monuments. Many of these marvelous areas in national forests are little known, and we had to do considerable research to find the most exciting and beautiful areas. I have tried to provide information for these areas in this book.

Since I am a professional botanist, my family and I spent considerable time in areas known as research natural areas which are part of a national network of ecological areas designated in perpetuity for research and education and/or to maintain biological diversity on National Forest System lands. Although many of the research natural areas are in remote areas that are not accessible to the ordinary person, others are more accessible. Special permission may be required to visit a few of them. Some of my favorites are included in this book.

When you begin your exploration of a national forest, you will need to obtain an up-to-date forest map showing where the major roads and back roads are located. These maps may be purchased from the district ranger stations or from the supervisor's office. While you are at the ranger stations, you can usually pick up several brochures describing trails and other points of interest. Whatever your interest is in the out-of-doors, you will undoubtedly be able to enjoy our national forests, which are truly a unique American treasure.

In 1984 I received a call from Alan Ternes, then the editor of *Natural History* magazine, published by the American Museum of Natural History in New York. He asked me whether I would be interested in writing a monthly column for the magazine about some areas in the national forests that I particularly liked. The areas didn't need to be particularly pretty, but they should have a biological or geological story to tell. My first article appeared in the November 1984 issue of *Natural History* magazine in my "This Land" column. The articles that I published in *Natural History* pertaining to national forests in the eastern United States accompany some of the chapters here as subsections, with minor revisions. I am grateful to *Natural History* magazine and its former editor, Ellen Goldensohn, for allowing me to republish these articles in this book.

I am indebted to Forest Service personnel, some of whom directed me to some little-known areas, and others who read drafts of the manuscripts of their particular national forest. Any errors that may have crept into the book are strictly my own. I am also grateful to Blake Edgar of the University of California Press, who suggested this series of books, and to Scott Norton, who has worked untiringly as my editor.

INTRODUCTION

During the rapid development of the United States after the American Revolution, and during most of the 1900s, many forests in the United States were logged, with the logging often followed by devastating fires; ranchers converted the prairies and the plains into vast pastures for livestock; sheep were allowed to venture onto heretofore undisturbed alpine areas; and great amounts of land were turned over in an attempt to find gold, silver, and other minerals.

In 1875, the American Forestry Association was born. This organization was asked by Secretary of the Interior Carl Schurz to try to change the concept that most people had about the wasting of our natural resources. One year later, the Division of Forestry was created within the Department of Agriculture. However, land fraud continued, with homesteaders asked by large lumber companies to buy land and then transfer the title of the land to the companies. In 1891, the American Forestry Association lobbied Congress to pass legislation that would allow forest reserves to be set aside and administered by the Department of the Interior, thus stopping wanton destruction of forest lands. President Benjamin Harrison established forest reserves totaling 13 million acres, the first being the Yellowstone Timberland Reserve, which later became the Shoshone and Teton national forests.

Gifford Pinchot was the founder of scientific forestry in the United States, and President Theodore Roosevelt named him chief of the Forest Service in 1898 because of his wide-ranging policy on the conservation of natural resources. Pinchot persuaded Congress to transfer the Forest Service to the Department of Agriculture, an event that transpired on February 1, 1905. He realized that the forest reserves were areas where timber production would be beneficial to the nation and where clear water, diverse wildlife, and scenic beauty could be maintained.

In 1964, the Wilderness Act was passed by Congress, authorizing the setting aside of vast areas that were still in pristine condition. Although the establishment of wilderness areas has preserved some of our most beautiful areas, it has also made these areas off-limits to anyone who is aged or has a physical disability, or who just cannot backpack for miles and miles into an area.

Slowly, as people from all walks of life began to use the national forests for recreational purposes, the U.S. Forest Service adopted the multiple-use concept, where timber production, wildlife management, conservation of plants and animals, preservation of clear water, maintenance of historic sites, and recreation could be accommodated in the national forests and enjoyed year-round. Although camping, picnicking, and scenic driving are the major recreational activities on national forest land, other activities include boating, swimming, fishing, hunting, whitewater rafting, horseback riding, bird-watching, nature study, photography, wilderness trekking, hang-gliding, rockhounding, and winter sports. Special areas are set aside for off-road vehicle activity.

Today, we have 155 national forests, although to save administrative costs, some have been combined. The U.S. Forest Service administers other areas as well, such as national grasslands, the Columbia River Gorge National Scenic Area, and the Lake Tahoe Basin Management Area.

NATIONAL FORESTS IN ALABAMA

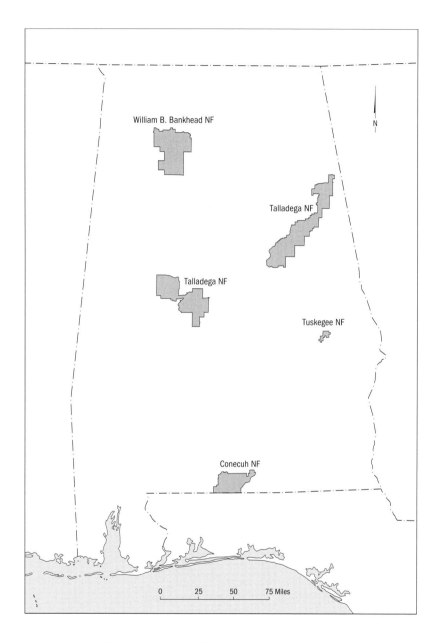

William B. Bankhead NF

Talladega NF

Talladega NF

Tuskegee NF

Conecuh NF

N

0   25   50   75 Miles

Four national forests are within the boundaries of Alabama and together are in the three physiographic provinces of the state. The Conecuh is in the Coastal Plain Province of Alabama, the Tuskegee is in the Piedmont Province, and the Talladega and William B. Bankhead are in the Mountain

Province. All four of the national forests are under the administration of one forest supervisor, whose office is at 2946 Chestnut Street, Montgomery, AL 36107. Alabama's national forests are in Region 8 of the United States Forest Service.

Conecuh National Forest

SIZE AND LOCATION: 83,898 acres along the southern border of Alabama, between Andalusia and Brewton. Major access routes are U.S. Highway 19 and State Routes 55 and 137. District Ranger Station: Andalusia. Forest Supervisor's Office: 2946 Chestnut Street, Montgomery, AL 36107, www.southern-region.fs.fed.us/alabama.

SPECIAL FACILITIES: Boat ramp; swimming beaches.

SPECIAL ATTRACTIONS: Open Pond Recreation Area; Blue Lake Recreation Area.

Although the Florida Panhandle separates the Conecuh National Forest from the Gulf of Mexico, the national forest is still in the Coastal Plain Province where boglike habitats alternate with stands of longleaf pine in a setting typical of the terrain adjacent to the Gulf Coast. The Conecuh National Forest is situated between the Conecuh River to the northwest and the Yellow River to the southeast. The Yellow River Basin includes a number of open-water ponds that may be used for swimming and fishing. The headwaters of the Blackwater River are also in the Conecuh National Forest.

The uplands in the national forest are dry and sandy, supporting fine forests of longleaf pine and a good variety of broad-leaved deciduous trees. Of unusual interest in the dry sandy woods are the endangered red-cockaded woodpecker and gopher frog. Scattered throughout the Conecuh National Forest are wetlands formed by sinkhole ponds, natural springs, and bottomland swamps. Swamp gum, tupelo gum, bald cypress, pumpkin ash, swamp cottonwood, overcup oak, swamp chestnut oak, and cherrybark oak are common trees in the bottomland swamps, while more boggy terrain is home to many carnivorous plants and several other rare plant species.

One of the best ways to experience what the Conecuh National Forest has to offer is to hike the Conecuh National Recreation Trail (commonly referred to as simply the Conecuh Trail), which was constructed during the last quarter of the 20th century by the Youth Conservation Corps. This trail meanders for 20 miles through longleaf pine forests, longleaf pine savannas, and bottomland hardwood forests; around several ponds, springs, and bogs; and across small streams. Foot bridges have been built over most of the streams. All of the trail is within the 22,500-acre Blue Spring Wildlife Management Area.

Several places in the Conecuh National Forest offer entry points to the Conecuh Trail. If you opt to start at the southern end of the trail, you should seek out the Open Pond Recreation Area at the end of County Road 24, about 1.5 miles east of State Route 137. The Recreation Area consists of approximately 280 acres around a large body of water known as Open Pond. The area offers camping facilities, picnic areas, boat ramps, and, of course, hiking. From the campgrounds one may hike the Lake Shore Trail that completely encircles Open Pond. Be on the lookout for southern bald eagles and ospreys in the sky and alligators on the wet ground. From the campgrounds you may also pick up the 4.5-mile southern loop of the Conecuh Trail that circles around Ditch Pond and Buck Pond, follows Five Runs Creek for 0.5 mile, passes Blue Spring, and returns to the Open Pond Recreation Area. Blue Spring is a large natural spring of clear blue water. In the ponds you may see one or more of the 10 species of bladderworts (*Utricularia*) that live in the Conecuh National Forest. These aquatic carnivorous plants have numerous bladders that are attached to a multibranched stem system below the water's surface. Equipped with sensitive hairs, the bladders invaginate quickly when the sensitive hairs are brushed by a minute aquatic organism, sucking the hapless creature into the bladder where its nitrogenous compounds are broken down by enzymes. Most of the bladderworts produce attractive yellow flowers, but a couple of them have purple flowers. Some of the bladderworts may survive for a while stranded on mud.

The more extensive northern part of the Conecuh Trail leaves Open Pond Recreation Area and passes a natural spring as it stays in low terrain for 3 miles until it comes to Blue Lake in the 305-acre Blue Lake Recreation Area. Featured attractions at Blue Lake are swimming and picnicking. From Blue Lake you may begin the northern loop of the Conecuh Trail, which is 11.5 miles long, ending again at Blue Lake. This part of the trail passes by a swampy area and crosses Camp Creek before circling below the two Nellie Ponds. From Nellie Ponds it is 1.1 miles to Gum Pond. The trail then meanders around the northern and western edges of a large, formidable swamp before passing the two Mossy Ponds and several smaller ponds. After crossing Moccasin Branch, the trail eventually returns to Blue Lake.

Along the shores of the ponds that have filled limestone sinks, look for gorgeous pink-flowering, willow-leaved meadow beauty and the yellow heads of the uncommon Conecuh yellow-eyed grass during summer and autumn. On terraces above the streams you may see the rare needle palm, named for the long, black spines usually found at the base of the plant.

The best places to see carnivorous plants and other rare species are in the bogs and wet pine savannas. In addition to five kinds of pitcher plants, four species of sundews, and three types of butterworts, all carnivorous, there are also spreading rose orchid, thistle-leaved aster, and a kind of turk's-cap lily known as the panhandle lily. If you decide to explore in the pitcher plant savannas, be on the lookout for eastern diamondback rattlesnakes, including some extremely large ones that will scare the pants off you.

Ponds bordered by bald cypresses and swamps containing bald cypresses are very scenic, with several species of bromeliads, including Spanish moss, clinging to the tree branches. Also in the swamps is a climbing member of the heath family, known as *Pieris*, and gorgeous white arum, with spoon-shaped white flowers and arrowhead-shaped leaves.

Other interesting areas in the Conecuh National Forest are Bear Bay, a dark swampy area south of Forest Road 305 a mile west of Otter Pond; Sandstone Hill, a high forested hill of sand just east of Forest Road 311; and Brook Hines Lake, the largest body of water in the Conecuh National Forest at the extreme southwestern corner of the national forest.

Two historic lookout towers are in the national forest, one near Open Pond and one near Parker Springs just north of Brook Hines Lake.

Conecuh Bogs

The primary activity in Alabama's Conecuh National Forest is timber production. The original tree cover was harvested long ago, and subsequent generations of trees have been planted and cut. Stands of slash pine and longleaf pine, arranged in densely crowded rows, cover much of the sandy uplands of the forest, interspersed with recently clear-cut tracts in which ragged tree stumps are surrounded by scrubby regrowth. But tucked away within this artificial and often unattractive landscape are more than two dozen pitcher plant bogs, ranging in size from a few square feet to a few acres, that still support a remarkable variety of plants and animals.

Pitcher plants can be found in boggy habitats all across the Coastal Plain of the southeastern United States, from the Carolinas and Florida to southern Louisiana and southeastern Texas. George Folkerts, a former student of mine and now a zoology professor at Auburn University, as well as an authority on carnivorous plants, guided me through the Conecuh National Forest. He noted that pitcher plants grow in 11 different kinds of wetlands—from river terraces to sphagnum mat bogs to savannas, swales, and seepage bogs. Although these habitats differ in their topography and source of water, they all have an acidic soil, saturated for at least a portion of the year, and depend on periodic fires to maintain their characteristic mix of vegetation.

The pitcher plant bogs in the Conecuh National Forest are seepage bogs created by rainwater percolating down through the sandy uplands and accumulating near the bottom of slopes. In some instances, a distinct shrub swamp community is found at the very bottom of the slope, where several inches of water stand throughout the year, while the pitcher plant bogs begin a little upslope, where the soil is saturated but not flooded. Rills and rivulets of cool, clear water, most less than a foot wide, form networks throughout the bogs.

The bogs contain at least 20 different species of carnivorous plants—plants that obtain some of their nutrition from the insects they trap. These include four kinds of sundews, three butterworts, nine bladderworts, and five types of pitcher plants. In addition, Conecuh bogs are home to a dozen species of wild orchids and a variety of sedges, including some known as beaked rushes.

Animal life is equally diverse. Cottonmouths often emerge from the shrub swamps to sun themselves in the open bogs. Diamondback rattlesnakes from the surrounding uplands frequently come to the bogs for water and then return to their drier habitat. The narrow trails used by armadillos that visit the bogs to drink can be seen. These nocturnal animals use the same trails so often that they wear them into deep ruts.

When Folkerts and I were in one of the bogs in April 1991, we saw a female crane skim just above one of the narrow rivulets and periodically dip her abdomen into the water to lay her eggs. On that spring day, caddis flies were swarming in the air. Folkerts and I paused to look at the emerging leaves of the orange macranthera, a beardtongue-like plant. During autumn, the gorgeous orange, tube-shaped flowers of this plant open to welcome the hummingbirds that are migrating through the area.

In one of the shallow pools we saw tadpoles that would mature into pine barrens tree frogs, one of the rarest amphibians in the United States. This species lives in two widely separated places, the pine barrens of New Jersey and the pitcher plant bogs of the Coastal Plain. Two species of fish, a shiner and a madtom, also live in the rills of the pitcher plant bogs.

Pitcher plants are noted for leaves that are modified into elongated tubes, or pitchers, typically hooded by their folded-over tips. But these perennial plants also produce long, flat leaves and a large flower that hangs from the tip of a leafless stalk.

Each pitcher collects water, which fills the bottom of the funnel and mixes with digestive enzymes that the plant secretes. To attract insects, the plant usually has nectar-producing glands on the hood as well as the rim of the pitcher, which often is shiny and slippery. An insect alighting on the rim tends to plummet into the watery abyss. Downward-pointing hairs that grow inside the tube prevent the victim from crawling out to safety.

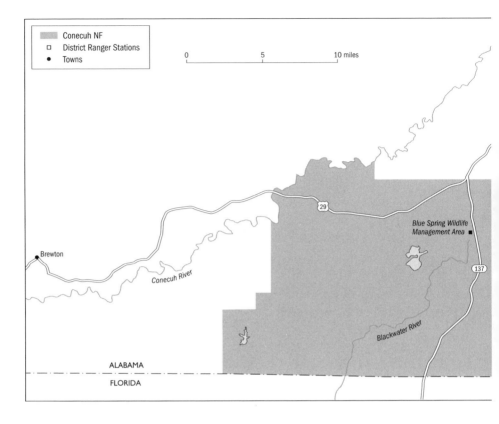

The pitcher's hood is arranged in a variety of positions, depending on the species, but never in a position that blocks the entrance to the funnel. Early naturalists speculated that the hood was hinged and that it automatically sealed off the opening when an insect fell inside, but this is not the case.

The most beautiful pitcher belongs to the white-topped pitcher plant. It grows up to 2 feet tall, and its upper end and hood are white with conspicuous green veins. These pitchers are commonly collected and sold for floral decoration. The Office of Scientific Authority of the United States Fish and Wildlife Service issues permits for this commerce, but many pitchers are harvested illegally every year, placing the plants in jeopardy of extinction.

The yellow pitcher plant is probably the most common and most sturdy of them all, with bright yellow or yellow-green pitchers up to 2.5 feet tall. Wherry's sweet pitcher plant has an erect pitcher usually less than 1 foot tall with a very short, wide, maroon-veined hood. The purple pitcher plant (pl. 1), the only one that is also found outside the Coastal Plain (it grows throughout the northeastern United States and westward to the Great Lakes), has a curved pitcher that more or less lies on its back. Its mouth is completely open because the hood is straight and does not arch over the tube.

The parrot pitcher plant is distinct. The variegated red-and-white pitchers are only 4 to 8 inches tall, and the hood is an inflated structure with a pointed, beaklike tip that superficially resembles the outline of a parrot's head. In spring, the pitcher stands erect and has a broad, green finlike "wing" that extends along the front of the pitcher from the base to the hood. This structure exposes more plant surface so that sunlight for photosynthesis reaches the surface. Pitchers that form later in the year essentially lie on their backs and lack the broad wing.

While many people think of pitcher plants as organisms that exploit insects, Folkerts has found that flies, mosquitoes, and other insects utilize pitcher plants for their own survival. The larvae of the mosquito *Wyeomia*, those of a midge, and those of several species of sarcophagid flies complete their larval development in the pitcher, feeding on microorganisms and insect corpses trapped in the pitcher's fluid. As much as half the prey trapped in the pitcher may be consumed by such larvae, which might seem to be detrimental to the pitcher plant. But Folkerts indicates that the larvae produce large quantities of nitrogen-rich wastes, which probably nourish pitcher plants.

Wasps, mites, and moths also feed on the tissues of the pitchers. According to Folkerts, the pitcher plant moth spends nearly its entire life in the pitcher. In spring, the female moth lays one to several eggs on the inside wall of a newly formed pitcher. The eggs develop into larvae that girdle the pitcher with a narrow feeding channel, which causes the upper portion to wilt and topple, closing off the opening. The larvae eat only the inner tissues of the pitcher, leaving the outer epidermis intact. Shortly before pupation, the larvae cut a tiny drainage hole in the pitcher. Until the females leave to lay their eggs, adults rarely stray from the pitcher, and copulation takes place there.

The reports of early European settlers show that pitcher plant bogs were once more common in the southeastern United States. The bogs were maintained by natural fires, which were started by lightning and spread unchecked over large areas. North American Indians also burned areas to drive game and for other reasons. Such fires killed or retarded the growth of the shrubs and trees that tend to invade open bog habitats. During the 1930s, however, the U.S. Forest Service, with its symbol of Smokey the Bear, campaigned against fires in both forested and unforested lands. As a result, fires were not allowed to burn even in open bogs, and dozens of woody species invaded them, wiping out some and drastically reducing the size of others.

At present, the Conecuh National Forest defends and enhances bog sites by protecting them from heavy fire equipment that is now prohibited from entering the bogs, the transition zone, or the immediate recharge area. Control burns in the bogs now occur periodically throughout the year. In addition, hand and light mechanical cutting of encroaching shrubs is being done.

Talladega National Forest

SIZE AND LOCATION: 389,834 acres in east-central and west-central Alabama, on either side of Interstate Highway 65. Major access routes are Interstate Highways 20 and 65; U.S. Highways 75, 82, and 431; and State Routes 9, 25, 48, 49, 77, 148, 183, and 219. District Ranger Stations: Brent, Heflin, and Talladega. Forest Supervisor's Office: 2496 Chestnut Street, Montgomery, AL 36107, www.southernregion.fs.fed.us/alabama.

SPECIAL FACILITIES: Boat ramps; swimming beaches; off-road vehicle areas.

SPECIAL ATTRACTIONS: Talladega Scenic Drive.

WILDERNESS AREAS: Cheaha (6,544 acres); Dugger Mountain (9,200 acres).

The Talladega National Forest consists of two separated geographic areas representing two different biological provinces. The larger portion of the Talladega National Forest that lies east of Birmingham is mountainous, being part of the southernmost extension of the Appalachian Mountains. A smaller district of the Talladega National Forest is west of Birmingham in a region of undulating hills of the Piedmont Province.

Hiking all or part of the Pinhoti National Recreation Trail will give the forest visitor a firsthand view of the mountainous district of the Talladega National Forest. The trail winds for 102 miles up and down mountain slopes, through valleys, past small settlements, and over several streams. Most of the vegetation types found in this part of the Appalachian Mountains may be encountered along the trail. On upper ridges are mixtures of blackjack oak, post oak, white oak, mockernut hickory, pignut hickory, and persimmon trees, while on the mountain slopes are basswood, red mulberry, and more oaks and hickories. The densely shaded forested coves are home to tulip poplars, American beech, Ohio buckeye, flowering dogwood, and magnolias.

The southern trailhead for the Pinhoti Trail is about 10 miles east of the town of Talladega near Clairmont Gap on Talladega Mountain. For a while the trail stays upland, closely paralleling Forest Road 600 and actually crossing it twice. After about 6 miles, the trail comes to Adams Gap and the edge of the Cheaha Wilderness. For the next 8 miles, the Pinhoti Trail winds through the wilderness. This is one of the more difficult parts of the trail since the terrain is often very steep and rocky.

At the center of the wilderness, the Pinhoti Trail crosses a main trail junction where the Pinhoti continues northward, Odum Trail heads southward, and Chinnabee Silent Trail goes off to the west. If you choose to take the Odum Trail, built by Boy Scouts in 1961, you will climb over Cedar Mountain and end up at the High Falls parking lot 4.7 miles away. The Chinnabee Trail goes westward for 6 miles to Lake Chinnabee Recreation Area, which features a campground and picnic area near the lake. Where the trail crosses Cheaha Creek, you will find a low but pretty waterfall. The Chinnabee Silent Trail was constructed by a Boy Scout troop from the Alabama Institutes for Deaf and Blind and is named for the legendary chief Chinnabee Selocta. A 1.5-mile trail encircles Lake Chinnabee.

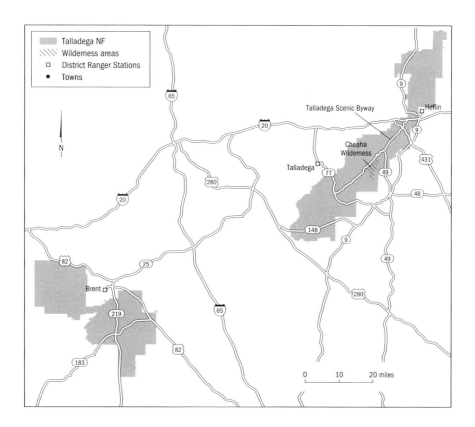

The Pinhoti Trail leaves the Cheaha Wilderness and passes through Cheaha State Park before reentering the Talladega National Forest and climbing up Pine Mountain. The trail then crosses Hillabee Creek and passes between Hillabee Lake to the west and the smaller Morgan Lake to the east. As the Pinhoti Trail crosses over Horseblock Mountain, it passes under Interstate Highway 20. On the approach to Flag Mountain is the Lower Shoal Shelter where hikers may rest for a while. After going around the eastern edge of scenic Highrock Lake, the Pinhoti Trail reaches Pine Glen Campground, which can also be reached by driving Forest Road 531. Highrock Lake is stocked with bream and largemouth bass. Near the campground is the historic Cole Cemetery and Shoal Creek Church, one of a few hand-hewn log churches left in Alabama, built between 1885 and 1890. Two miles north of the Pine Glen Campground is Sweetwater Lake where the Pinhoti Trail follows along the western edge of the lake. Just beyond the northern end of Sweetwater Lake is the Laurel Shelter. The last 3.2 miles connect Laurel Shelter with Coleman Lake, which features a fine campground.

From the Lower Shoal Shelter to Coleman Lake Campground, the Pinhoti Trail is in the Choccolocco Game Management Area. This 40,000-acre region is penetrated by numerous streams and back roads. The Talladega National Forest extends for an additional 5 miles north of the Choccolocco Game Management Area.

Between the towns of Anniston and Piedmont is rugged Dugger Mountain, named for Civil War veteran J. Taylor Dugger, who at one time owned and preserved the mountain. At 2,140 feet, it is the second-highest point in Alabama. Most of this steep-sided mountain is in the Dugger Mountain Wilderness. Large boulders nestled beneath pine trees are on the top of the mountain, with huge American beech trees at lower elevations. The Pinhoti Trail passes through the wilderness for 7.7 miles.

If you are not up to hiking, you may still grasp the character of this part of the national forest by driving the Talladega Scenic Drive (State Route 281), a 23-mile road through the prettiest part of the national forest. It begins about 2 miles west of Heflin and heads south over Horseblock Mountain.

At the southwestern end of this part of the Talladega National Forest is the Hollins Game Management Area, where you might get a glimpse of wild turkeys. This area begins 3 miles west of Sylacauga on State Route 148. Two miles north of the Hollins Game Management Area is the Horn Mountain Lookout Tower.

The Oakmulgee District of the Talladega National Forest lies west of Interstate Highway 65 and south of Interstate Highway 59/20, in the upper coastal plain of Alabama where low rolling hills and crooked creeks dominate. Most of the northwestern section of this district consists of the Oakmulgee Wildlife Management Area, which is just south of U.S. Highway 82 midway between Tuscaloosa and Centreville. Numerous roads criss-cross the area; while driving them, particularly near dusk, you may spot white-tailed deer, squirrels, foxes, skunks, beavers, rabbits, mink, weasels, possibly bobcats, and the state's largest population of red-cockaded woodpeckers. The Pondville Lookout Tower is 0.5 mile north of State Route 25 just west of County Road 44.

Just outside the southwestern corner of the management area is Payne Lake Recreation Area, which has two campgrounds, picnic areas, boat ramps, swimming beaches, and a nature trail. The trail, just beyond the northern end of Payne Lake, is accessed along County Road 49. This nature trail provides a good cross section of this part of the Piedmont as it goes from a swamp forest with bald cypress to a bottomland forest with sweet bay and sweet gum, through mesic woods, dry woods, and, finally, to a very arid ridgetop. Common trees seen along the route include tulip poplar, American beech, flowering dogwood, and several kinds of oaks, hickories, and pines. Nearby is the historic Payne Lake Lookout Tower.

The remainder of the Oakmulgee District, lying east of the wildlife management area, consists of a mixture of private land and national forest land. The Cahaba and Perry Mountain lookout towers are in this region. An interesting swampy woods occurs along Buck Creek, beginning near the southern end of Forest Road 424-D. Several historic cemeteries are in the area.

Tuskegee National Forest

SIZE AND LOCATION: 11,252 acres in east-central Alabama, between Columbus and Montgomery. Major access routes are Interstate Highway 85, U.S. Highways 29 and 80, and State Route 186. District Ranger Station: Tuskegee. Forest Supervisor's Office: 1946 Chestnut Street, Montgomery, AL 36107, www.southernregion.fs.fed.us/alabama.

SPECIAL FACILITIES: Boat ramp; viewing platform.

SPECIAL ATTRACTIONS: Taska Recreation Area.

When the United States Forest Service acquired the property in 1959 for the creation of the Tuskegee National Forest, it purchased land that was worn-out farmland and forested areas that had been cleared time and again by the previous owners. With good forest management techniques, the Tuskegee

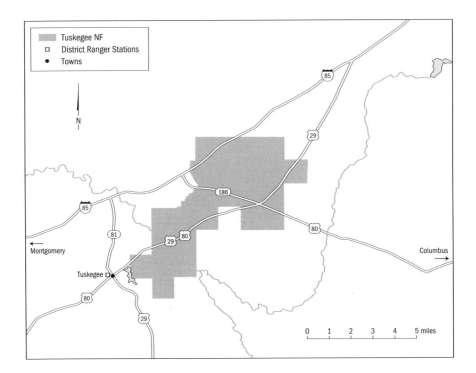

National Forest now consists of secondary forests where visitors may hike, camp, picnic, fish, canoe, and simply commune with nature.

The Tuskegee National Forest lies in Alabama's Piedmont Province, the middle province between the Coastal Plain Province to the south and the Mountain Province to the north and west. The area consists of undulating terrain interspersed by several creeks, including the Choctafaula and Uphapee.

For hikers, the famous Bartram Trail crosses the width of the Tuskegee National Forest, beginning at the junction of County Road 53 and Forest Road 913 a short distance north of the Tsinia Wildlife Viewing Area, and ending along U.S. Highway 29 less than 1 mile west of the intersection of U.S. Highways 29 and 80. William Bartram was a naturalist who wandered through this area during the 1770s. Now a National Recreation Trail, the Bartram Trail traverses through 8.5 miles of national forest land. Oaks, hickories, ashes, and elms are plentiful along the trail, and during late April and early May, the blossoms of magnolias, flowering dogwoods, and redbuds add color to the forest above a rich ground layer of violets, wild ginger, trilliums, wild geraniums (fig. 1), buttercups, and many other wildflowers. If you don't want to hike the entire 8.5 miles, you may join the Bartram Trail where it crosses forest roads at six places across the national forest.

Less than 1 mile south of the southern trailhead of the Bartram Trail is the Tsinia Wildlife Viewing Area, a 125-acre tract once used by the Creek Indians. Tsinia features a short boardwalk and platform where you may observe the surrounding vegetation, an ample diversity of bird life including wild turkeys, and white-tailed deer, squirrels, rabbits, raccoons, and possibly skunks. Small ponds dot the area, but fishing is not permitted here. The Tsinia Wildlife Viewing Area is about 2 miles northeast of Tuskegee along U.S. Highway 29/80.

By continuing to drive northeast along U.S. Highway 29/80, you will come to the Taska Recreation Area. Picnic tables with grills are nestled beneath the canopy of pine trees. Nearby is the Pleasant Hill Lookout Tower.

Fishermen will find two ponds at the northern end of the Tuskegee National Forest along County Road 54. The smaller Chutkee Okhussee Pond and the larger Thloko Okhussee Pond are waiting for fishermen to try their luck at catching largemouth bass, catfish, and bream. Anglers may also cast into Choctafaula Creek, which meanders across the northwestern corner of the national forest, and Uphapee Creek, which cuts across the southwestern corner.

For hunters wishing to hone their skills, the Uchee Firing Range is available off Forest Road 910 a little west of the fishing ponds. It is accessible to visitors with disabilities.

Figure 1. Wild geranium, Tuskegee National Forest (Alabama).

Additional attractions include more than 14 miles of horseback riding on the Bold Destiny/Bedford V. Cash Memorial Horse Trail, with a trailhead at the intersection of Forest Roads 905 and 906. The 4-mile Pleasant Hill Trail, with trailheads at the intersection of State Highway 186 and Forest Road 900, and at the intersection of U.S. Highway 29 and Forest Road 930, is worth taking.

William B. Bankhead National Forest

SIZE AND LOCATION: 181,156 acres in northwestern Alabama. Major access routes are U.S. Highway 278 and State Routes 33 and 195. District Ranger Station: Double Springs. Forest Supervisor's Office: 2946 Chestnut Street, Montgomery, AL 36107, www.southernregion.fs.fed.us/alabama.

SPECIAL FACILITIES: Boat ramps; swimming beaches; horse trails.

SPECIAL ATTRACTIONS: Lake Lewis Smith; Clear Creek Recreation Area; Houston Recreation Area; Corinth Recreation Area.

WILDERNESS AREA: Sipsey (12,726 acres).

The Sipsey Wilderness encompasses the most spectacular areas in the William B. Bankhead National Forest. Steep-walled, deep gorges have been carved by the Sipsey River and its tributaries Quillan Creek, Hubbard Creek, Thompson Fork, Bee Branch, Braziel Creek, Hagood Creek, and Borden Creek. In places, the gorges are more than 75 feet deep, surrounded by vertical walls. The gorges are often so narrow that sunlight sometimes seldom reaches the forest floor. In these canyons are cove forests, similar to the Appalachian cove forests in North Carolina and Georgia. Some of the largest trees in Alabama grow here, particularly in the area where Bee Branch empties into the Sipsey River.

Because of its geographic position, the vegetation in the Sipsey Wilderness has representatives of three biological provinces: the Cumberland Plateau, the Appalachian Plateau, and the Coastal Plain. Many species in the Sipsey Wilderness are near the southern limit of their ranges, including hemlock, sweet birch, mountain holly, bush honeysuckle, ferns such as silvery spleenwort, Goldie's fern, mountain spleenwort, and Virginia polypody, and wildflowers that include yellow trillium, dutchman's breeches, brook saxifrage, rattlesnake plantain orchid, blue cohosh, mountain anemone, showy orchis, and spikenard.

The cliffs along Bee Branch comprise one of only a very few places in North America where two kinds of filmy ferns grow near each other, both discovered in Alabama by Judge Thomas M. Peters in 1853. Trudell's spleenwort is another rare fern here that is known in only a few other places in the southeastern United States.

Rare woody plants that occur in the Sipsey Wilderness are two species of wild camellias of the genus *Stewartia* and three kinds of magnolias, including the bigleaf magnolia.

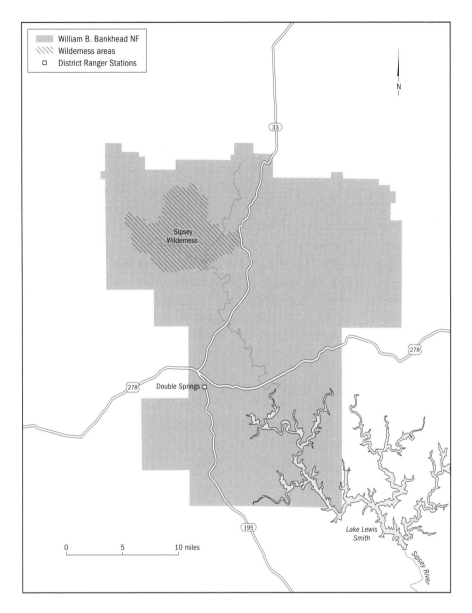

Because of the intense shade, the gorges are cool in the summer, a feature conducive for the growth of plants from more northern latitudes.

Hiking into the gorges, you will encounter forests dominated by hemlock, American beech, and tulip poplar (pl. 2), although other commonly occurring species such as white oak, rock chestnut oak, American holly, sweet gum, sugar maple, hop hornbeam, mockernut hickory, and pignut hickory are found here as well. A well-developed shrub layer includes strawberry bush, mapleleaf viburnum, mountain laurel, large-flowered snowbell bush, oakleaf hydrangea, and spicebush. Wildflowers in the gorges are partridge berry, hepatica, foamflower, cucumber root, wild geranium, rue anemone, Jack in the pulpit, green violet, and yellow trout lily.

The best route to see the big trees in the Bee Branch gorge is from the north. After parking your vehicle at the end of the road at the historic Gum Pond Cemetery, hike a little more than 4 miles to reach the most scenic and vegetatively diverse part of the area.

The western side of the Sipsey Wilderness is accessed off Forest Road 210 near Kinlock Springs, while the southern edge of the wilderness may be reached from the Sipsey River Campground along County Road 60. Several old cemeteries are within the wilderness.

Three miles east of the Sipsey Wilderness along Forest Road 254 are the historic Pine Torch Church and Cemetery. Built between 1850 and 1855, the church was named because pine torches were burned for light. The church is no longer used on a regular basis.

Historic lookout towers are the Central Lookout near the Black Warrior Work Center along State Route 33, the Moreland Lookout just east of Forest Road 121-B, and the Black Pond Lookout along County Road 8. A unique geological feature is a natural sandstone bridge near County Road 63 about 1.5 miles north of U.S. Highway 278. The Owl Creek Horse Camp, located in the northeastern corner of the national forest along Forest Road 262, is popular among equestrians. Swimming is permitted at Clear Creek, Houston, and Corinth recreation areas.

Bee Branch

Carving deep canyons through Alabama's William B. Bankhead National Forest, Bee Branch and other tributaries enter the wild and clear Sipsey Fork at the northern edge of the Black Warrior River basin. Between the canyons, sandstone ridges support trees—such as sourwood, flowering dogwood, persimmon, sassafras, shortleaf pine, Virginia pine, and a variety of oaks— that tolerate relatively dry, open conditions. Preferring the steep slopes are red maple, American holly, mountain laurel, silverbell, and bigleaf magnolia. But the cooler, shaded ravines harbor the most unusual assortment of plants and animals, including many that are at the southernmost limit of their range.

The most conspicuous tree in these canyons is the eastern hemlock, which grows naturally from southern Canada down through the Appalachians, with western extensions to Minnesota and Indiana. A stately tree with short, flat, deep green needles, the eastern hemlock thrives at the bottom of rich, dark canyons, where abundant trees usually close the canopy. Since hemlock seedlings do not do well in the deep humus that accumulates on the forest floor, however, most take hold near streams, where rushing waters wash away the humus before it can build up.

This region is also the southwestern-most location of the sweet birch. Both the birch and the hemlock probably migrated to the area when it was cooler, during times of Ice Age glaciation, but were then left behind in isolated pockets as the glaciers retreated northward. The beech tree may also have shared this fate. Other trees found in the canyons, including the giant tulip poplars that give the forests a majestic, cathedral-like appearance, fall well within their normal ranges.

Many wildflowers in the canyons are also pushing at their southern limits, including blue cohosh, rattlesnake plantain orchid, and barren strawberry. Among the animals at the southern edge of their range are the colorful red milk snake and the shade-loving seal salamander. Other unexpected animals are the noisy barking tree frog, at the northwestern limit of its range, and the green anole, a lizard, at the northern edge of its range. The flattened musk turtle, on the other hand, is known only from the Black Warrior River basin.

The region is also famous for its diversity of ferns and fern-related plants growing in the heavily shaded canyons and on their wet, moss-covered sandstone walls. The glade fern is one of those at the southern edge of its range. More unusual are two kinds of filmy ferns, first discovered here in 1853 by Judge Thomas M. Peters. One, the common filmy fern, has feathery leaves 4 to 8 inches long and grows far back under rock overhangs out of direct sunlight. Finding it usually requires crawling on hands and knees. It has been observed in most of the states in the Southeast and as far north as West Virginia, southern Ohio, and southern Illinois. The other, much rarer species, known as Peters's filmy fern, has undivided leaves only 0.25 inch long. It is known only from a few places in Alabama, Florida, Georgia, Tennessee, and the Carolinas, where it grows on wet rocks.

Filmy ferns, whose 450 species are mostly tropical, are so different from other ferns that they are placed in their own botanical family. Like all ferns, they reproduce with spores, not seeds; but instead of growing upright, filmy ferns send threadlike, sometimes branched stems creeping over the surface of wet rocks or moist soil. The delicate filmy leaves arise from buds that form along the slender stems. The leaves are only one cell layer thick, so thin that one can read newsprint through them. A sparse network of dark-colored veins carries water and nutrients to the leaf cells.

During the summer, short, black, slender bristles appear along the edges of the leaves, accounting for the filmy fern's alternate common name, the bristle fern. Spore-bearing structures form at the base of each bristle. After the spores fall onto the moist rocks or soil, they germinate into minute, branched filaments that resemble certain types of algae. These filaments bear minuscule sex organs that produce sperm and eggs. If a sperm unites with an egg, the fertilized egg will develop into the more familiar, spore-producing plant, and the cycle will start over again.

NATIONAL FORESTS IN FLORIDA

Three national forests in Florida have developed recreational facilities. The Ocala, the southernmost national forest in Florida, is the only national forest in the United States with an abundance of subtropical vegetation. The Apalachicola National Forest, in the Florida Panhandle, has several unusual botanical and geological areas. North of Lake City is the Osceola National Forest.

A fourth national forest, the Choctawhatchee, was established in 1908 with headquarters at DeFuniak Springs and later in Pensacola. During World War II, its land was transferred to the War Department for military purposes. The original forest supervisor's headquarters was at Camp Pinchot, which is now a part of Eglin Air Force Base. Currently, the Choctawhatchee consists of a few isolated parcels of longleaf pine, wiregrass, saw palmetto, and turkey oak communities; it has no facilities. It is under the auspices of the forest supervisor of the national forests in Florida in Tallahassee. In fact, all four of the national forests in Florida are under the management of one Forest Service supervisor, whose office is at 325 John Knox Road, Suite F-100, Tallahassee, FL 32303. The national forests in Florida are in Region 8 of the United States Forest Service.

Apalachicola National Forest

SIZE AND LOCATION: 569,804 acres in the Florida Panhandle, immediately southwest of Tallahassee. Major access routes are Interstate Highway 10; U.S. Highways 98 and 319; and State Routes 12, 20, 22, 65, 67, 267, and 373. District Ranger Stations: Bristol and Crawfordville. Forest Supervisor's Office: 325 John Knox Road, Suite F-100, Tallahassee, FL 32303, www.southernregion.fs.fed.us/florida.

SPECIAL FACILITIES: Boat ramps; swimming beaches; horse trails; bicycle trails.

SPECIAL ATTRACTIONS: Rock Bluff Scenic Area; Morrison Hammock Scenic Area; Leon Sinks Geological Area; Apalachee Savannahs Scenic Byway; Fort Gadsden Historic Site; Silver Lake Recreation Area.

WILDERNESS AREAS: Bradwell Bay (24,602 acres); Mud Swamp/New River (8,090 acres).

The Apalachicola National Forest is a marvelous assemblage of pine flatwoods, savannas, wetlands, wilderness, and some surprises. The Apalachicola River is outside the western boundary of the national forest, and the Ochlockonee River bisects the national forest into two nearly equal-sized western and eastern units. Except in the wilderness areas, enough highways and forest roads are available to access most parts of the Apalachicola National Forest.

The Florida National Scenic Trail crosses the entire width of the national forest, providing a way to observe all of the vegetation types that occur within the national forest. The eastern trailhead is along U.S. Highway 319, just west of the junction of this highway with U.S. Highway 98. The trail then crosses the Sopchoppy River and follows this river for several miles before crossing the Bradwell Bay Wilderness. There are also three trailheads along Forest Road 329 and one trailhead along Forest Road 314. After entering the wilderness, the Florida National Scenic Trail crosses the Ochlockonee River on Forest Highway 13 and then meanders through the northwestern corner of the Apalachicola National Forest, passing along the east side of Camel Lake before leaving the national forest at State Route 12. This interesting trail winds through forests of longleaf pine, savannas, and wetlands, with occasional bridges provided in the wettest areas.

The Apalachee Savannahs Scenic Byway is a great way to get a firsthand view of the wonderful longleaf pine savannas that occur just east of the Apalachicola River. The northern few miles of the scenic byway are on State Route 12 south of Bristol, but the road soon changes to County Road 379 to Sumatra. From Sumatra to the southern end of the scenic byway, the byway is State Route 65.

The savannas have scattered longleaf pines above an open, parklike understory dominated by wiregrass. Palmettos occur here and there. In the often wet and sandy soil is one of the most botanically diverse habitats in the United States. In addition to winding through the savannas, the scenic byway also passes magnolia bays, depressions containing bald cypresses, and ridges with scrub oaks. If you explore the savannas along the scenic byway, you will encounter many plants that live only in these savannas. Pitcher plants are frequent, and ferns are plentiful, including cinnamon fern, royal fern, Virginia

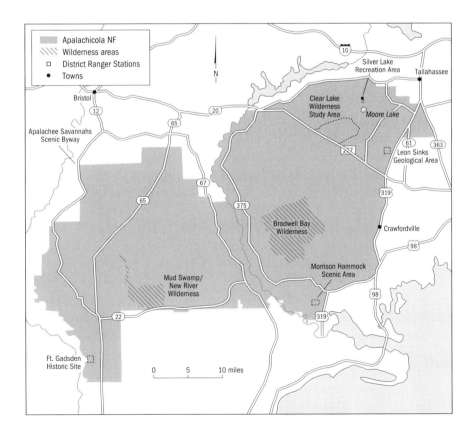

Figure 2. Fort Gadsden, Apalachicola National Forest (Florida).

chain fern, and netted chain fern. One of the finest longleaf pine savannas is at Post Office Bay (discussed later) near the village of Sumatra.

Several side roads to the west of the scenic byway lead to the Apalachicola River or some of its tributaries. Boat ramps are at Big Gully Landing, White Oak Landing, Cotton Landing, and Hickory Landing. There is also a campground at Hickory Landing. Forest Road 129 goes west to the Fort Gadsden Historic Site (fig. 2) on the east side of the Apalachicola River. In 1814, British colonel Edward Nicolls constructed a fort overlooking the Apalachicola River, naming it British Fort. Its purpose was to establish a stronghold against the Spaniards. At the fort, the British trained 3,000 Indians and 300 African slaves to protect the fort and the area adjacent to the Apalachicola River. On July 17, 1816, Colonel Duncan Clement and his party of 116 soldiers and Major William McIntosh and his company of 150 Lower Creek Indians surrounded British Fort. It is reported that one huge shot from a gunboat landed in the fort, blowing it to pieces and killing all but 33 of the 300 people who were in the fort. In April 1818, during the First Seminole War, Andrew Jackson ordered Lieutenant James Gadsden to build a new fort on the site of British Fort to be used as a supply base for Jackson's forces. Jackson named the fort in Gadsden's honor, and Gadsden maintained the fort until 1821 when Florida was ceded to the United States by Spain. The fort was unused until 1862 when Confederate troops utilized it during the Civil War. Fort Gadsden is surrounded mostly by longleaf pine forests, but an interesting cypress dome is also nearby.

From the southern end of Lake Talquin, the Ochlockonee River enters the Apalachicola National Forest and flows southward through the heart of the national forest. It is a popular river for canoeing, and the Forest Service recommends spending 3 days to do the entire length of the river. The river flows through woods that crowd the river's corridor. One of the best places to observe the vegetation along the river is at the Rock Bluff Scenic Area about 2 miles south of State Route 20. Forest Road 390 actually penetrates the scenic area. County Highway 375 follows the course of the river's eastern side. There are boat accesses at Pine Creek Landing, Revell Landing, Mack Landing, Hitchcock Lake, and Wood Lake, with campgrounds at some of these areas.

Northeast of the Ochlockonee River near where the river leaves the Apalachicola National Forest is Morrison Hammock Scenic Area, one of the nicest remaining hammocks in the national forest. The scenic area is located just east of County Highway 399. This pristine 300-acre area consists of hardwood forests interspersed with large specimens of spruce pines and loblolly pines.

One of the more unusual areas in any of the national forests is the Leon Sinks Geological Area (pl. 3), where numerous limestone sinks, caverns, and natural bridges have been formed in what is known as the Woodville Karst Plain. Sandhills with scrub oaks, longleaf pine forests, and wet hammocks surround the geological area. The sinks are about 7 miles south of Tallahassee, adjacent to U.S. Highway 319. One mile south of Leon Sinks, on the east side of U.S. Highway 319, is a smaller area of sinks known as the River Sinks Geological Area, although there is no public access at this time at River Sinks.

Bradwell Bay Wilderness is located near the center of the Apalachicola National Forest. It is entirely surrounded by forest roads. Much of the wilderness is in constant shallow standing water swampland and consequently is not easy to navigate. Black titi is the dominant shrub in the wilderness under a canopy dominated by pond pines or slash pines. Several other shrubs are common, including swamp cyrilla, little-leaf cyrilla, myrtleleaf holly, fetterbush, large gallberry, sweet pepperbush, staggerbush, and scentless bayberry.

The northeast corner of the Apalachicola National Forest, between County Highway 267 and Tallahassee, is criss-crossed by numerous roads. Several lakes of varying sizes are in this area, as well as the Munson Hills Off-Road Bicycle Trail and the Vinzant Horse Trail. The Munson Hills Off-Road Bicycle Trail passes through a series of sand dunes. Longleaf pines provide shade along much of the trail, and the trail often borders ponds and a variety of wetlands. In the valleys between the dune ridges are hammocks where wild black cherry, sassafras, and various species of oaks are common. The Vinzant Horse Trail winds for 30 miles through pine forests, bays, and swamps and across streams. Adjacent to County Highway 267 is a wild area known as the Clear Lake Wilderness Study Area, with Clear Lake at the eastern edge of the area.

Mud Swamp/New River Wilderness is in the southwestern part of the Apalachicola National Forest, covering 8,000 acres of black titi swamp. The New River that flows through the wilderness, however, is lined by beautiful Atlantic white cedars before the river enters the equally wild and swampy Tate's Hell Swamp. This is one of the wildest and most perilous wilderness areas in the country. It has no trails, and most of it is covered by waist-deep water. Alligators, cottonmouths, and hordes of mosquitoes are just some of the hazards that visitors should be aware of if entering the wilderness.

Silver Lake Recreation Area is very popular since it is a mere 8 miles west of Tallahassee via State Route 20. The spring-fed lake is surrounded by scenic bald cypresses with wispy Spanish moss hanging from their branches. There is a swimming area with a white sand beach and picnic area. One mile south of Silver Lake is Moore Lake, where you can observe a high pineland community. This dry sandy region is dominated by longleaf pine with a thick understory of turkey oak. Also beneath the pines are live oaks, southern red oaks, persimmons, black gums, and flowering dogwoods. Bluejack oaks occur occasionally. A dense growth of wiregrass covers the ground in most places, and bracken fern is common. Wildflowers in the high pineland include golden aster, rose rush, blackroot, elephantsfoot, blazing star, greeneyes, butterfly pea, partridge pea, goat's rue, colicroot, and rattlesnake-master.

Occasional large sandy depressions near Moore Lake may support a woods of mockernut hickory, pignut hickory, persimmon, southern red oak, laurel oak, the evergreen southern magnolia, basswood, sweetbay, black gum, and even American beech. Shrubs in these depressions include farkleberry, horse sugar, and, where water sometimes stands, buttonbush. Partridge berry often crawls on the ground.

On the east side of State Route 373 about 10 miles southwest of Tallahassee is Trout Pond Recreation Area. A picnic area with a fishing pier accessible to visitors with disabilities is available at the lake.

Just south of the Clear Lake Wilderness Study Area, on either side of State Route 267, is a variety of plant communities within 2 miles of each other. Less than 1 mile southeast of Forest Road 307, on the east side of State Route 267, is a fine black gum swamp. This swamp is often inundated by up to 2 feet of water, although several drier pockets of soil occur in the swamp. Black gum dominates the swamp, with a border of large slash pines. Shrubs found in this swamp are Virginia sweetspire and sweet pepperbush. Less than 1 mile farther south on State Route 267, immediately after crossing Forest Road 369, is a large black titi swamp on the southwest side of the highway. This swamp merges into a pine flatwoods community at its southeastern edge. This is one of the most common plant communities in the Apalachicola

National Forest. Longleaf, slash, and occasional pond pines stand above a dense shrubby thicket of saw palmettos. Other shrubs occurring with the saw palmettos include gallberry, running oak, little-leaved blueberry, dwarf huckleberry, dangleberry, dwarf live oak, and wax myrtle. Attractive wildflowers are meadow beauty, crow poison, several species of milkworts, yellow colicroot, rose mallow, and several species of St. John's wort. The curious little bog buttons is frequent in moist depressions. Across State Route 267 is an extensive bald cypress swamp.

On the far western side of the Apalachicola National Forest, Forest Road 115 toward White Oak Landing passes fine examples of a river swamp on the north side of the road and a loblolly pine forest on the south side. The river swamps occur along permanent rivers and streams, and a rich assemblage of trees is present: ogeechee lime, red maple, tulip poplar, black gum, water oak, sweetbay, swamp bay, sweet gum, loblolly pine, and bald cypress. The dense understory contains viburnum, pinckneya, swamp cyrilla, sweet pepperbush, swamp azalea, and black titi. In floodplains are American elm, swamp cottonwood, pop ash, water hickory, black willow, and diamondleaf oak. The loblolly pine forest has dense pines above an entanglement of greenbriers, muscadine, and Virginia creeper.

A habitat known as a *bay swamp* occurs in the Apalachicola National Forest. Sweetbay is a dominant tree, with swamp bay and loblolly bay occurring in lesser numbers. A well-populated understory includes giant cane, wax myrtle, swamp cyrilla, sweet pepperbush, Virginia sweetspire, swamp azalea, fetterbush, large gallberry, myrtleleaf holly, and red chokeberry. Mounds of sphagnum appear throughout the bay forest, and in certain places, water stands for much of the year.

Fort Gadsden Dome

Built by the British to recruit blacks and Indians to fight against the United States in the War of 1812, Fort Gadsden was abandoned after sustaining just one devastating blast. Today a replica of the wooden fort stands at the historic site, which falls within the Apalachicola National Forest in the Florida Panhandle. The surrounding terrain consists primarily of sandy flatwoods, with stands of 60-foot-high longleaf pines seeming to stretch endlessly toward the Gulf of Mexico, some 20 miles south. Here and there, however, shallow depressions filled with standing water support shorter, dome-shaped patches of pond cypress trees and other vegetation. This type of wetland, known as a *cypress dome*, is found only in the Coastal Plain of the southeastern United States. One easily accessible dome is 3 miles southeast of Fort Gadsden, off Forest Highway 143.

Cypress domes have a rounded profile, with trees rising 30 feet or so at the center, surrounded by lower trees and shrubs at the periphery. Covering a few acres, they arise where the terrain is perhaps 5 feet lower than the surroundings. Water seeps through the sandy soil from the higher ground but can drain only slowly through the nearly impervious, mucky floor of the dome. As a result, water usually stands in the dome for most or all of the year. At the center of the dome, where the ground is lowest, the water may be 3 to 4 feet deep.

Pond cypresses are gray-barked, gnarled trees that often have an irregular, flattened top. Although some botanists consider them the same species as the more common bald cypress, close observation of the two reveals their distinctness. According to the late Robert K. Godfrey, an authority on the flora of the Florida Panhandle, the most obvious difference is in the orientation of the leaves. Most of the pond cypress leaves are twisted so that they stand erect along the upper side of the twigs, creating a featherlike effect. By contrast, the leaves of bald cypress are arranged in a spiral all around the twigs.

In addition, pond cypress leaves are gradually tapered and rigid, while bald cypress leaves are soft and flat and abruptly tapered at the tip. The bark on older pond cypress trees breaks up into vertical, rectangular plates about 1 inch thick, whereas the bark of the bald cypress shreds into thin strips. Both kinds of trees produce woody "knees" when subjected to standing water over time, but those of the pond cypress are shorter and nearly columnar, with thick bark over their tips, while those of the bald cypress are longer and more conical, with thin bark at the tip.

The pond cypress trees are the ones that make it easy to spot a dome while traveling through the Apalachicola National Forest, but many other plants have adapted to living in and around these habitats. A few other trees that are at home in standing water are red maple, pop ash, and swamp black gum (another controversial species, since a number of botanists do not regard it as distinct from the upland black gum).

The climbing pieris, a woody vine in the heath family, germinates in the mucky soil of some of the domes. The stem grows upward beneath the outer bark of a tree, often a pond cypress; then, 2 feet or more above the water, it bursts through the bark and arches out, bearing green, leathery leaves. Eventually it produces white, bell-shaped flowers and spherical, dry seed capsules.

Two species of St. John's wort—shrubs with narrow, needlelike leaves and reddish, fibrous bark—are found in cypress domes. Both may grow 6 feet tall. The clustered St. John's wort prefers the shallower water at the periphery of the dome; if the water level fluctuates a lot, the shrub may put out slender, woody prop roots, which provide additional anchorage, from the lower part of its stem. Chapman's St. John's wort grows near the center of the dome if the water is deep enough. Its bark consists of soft, spongy cork 1 to 2 inches thick, which protects it well from fire.

Myrtleleaf holly, a shrub with short, narrow, leathery leaves and tiny red berries, may attain the stature of a small tree while growing in the standing water of a cypress dome. Four other shrubs, all with clusters of white flowers, are found at the periphery. These are sweet pepperbush (with leaves that resemble those of an alder), titi, black titi, and storax. Small, shrubby, blue-berried hollies, known as gallberries or inkberries, grow just outside the borders of the dome. Diverse wildflowers and sedges are also distributed throughout the dome, their location dependent on water depth or the degree of saturation of the soil.

Carnivorous plants are often present. The tall yellow pitcher plant grows in shallow water, while the parrot pitcher plant, whose insect-trapping leaves lie flat on their backs, lives in heavily saturated soil along the edge. Commonly growing along with the parrot pitcher plant are two sundews, whose leaves have sticky hairs that glisten in the sunlight and trap tiny insects. One of the sundews has a cluster of leaves, usually red, that consist of an elongated leaf stalk and a small round blade. The other, known locally as sunthreads, has slender, gray, erect leaves up to 8 inches long and 0.125 inch wide.

Several kinds of butterworts grow in the saturated soil outside the cypress domes. These insect-trapping plants with solitary flowers have a small cluster of leathery leaves whose edges are rolled upward. Some of the leaf cells secrete an oily substance that gives off a funguslike odor apparently attractive to gnats. The insects that investigate the smell become mired in the plant's secretion and drown. The leaves then roll farther upward and inward to form a shallow cup into which digestive enzymes are secreted. Any object containing nitrogen appears to trigger the leaf to secrete the digestive enzymes. As a result, the butterwort will digest not only gnats but also pieces of leaves and small seeds.

Leon Sinks

The low hills of sandy, red clay that surround Tallahassee, Florida, drop off to the delta plains a few miles south of the city. To the casual observer, the plains, which slope gently for almost 20 miles to the Gulf of Mexico, seem a monotonous sea of sand that supports large numbers of longleaf pines and little else. But some surprises lurk beneath the surface. Under the loose quartz sands is a bed of easily dissolved limestone. The abundant rains that fall in this humid, subtropical region (an average of 57 inches annually) trickle through the permeable sands into this underlying layer. Percolating groundwater dissolves the limestone, rapidly lowering parts of the terrain. The resultant depressions, or *sinks*, pock the landscape. Shallower sinks are filled with sand, but deeper ones often accumulate water.

A cluster of water-filled sinks perforate the sandy plains in the Apalachicola National Forest at the southern edge of Leon County; these are known collectively as the Leon Sinks. A circular route that follows an old sand road for about 1.5 miles passes by the five major sinks in the Leon Sinks Geological Area. Each sink has its own distinctive feature. Hammock Sink (pl. 4) is a large, circular depression that drops 65 feet from the surrounding pinelands. Its basin, filled with clear, turquoise blue water, reflects the silhouettes of some of the surrounding trees. Black Sink is a funnel-shaped depression that drops into a deep, dark abyss. Gopher Sink, a cavernous hole lined with rocks, takes its name from its vague resemblance to a den inhabited by gopher tortoises. Natural Bridge Sink includes a tree-covered limestone bridge. Fisher Creek runs beneath the bridge and disappears underground, only to resurface 150 feet away.

Dismal Sink is the most impressive of the Leon Sinks. Two hundred feet from rim to rim, it drops abruptly 75 feet to a dark pool of unknown depth. For 20 feet above the water surface, the sheer, dripping walls are of solid rock. Handsome ferns, whose roots somehow have found narrow crevices in the rock, display their feathery fronds above the water. There is no safe way to descend into Dismal Sink, and extreme care should be taken even when walking near its rim.

The abundant water in the sinks nourishes a community of plants that contrasts with the surrounding drier pinelands. Where the steep slopes of the sink are permanently moist, large specimens of laurel oak, water oak, white oak, southern magnolia, sweet gum, American beech, basswood, sweet bay, and black gum may be found. On somewhat drier slopes, the dominant trees are mockernut and pignut hickories, red oak, and persimmon. These species, collectively termed *southern mixed hardwoods*, are found over much of the southeastern United States but are expected in the sandy coastal plains of the Florida Panhandle. Beneath the hardwoods is a layer of shrubs that consists of two gnarled members of the heath family, farkleberry and squaw huckleberry; a yellow-flowered shrub known as horse sugar; and small, symmetrical dogwoods. An entanglement of creepers often makes hiking difficult, while at ground level are wildflowers not found in the sandy surroundings.

The hike from one sink to the next takes the visitor through a different world of sand and sand-tolerant species. The relatively nonorganic, unconsolidated quartz sand primarily supports a sparse forest of longleaf pine and blackjack and turkey oaks. The pines may approach 100 feet in height, but the oaks grow more slowly and seldom reach 20 feet. The herbaceous undergrowth is sparse, and large patches of sand remain exposed to the sun. Colonies of bracken fern, which can prosper under adverse conditions, give a false hope that the sand community can support lush vegetation.

Longleaf pine, whose mature needles are up to 15 inches long, is well adapted to grow in the often-deep sand of the Coastal Plain. It is found from North Carolina, around peninsular Florida, and over to eastern Texas, skipping only the Florida Keys, the Everglades, and the bottomland woods adjacent to the Mississippi River. For the first 6 years of its life, while its roots are developing, the young longleaf pine looks like a mounded clump of bright green grass. Its main bud remains level with the ground, where it forms long needles in dense clusters. When the little tree is about 7 years old, however, it loses its grasslike appearance and begins a rapid upward growth, which continues until the tree begins to decline in vigor after 300 or 400 years. Under optimal growing conditions a longleaf pine may attain a height of 120 feet and a trunk diameter of nearly 3 feet.

Because it cannot tolerate shade, the longleaf pine does not withstand competition from other trees, including others of its kind. The roots of a mature longleaf pine radiate for a distance of 50 to 100 feet beneath the tree, and seedling longleaf pines that germinate within this radius survive only a short time. As a result, trees in a stand of longleaf pines are widely spaced. Considerable sunlight penetrates to the sandy ground between the trees, providing an ideal situation for the growth of wiregrass.

The wiregrass and the millions of pine needles that fall to the sand provide rapid-burning fuel for surface fires. Lightning frequently ignites such fires, which burn over wide areas of the pine community. One reason for the longleaf pine's success is that it is the most fire-resistant tree in the eastern United States. In the early growth stage, the dense cluster of needles protects the terminal bud from fire damage. Once the thick, flaky bark has formed on the trunk, the tree resists hot fires as if it had a covering of asbestos. In an experiment at Florida State University, a slab of bark from a mature longleaf pine was exposed to an acetylene torch for 1 minute. A thermometer placed on the side away from the flame registered only a 3-degree rise in temperature, with little damage to the bark.

The branches of the longleaf pine, unlike those of many other pines, are widely spaced, often giving the tree an appearance of spindliness. If the lower branches of a longleaf pine should ignite, the fire usually cannot advance to the branches above. The wide separation of the trees makes it unlikely that fire could spread from the crown of one tree to another.

The success of the longleaf pine actually depends on fire because periodic fires kill young saplings of other trees that might eventually provide competition. In addition, longleaf pine cones fall to the sandy ground unopened, and only a searing fire will cause them to open. In this way, seeds are liberated under the most favorable conditions. The tree's dependence on fire was suggested as early as 1847 by British geologist Sir Charles Lyell, who observed this species on a visit to Alabama. Several ecologists believe that if fire were kept out of a longleaf pine stand long enough, many of the trees that now only line the limestone sinks would germinate there and eventually choke out the shade-intolerant pines, ultimately forming a continuous community of southern mixed hardwood species.

Post Office Bay

"These are the crown jewels of the pineland wet savannas in North America," remarked Steve Orzell as he, Edwin Bridges, and I drove through the Florida Panhandle southwest of Tallahassee, observing large expanses of grassland that intrude like fingers into the forests of longleaf pines. Orzell and Bridges, a pair of dedicated Florida botanists, had been studying the ecology and plant diversity of these wet savannas and were eager to introduce me to the different types found in a part of the Apalachicola National Forest known as Post Office Bay.

The savannas and their surroundings appear nearly flat, but slight differences in elevation enable distinct plant communities to develop. At the higher and drier end of the spectrum is a forest dominated by longleaf pines, with a dense, often-impenetrable undergrowth of saw palmettos. As the elevation drops ever so slightly (the gradient is seldom more than 6 feet over a distance of 600 feet), the pines suddenly drop out and the wet savanna begins, dominated by grasses and sedges growing in saturated or mucky soils. Deeper into the savanna, more surface water accumulates; when it reaches a depth of perhaps 8 to 12 inches, the savanna ends and a swampy zone of 2- to 3-foot-tall, water-tolerant shrubs appears.

Where the standing water is between 1 and 3 feet deep are stands of ascending pond cypresses, known as *cypress domes*. The border of a cypress dome consists of two species of St. John's worts, 6-foot-tall shrubs with needlelike leaves. One of these, which grows in slightly deeper water than the other, has an inch-thick layer of tightly compressed, papery, reddish bark that protects the shrubs from fire. A shrubby black gum that is found nowhere else sometimes grows in the cypress dome.

Although wiregrass, a very narrow-leaved grass, dominates all the savanna areas, there are actually two types of wet savannas. Botanist Andre Clewell noted this some years ago, calling them the Verbesina Phase Savanna and the Pleea Phase Savanna, based on the characteristic flower that grows in each.

The Verbesina Phase Savanna occupies soils topped by as much as 8 inches of clay. The surface consists of a hardpan, or hardened layer, that is nearly impermeable to water. Water stands on the hardpan throughout the wetter seasons of the year, mixing with the clay to form a black, sticky muck. The *Verbesina*, or crownbeard, which gives this savanna its name, is unique to the Florida Panhandle. Its numerous round heads of yellow, tubular flowers grow at the tips of 3-foot-long branches. Carnivorous plants also prosper here: sundews (a tiny one with yellow flowers and another that grows up to 10 inches tall with inch-wide purple flowers); the tall yellow pitcher plant and the lower-growing red pitcher plant; and two kinds of butterworts, plants that entrap insects in their leaves. Other plants are two types of lady's hatpins, with their solitary white flowering heads perched atop needlelike stems, and a diversity of sedges known as beaked rushes and nut rushes.

Rarities found in the Verbesina Phase Savanna include the grass-leaved blackeyed Susan, which has maroon flower heads; a large, blue-flowered skullcap of the mint family; and a foot-tall sunflower whose only leaves are crowded at the base of the plant.

The round-toothed milkwort is a plant whose seeds are produced from underground flowers; one advantage of this form of reproduction is that the flowers and seeds are safe from fire.

The Pleea Phase Savanna has soil that is a sandy loam instead of clay. Although the water from the adjacent longleaf pine forests continually drains through the sand, the water table is high: if you press your foot down hard, water will gurgle up a few inches away. *Pleea*, or rush featherling, is always common here. A member of the lily family, it has small white flowers, and its leaves resemble those of an iris. Species associated with it include pitcher plants, sundews, and bladderworts. Also carnivorous, bladderworts have small bladders; attached to an intricate stem system below the water, these bladders can suck in microscopic insects.

The wet savannas of Post Office Bay are not a stable community. Their health and vigor often depend on the amount of rainfall during the year. Periods of drought lower the water table, with the result that some species fail to flower or even to come up. Fires are also essential to the savanna ecology. The savannas in Apalachicola National Forest are now deliberately burned every 3 to 5 years, usually in April after plant growth has resumed. The fires are often set to include the adjacent longleaf pine forests and allowed to spread across the savannas, burning out naturally in the deeper water.

In the longleaf pine forests, fire knocks back the dense growth of saw palmettos. During the first year following a burn, wiregrass takes over beneath the pines, forming a continuous, light green cover. Eventually the saw palmettos regain their prominence in the understory.

April fires burn well in the wet savannas despite the seepage water overlying the clay and saturating the sandy loam. Fire kills such woody plants as titi and inkberry, which otherwise tend to displace more fragile species. An April burn compresses the flowering season, with species that normally bloom in April and May delaying their flowering until summer and autumn, joining the species that normally bloom then. With more than 100 species of plants bursting into flower, these areas provide a profusion of color. Wiregrass, the dominant species of wet savannas, blooms vigorously following a burn but otherwise may not.

As it burns toward the deeper waters of a cypress dome, the fire begins to die out. The shrubby St. John's worts, with their thick layers of bark, remain unharmed, but the cypresses on the periphery will be singed and their growth retarded for a while. The cypress trees in the center of the dome usually remain unscathed. The resultant layered effect—with 6-foot St. John's worts, stunted 10- to 12-foot cypresses, and 40- to 50-foot cypresses—gives the cypress dome its rounded shape.

Rock Bluff

Originating in southern Georgia, the Ochlockonee River twists and turns through the Apalachicola National Forest in the Florida Panhandle and empties into the Gulf of Mexico. The region is primarily flat, with stretches of longleaf pine growing in dry sand above an understory of saw palmetto. But scattered throughout are other habitats—forests of oaks or magnolias, swamps of bald cypress or gum, pond cypress domes, wet savannas, and pitcher plant bogs. A variety of vegetation appears at Rock Bluff, a sandy ridge that runs for a half mile along the east side of the Ochlockonee River. North to south, the ridge rises from 30 to nearly 70 feet, its western side sloping ever more precipitously.

The dry ridgetop, whose lower end is accessible from a small parking area off Forest Highway 390, is massive sandstone covered with sand, which in places is several inches deep. This nutrient-poor soil supports longleaf pine, turkey oak, saw palmetto, bracken fern, and little else. A little downslope toward the river, where there is a measure of shade and some more moisture, the vegetation is dominated by upland laurel oak, sometimes called Darlington oak.

Although the distinction is not commonly made, botanists for years have noticed that some laurel oaks live in dry, sandy, upland habitats, while others live in valley bottoms in more moist, shaded areas. Florida botanist Robert K. Godfrey argued persuasively that two species should be recognized. Among their differences, the upland laurel oak (*Quercus hemisphericus*) has leaves whose lower surface is completely smooth, while the swamp laurel oak (*Q. laurifolia*) has matted hairs on the undersurface of its leaves. Although the leaves of the two species are elliptical, the swamp laurel oak, also known as diamondleaf oak, always has at least a few that are shaped like narrow diamonds.

The upland laurel oaks grow loosely spaced, allowing sunlight to reach the forest floor. The sandy soil is acidic, supporting several members of the acid-loving heath family—farkleberry, deerberry, and Elliott's blueberry. All these shrubs or small trees are kinds of wild blueberries.

The forest elsewhere on the western slopes is a combination of hardwoods and pines, actually an extension of the mixed hardwood forests of the southern Appalachians. Among the species are white oak, basswood, American beech, spruce pine (with short, slender, very flexible needles), and loblolly pine (with larger, thicker, straighter needles).

American holly, with its glossy, spiny-toothed, evergreen leaves and shiny red berries, adds color to the forest during the winter, while red buckeye, flowering dogwood, redbud, and silverbell all bloom in April. The silverbell's 1.5-inch-long, bell-shaped flowers are much larger than those of other silverbell trees in the southern United States. In Godfrey's view, this is a distinct variety limited to southeastern Alabama, southwestern Georgia, and the Florida Panhandle west of Tallahassee.

Also flowering in April is the sweetleaf, or horse sugar tree, with its inch-round clusters of fuzzy, pale yellow flowers. The leaf of this species tastes sweet, particularly near the central vein, owing to sugar that the leaf cells manufacture and store. A little later in the spring, the large, white flowers of the shrubby snowbell bush, a first cousin to the silverbell, brighten the mid-layer in the forest.

A tree found on the forest slopes is tulip poplar, a member of the magnolia family, whose creamy yellow flowers resemble tulip blossoms. The genus *Magnolia* itself boasts the most spectacular blossoms. Rarest of those at Rock Bluff is the pyramidal magnolia, considered to be endangered in Florida. The white flowers of this deciduous tree stand erect when they bloom during the first half of April. Southern magnolia, with its large, glossy, leathery, evergreen leaves, blooms in May with the largest flowers of all magnolias—up to 10 inches across. Also flowering in May is the sweet bay magnolia, a tree with narrow, white-backed leaves and a rich, creamy flower. The leaves of the sweet bay are evergreen in the warm parts of its range, such as Florida and elsewhere in the South, but may be shed farther north, where colder winters prevail.

The base of Rock Bluff lies about 100 feet from the Ochlockonee River, whose lazy waters stretch about 200 feet wide. Blocking access to the river's edge at this location is a streamlike feature, actually an old oxbow of the river. The low terrain leading up to the oxbow is a bog, fed by cool, clear spring water that seeps slowly from the base of the slopes throughout the year. The water may be more than a foot deep in places.

Trees shade the bog during the growing season—sweet gums, river birches, swamp bays, swamp chestnut oaks, and bald cypresses. One mammoth bald cypress, whose top was knocked off during a past lightning storm, has a trunk 12 feet thick at shoulder height. Wetland vines that climb the trees include climbing hydrangea, cross vine, and bamboo vine.

Colorful wetland shrubs in and near the seepage bog include dog hobble, titi, and Virginia willow, all with white flowers; deciduous holly with red berries; and two kinds of viburnum with blue berries. Among the wildflowers are yellow stargrass of the iris family, brookweed of the primrose family, and Indian plantain of the aster family, as well as St. John's wort and lizard's tail. Two rarities are the needle palm and the camellia-like silk bay shrub.

At least 12 kinds of wetland sedges grow in the bog, along with a few grasses and rushes. Scattered throughout are patches of sphagnum moss and several attractive ferns, from the cinnamon fern with 4-foot-long fronds, to royal fern, southern lady fern, southern shield fern, sensitive fern, and netted chain fern. Maintained by the reliable trickle of water, these wetland plants add an extra measure of diversity to the varied vegetation at Rock Bluff.

Ocala National Forest

SIZE AND LOCATION: 430,000 acres in central Florida, east of Ocala. Major access routes are State Routes 19, 40, 42, 314, 314A, 445, and 445A. District Ranger Stations: Silver Springs and Umatilla. Forest Supervisor's Office: 325 John Knox Road, Suite F-100, Tallahassee, FL 32303, www.southernregion.fs .fed.us/florida.

SPECIAL FACILITIES: Boat ramps; swimming beaches; mountain bicycle trail; horse trails; off-road vehicle trails; visitor centers.

SPECIAL ATTRACTIONS: Juniper Springs; Alexander Springs; Salt Springs; Silver Glen Springs.

WILDERNESS AREAS: Alexander Springs (7,887 acres); Billies Bay (3,063 acres); Juniper Prairie (14,294 acres); Little Lake George (2,832 acres).

The Ocala National Forest, located east of central Florida's rapidly growing city of Ocala, stands unique among the national forests in the United States.

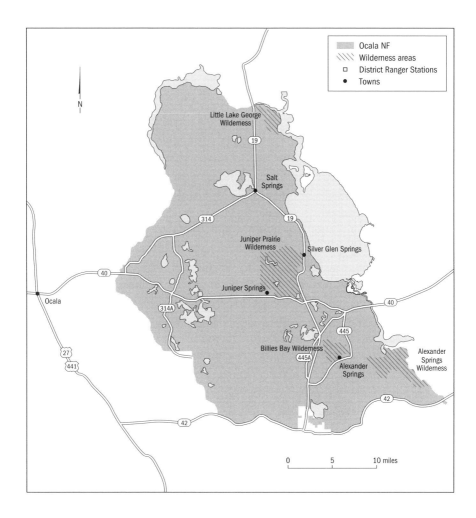

Combining subtropical and temperate elements, the Ocala National Forest, sprawling over nearly 430,000 acres, is the southernmost national forest in the contiguous United States, located entirely below the 30th parallel.

The diversity of features in the Ocala National Forest is greatly responsible for the variety of life found within its boundaries. Clear springs, subtropical vegetation, a huge variety of animal life, mystically haunting creeks, and more than 600 lakes are only some of the enticements that beckon the visitor.

Recreation of all types can be found, from a canoe float trip down Juniper Creek, to scuba diving at Alexander Springs and swimming at Juniper Springs, to hiking the Florida National Scenic Trail. Fishing is always rewarding, and hunting in season is available, too. Several campgrounds and picnic areas are scattered throughout the forest.

State Route 40 bisects the Ocala National Forest west to east, while State Route 19 bisects it north to south. State Route 42 forms much of the southern border of the national forest. Numerous county and forest roads, some of them in deep sand, cross parts of the national forest. Care should be taken when driving on the deep sand roads.

The western edge of the Ocala National Forest is about 10 miles east of Ocala. Entering this part of the forest via State Route 40, you come to the Ocklawaha Visitor Center shortly after crossing the Ocklawaha River. The Ocklawaha River flows for about 70 miles, more or less forming the western and northern boundaries of the Ocala National Forest. The river is fed by a chain of connected lakes including Griffin, Eustis, Harris, and Apopka; crystal-clear water from Silver Springs; and overflows of Orange Lake. Fishing for channel catfish, panfish, and bass is usually rewarding in this river.

During heavy rains, the river often overflows into the adjacent floodplains where swamps and mesic hammocks are overgrown with lush vegetation. The swampy woods are dominated by bald cypress, tupelo gum, red maple, pop ash, cabbage palm, water locust, and water hickory, with a shrub layer of palmetto, wax myrtle, Virginia sweetspire, dahoon, buckthorn, and buttonbush. The mesic hammocks are dominated by southern magnolia, associated with American holly, laurel oak, water oak, live oak, and loblolly pine, beneath which grow saw palmetto, rusty lyonia, staggerbush, and American beautyberry. At a few places near the Ocklawaha River are rich areas of hardwood trees that include Shumard oak, cherrybark oak, pignut hickory, mockernut hickory, sugarberry, tulip poplar, and wild black cherry. Beneath these trees are a shrubby viburnum, strawberry bush, crossvine, and bristly greenbrier.

A beautiful corridor of the Ocklawaha River is the 17-mile segment between the Delks Bluff Bridge and Eureka. Between Eureka and the Rodman Lake Dam, the river is bordered by swampy woods that are frequently inundated. The 9-mile stretch of the Ocklawaha River between the Rodman Dam and the St. John's River goes through undisturbed swamps.

Immediately south of the Ocklawaha River where it enters the St. John's River is the Little Lake George Wilderness where nearly 3,000 acres of the St. John's Flatwoods are managed as a wilderness area. Most of the wilderness is seasonally flooded, and the vegetation is so dense as to appear junglelike. Slash, pond, and loblolly pines occur above an understory thicket of palmetto, gallberry, and wax myrtle. Sandhill cranes inhabit the wilderness.

Three miles south of the wilderness is Cat Head Pond, a small but deep pond along County Road 43. A Civil War battle was fought nearby, and legend has it that the winner of the battle dumped the loser's cannon and sidearms into the pond.

By driving Forest Road 77-1, which follows the southern side of the

Ocklawaha River, about 1.5 miles southeast of Rodman Dam, you come to Davenport Landing. This was a small town with only a few houses, from about the 1860s, where riverboats stopped to get fuel.

After entering the national forest's western side on State Route 40, the highway is surrounded on both sides by numerous lakes for the next 10 miles. North of the highway is a good example of the Ocklawaha Flatwoods community where slash, pond, and loblolly pines are common. In moist areas of the flatwoods are oak hammocks.

South of State Route 40 and Church Lake is an area of undulating sand hills with large level areas supporting marshes and prairies. This region is punctuated by small lakes and a few isolated sinkholes. The lakes have clear water and sandy bottoms. The Florida and greater sandhill cranes occur throughout these sandy hills, particularly around Church Lake, Sherrill Mill Pond, and, somewhat farther south, Mud Prairie Lake. The endangered red-cockaded woodpecker is found here as well. Longleaf pine, slash pine, and turkey oak are the dominant woody plants.

County Road 145AV north from State Route 40 goes to Lake Charles. This 360-acre lake is surrounded by a fringe of titi swamp, slash pine, and bald cypress. Two and one-half miles southeast of Lake Charles is Redwater Lake. On all sides of Lake Charles and Redwater Lake is an extensive area of swamps, hammocks, and lakes. Ospreys nest at both lakes. The lakes in this scenic area have thick, mucky bottoms and dark brown water stained with tannic acid from oak leaves. Much of the vegetation in this region is slash pine flatwoods with a dense understory of gallberry. By continuing north past Lake Charles, you will come to Fore Lake and Lake Shore campgrounds in a few miles.

Mud Lake lies 1 mile north of State Route 314 and 2 miles north of Lake Eaton. Eaton Creek flows north from Mud Lake, and Mason Bay Springs is immediately west of Mud Lake. Mud Lake spans 500 acres and has an average depth of 20 inches. Situated in a large hardwood swamp, it is unique in having unusual organic ooze on its bottom. It is said that only four lakes in the world possess this type of organic ooze: two in Central Africa, Saddlebags Lake north of Lake Okeechobee, and Mud Lake. The oozy bottom of Mud Lake consists of a layer of algae with undecayed insect fecal pellets. The ooze accumulates very slowly and, at a depth of 3 feet, is estimated to be about 2,000 years old. The alkaline waters of the lake support a varied fish population that includes gizzard shad, shellcracker, Seminole killifish, bluegill, bass, and crappie. The very small amount of bacteria in the lake accounts for the slow decay rate of dead fish and other organic matter.

South of Mud Lake is Lake Eaton with a nearby Indian mound. East of the lake is the Lake Eaton Sinkhole Trail. The trail winds for 0.5 mile through sand pines and scrub oaks to the huge sinkhole. The dry sinkhole is 450 feet wide with a maximum depth of 80 feet. An observation deck allows you to look into it, and stairs lead to its bottom. At the base of the stairs is an oak hammock with live oak, magnolia, flowering dogwood (pl. 5), loblolly pine, and palmettos. A 1.2-mile trail also leads from the parking area to the sinkhole.

Across the road from the Lake Eaton Sinkhole Trail is the Lake Eaton Loop Trail. It is 0.8 mile from the parking area along Forest Road 79 to Lake Eaton, through sand pine and scrub oak. As the trail decreases in elevation near the lake, a hardwood forest is encountered with red maple, water oak, laurel oak, loblolly bay, bald cypress, wax myrtle, and buttonbush the common woody species. Lake Eaton is located in a region of shallow swamps and wet prairies, while longleaf pine, sand pine, and turkey oak occur on higher land.

State Route 40 continues eastward, with Owens Lake, Tobe Lake, and Mill Dam Lake to the north of the highway and North Lake, Lake Bryant, and Halfmoon Lake to the south. These lakes are associated with wet prairies, sand pine flatwoods, longleaf pine uplands, and marshes. Hardwood hammocks occur around Halfmoon Lake. Mill Dam Lake has a picnic facility, boat ramp, and group camp. Halfmoon Lake has a boat ramp.

The lakes end abruptly east of Mill Dam Lake and Halfmoon Lake, and State Route 40 enters the vast area of the Ocala National Forest known as the Big Scrub. The Big Scrub stretches north all the way to the Ocklawaha River, south to the southern border of the Ocala National Forest near Nicotoon Lake, and east to State Route 19. The 191,000-acre Big Scrub is a sea of sand pine, with an abundance of scrub oaks and shrubs. Sand pines live for about 60 years, and extreme heat is necessary to open the pine cones and to release the seeds. Natural fires provide the heat and prepare the soil for seed germination. Without fire, the sand pines would be replaced by dwarf oaks and other species. Although sand pine is by far the most dominant tree in the Big Scrub, this area also contains a mixture of evergreen oaks, particularly myrtle oak, live oak, and Chapman's oak. Smooth palmetto and saw palmetto are common. The heath family is well represented by rusty lyonia, staggerbush, little-leaved blueberry, and a dwarf huckleberry, while the holly family is represented by Carolina holly, scrub holly, and gallberry. Rosemary, garbera, and sweet olive are locally abundant. The endangered Florida scrub jay lives in the Big Scrub. North of State Route 40, Forest Roads 67, 75, 88, and 97 pass through the heart of the Big Scrub, while south of State Route 40, Forest Roads 566, 573, 588, and 595 provide ample opportunity to observe this vast community. Big Scrub Campground is at the junction of Forest Roads 573 and 588.

North of State Route 40, the continuous area of Big Scrub is occasionally interrupted by stands of longleaf pine. These pines make up what are known as *longleaf islands*, referring to islands of longleaf pine in a sea of sand pine. West of Lake Delancy and north of Lake Kerr is a large area known as Riverside Island. Southeast of Lake Delancy is Norwalk Island, and adjacent to the west side of Lake Kerr is Kerr Island. Between Lake Kerr and Cat Head Pond is Syracuse Island, while Salt Springs Island is south of Lake Kerr. Most of these stands of longleaf pine are nearly 100 years old and have a parklike appearance. Red-cockaded woodpeckers are often seen here.

Two longleaf islands are historically significant. Hughes Island is 5 miles south of Lake Kerr, and Pat's Island is 8 miles southeast of Lake Kerr and 3 miles west of Lake George. Hughes Island consists of 880 acres of longleaf pines. Named for Hughes Stanaland, an early settler of the area, Hughes Island features an Indian burial ground, as well as the world's largest loblolly bay tree. Forest Roads 10A and 10B encircle Hughes Island, while Forest Road 10, the Old Grahamsville Road, passes through its center.

Pat's Island, at the northern end of Juniper Prairie Wilderness, is well-known because it was the setting for the 1938 Pulitzer Prize–winning novel *The Yearling* by Marjorie Kinnan Rawlings. Pat's Island consists of 900 acres, mostly stands of longleaf pine, but with occasional xeric, hydric, and mesic hammocks and sinkholes. The island is named for Patrick Smith, the first postmaster of a small settlement on Pat's Island that included a church and school. In 1872, the Reuben Long family moved to Pat's Island, and it was the descendants of this family, Calvin and Mary Long, with whom Rawlings lived in October 1933 and whose local stories she included in her epic book. The Long Cemetery is on Pat's Island. Forest Road 10 passes within 0.5 mile of the north end of the island. Another way to reach Pat's Island is to take the Yearling Trail that begins at State Route 19 across from the entrance to Silver Glen Springs and proceeds for 2 miles to Pat's Island. The Florida National Scenic Trail also passes through the center of Pat's Island from north to south.

Juniper Prairie Wilderness consists of 14,294 acres, including Pat's Island. Juniper Creek passes through the wilderness, as does the Florida National Scenic Trail. State Route 19 forms the eastern border of the wilderness, and State Route 40 the southern edge. Just outside the southwestern corner of Juniper Prairie Wilderness is the Juniper Springs Recreation Area. Juniper Springs is one of several springs in the Ocala National Forest. Along with nearby Fern Hammock Springs, it produces about 13 million gallons of 72-degree Fahrenheit spring water each day. You may walk into the old Mill House and look at a few exhibits and wildlife photos. There are several camping loops, and swimming is permitted in the spring's pool. A nature trail follows the south side of Juniper Run, connecting the Juniper Springs pool with Fern Hammock Springs. A very popular canoe run begins at the Juniper

Springs Recreation Area and follows Juniper Creek to State Route 19, where there is a wayside park and boat ramp. The distance of the canoe run is about 7 miles, and the route winds through subtropical vegetation and is often lined by bald cypress trees. About midway between Juniper Springs and the wayside park, at a sharp curve of Juniper Creek, is a steep sand bank. On top of the bank are various earthworks among the pines and palmettos that collectively are referred to as Machine Gun Nest Hill.

Numerous boils are present in the Juniper Springs area. Where the aquifer comes near the surface, boils appear. The boil may be so intense that it can sometimes be heard. A deep boil under the Fern Hammock Bridge makes a faint sound. The boils bubble through the sand, and the swirling white sands stand out vividly against the creek's darker bottom.

Just north of the wayside park on State Route 19 is Sweetwater Springs where the Forest Service has cabins for rent. On the east side of State Route 19 and immediately south of Sweetwater Springs is Morman Branch. Forest Road 71, a deep-sand Jeep trail, crosses the creek and winds through the area. Morman Branch is significant for its stands of Atlantic white cedar and the presence of a bay swamp forest. Rare plants include star anise, Florida willow, large-leaved grass-of-Parnassus, and Chapman's sedge. In addition, loblolly bay, needle palm, climbing hydrangea, swamp bay, leucothoe, royal fern, cinnamon fern, and many other species of ferns, shrubs, and wildflowers are present.

Two miles northwest of Pat's Island and accessible via Forest Roads 90 and 65 is Hopkins Prairie. This wet prairie is nearly 3 miles long and up to 0.5 mile wide. Numerous wetland species of plants grow here. The vegetation at Hopkins Prairie is in broad zones that correlate with the amount of inundation and whether the substrate is sand or peat. In the wetter areas with standing water are floating hearts and fringed yellow-eyed grass. The zone around this area is sandy peat that is periodically inundated. This zone contains Tracy's beaked rush, Small's yellow-eyed grass, and flat pipewort. Surrounding this zone is a wet, spongy humic layer over sand with substantial growth of maidencane grass and flat pipewort. Beyond this zone, in dry, exposed white sand, are horned bladderwort, Elliott's yellow-eyed grass, shrubby St. John's wort, and bog buttons. At the drier, outer zone is a rim of live oaks draped with Spanish moss, with saw palmetto and large-fruited beaked rush in the ground layer. Many birds visit Hopkins Prairie.

Completely bordered by State Routes 314 and 19 and County Road 316, Lake Kerr is the largest lake within the Ocala National Forest. Most of the area immediately surrounding the lake is privately owned, but the Forest Service owns Kauffman's Island, a peninsula that extends into the north side of the lake. The peninsula has small ponds and a scenic hardwood hammock.

Across State Route 19 from Lake Kerr is the Salt Springs Visitor Center, adjacent to Salt Springs Campground. Salt Springs Run is a clear stream from the springs that flows for 5 miles before emptying into Lake George. From a trailhead on State Route 19 is the Salt Springs Trail that winds for 1 mile to an observation platform at the edge of Salt Springs Run.

North of Lake Kerr is Lake Delancy, where the Forest Service has a campground near the northwestern corner of the area. The campground is situated beneath sand and longleaf pines. The lake itself is surrounded by stands of slash pine and an oak hammock. An extensive wetland known as Buckskin Prairie is 1 mile to the east of Lake Delancy.

After State Route 40 crosses State Route 19, it passes along the north side of Wildcat Lake where there is a boat ramp, swimming area, and picnic area. The lake is bordered by slash pine and several species of oaks. The highway then continues eastward to the town of Astor before leaving the Ocala National Forest as it crosses the St. John's River. South of Wildcat Lake is Grasshopper Lake, where the Forest Service maintains a boat ramp.

By driving State Route 19 south from its junction with State Route 40, you will pass several lakes on the west side of the road and Billies Bay Wilderness on the east side. Beakman Lake, Chain O'Lakes, Farles Prairie, and Buck Lake each has a boat ramp, with the last two lakes also having a campground. The Farles Prairie area has several interconnected sand bottom lakes, with most of the lakes lined with live oak, slash pine, and pond pine. The area is a typical sand pine habitat.

Billies Bay Wilderness is an exceptionally wild area of swamps, hammocks, and flatwoods. Nine Mile Creek and Billies Bay Branch wind through the wilderness. Dense hardwood forests create a junglelike appearance in the wilderness. Just outside the southern end of the wilderness is remarkable Alexander Springs, which puts forth 78 million gallons of crystal-clear water daily. The water is excellent for snorkeling and scuba diving. Surrounding the springs are palm hammocks, hardwood swamps, and sand pine ridges. Timucuan Indian Trail forms a loop from the campground around the east side of Alexander Springs. Much of this trail is on boardwalk. Access to the Alexander Springs Recreation Area is via State Route 445.

Alexander Springs gives rise to Alexander Springs Creek, which follows a circuitous route, eventually entering the Alexander Springs Wilderness and emptying into the St. John's River. Get-Out-Creek is a mystic black waterway in the wilderness. This fascinating creek is lined by cabbage palms and Spanish moss–draped live oaks. Also in the wilderness is Kimball Island. This area of 1,000 acres exhibits good examples of hardwood hammocks and hardwood river swamps. An Indian shell midden, native sour oranges, and a few remaining orange and grapefruit trees of late 19th-century groves are present.

North of the Alexander Springs Wilderness, between Stagger Mud Lake and Lake Dexter, are the Bowers Bluff middens. This archaeological area consists of three freshwater snail shell middens, the largest approximately 380 feet long, 100 feet wide, and 15 feet high. One-third mile to its north is another midden, this one 150 feet long, 80 feet wide, and 13 feet high. To the west is a third midden that is 100 feet long and 35 feet wide. Archaeological evidence suggests that the middens date to as early as 3,000 to 2,000 B.C.

Heading south, State Route 19 crosses the Florida National Scenic Trail, where there is trailhead parking on the west side. Two miles southwest of the trail crossing is the small water-filled Blue Sink. Just before State Route 19 exits the southern boundary of the Ocala National Forest, it comes to the Pittman Visitor Center. East of the visitor center is large Lake Dorr, surrounded by longleaf pines mixed with various hardwood species. The lake has a campground and two boat ramps along State Route 19. The Forest Service also rents a cabin at Lake Dorr.

Northeast of Lake Dorr and east of Forest Road 572-1 is Disappearing Creek. Here is a ravine about 20 feet deep, nearly 1,000 feet long, and 200 feet wide, surrounded by longleaf pines. The depression was formed by numerous small springs that flow from the banks, join together, and then disappear at the eastern end of the ravine. A mesic hammock is in the ravine, with fine specimens of magnolia, laurel oak, water oak, wild black cherry, basswood, and cabbage palm.

State Route 42 follows the southern boundary of the Ocala National Forest. Four miles east of State Route 19, State Route 42 crosses Blackwater Swamp. Although some of Blackwater Swamp was logged in the past, about 200 acres still exist of old-growth bald cypress and tupelo gum, with trees more than 150 years old. After another mile on State Route 42, Forest Road 536 goes north to Clearwater Lake and Campground. The Florida National Scenic Trail is immediately east of the campground. Forest Road 538-1 north of State Route 42 passes the historic Paisley Lookout Tower. The Paisley Woods Bicycle Trail begins near Clearwater Lake and traverses live oak domes, grassy prairies, and pine stands for 22 miles.

At the extreme southeast corner of the Ocala National Forest is the St. Francis Historical Area. St. Francis in 1887 was a thriving pioneer town on the west bank of the St. John's River. It supported a newspaper, a general store, a hotel, a warehouse, a sanitarium and health resort, numerous homes, and huge citrus groves. The coming of railroads and a disastrous freeze in 1894 brought an end to the town. Today the Forest Service has two trails that interpret the area, one covering 1.8 miles and the other 7 miles. For a while the trail follows the old Paisley–St. Francis Wagon Road around an oak hammock. The long trail then passes a bayhead swamp, a pine flatwoods, and a hardwood river swamp. The short trail passes a natural spring.

For horse riders, the 100 Mile Horse Trail is ideal as it covers many miles of flatwoods and prairies.

Alexander Springs

Issuing clear water at the rate of 78 million gallons a day from a large limestone cavern 25 feet belowground, Alexander Springs has the greatest flow of any natural spring on United States government land. The 300-foot-diameter pool it supplies is the main attraction of a recreation area in the Ocala National Forest, in the center of the Florida peninsula. At a constant 74 degrees Fahrenheit, the pool is a favorite of swimmers and scuba divers. The crowds do little harm to the environment, although they stir up sand on the adjacent beach, sending it into the rolling water.

The cavernous limestone that conducts water to Alexander Springs was laid down by a process that began several million years ago, when a warm, shallow sea covered what is now peninsular Florida. When small invertebrate animals and plants died, their hard parts accumulated on the sea floor, forming a layer of limestone bedrock several thousand feet deep, over which additional layers of sandy limestone and sand were deposited. After a fall in sea level, the present peninsula emerged as dry land, and the deposits became subject to erosion. Today the Ocala area is a rolling landscape that consists mainly of exposed sand, with a few limestone outcrops.

The pool created by Alexander Springs lies in a shallow basin. Among the trees that grow nearby in the sandy soil are cabbage palms, water oaks, and willow oaks draped with Spanish moss. Probably no oak species in the country has as many leaf shapes as the water oak, varying from highly lobed to not lobed at all. Willow oak (fig. 3), except for its acorns, looks little like an oak, because its leaves are extremely narrow, toothless, and lobeless.

In the deeper parts of the spring pool, threadlike stems of naiad, a flowering plant with microscopic flowers, form mats just beneath the surface. The water along the shoreline is clogged with water hyacinth, whose gorgeous pink flowers can scarcely compensate for its aggressive nature. A native of South America, it was introduced as a novelty at the 1892 Louisiana Exposition, from which it quickly spread across the South. But the water hyacinth also has its share of virtues. It is able to absorb toxic chemicals from polluted water, and its massive rhizomes provide a perfect home for the inch-long apple snail with its brown, paper-thin shell. At Alexander Springs, the apple snail is the only source of food for the limpkin, a brown, awkward-looking, 2-foot-tall bird that tiptoes on spindly legs along the water's edge.

The marshy soil adjacent to the water hyacinth is home to several small wetland species of flowering plants, including pennywort, whose rounded leaves disguise its membership in the carrot family; the nodding frog-fruit of the verbena family; and the weak-stemmed bedstraw, with whorls of narrow leaves up and down its stem. Growing as a web over much of the vegetation is the slender, viny Florida vetch, whose pale lavender, sweetpea-shaped flowers appear in May and June.

From the pool, water channels into Alexander Springs Creek, which flows through a cabbage palm–hardwood "hammock," grassy wetland flats, and a shady hardwood forest, to empty into the St. John's River 16 miles to the east. A *hammock* (or *hummock*) is a slightly elevated tract of forest in a marshy area. The one through which the creek flows is covered with an unusual mix of temperate and subtropical vegetation. Sixty-foot-tall cabbage palms from the tropics stand beside red maples, commonly associated with New England, and sweet bay magnolias and Carolina ashes from the Appalachians. In the shrub layers, saw palmettos of the tropics live with wahoo of the eastern United States and leatherleaf leucothoe of the Appalachians. Virginia creeper and poison ivy, so common in the Northeast and Midwest, climb on the same trees as do the Florida wild yam and the climbing hydrangea of the Southeast.

Figure 3. Willow oak, Ocala National Forest (Florida).

Ferns, many with 3-foot-long leaves, create impenetrable thickets in places. Common temperate species are royal fern, cinnamon fern, netted chain fern, and marsh fern, while golden polypody is subtropical. The small gray polypody, which has a broad range, often grows as an epiphyte (air plant), blanketing the upper side of tree branches. Shallow depressions in the cabbage palm–hardwood hammocks periodically retain water, supporting such wetland plants as arrowhead, lizard's tail, and an assortment of grasses, sedges, and rushes. Most are temperate, but at least one, Florida brookweed, is a subtropical species. To protect both the ecosystem and the visitor, the Forest Service has constructed a trail, much of it on elevated boardwalks, throughout this jungle.

One unusual plant in the hammock, whose divided leaves suggest a cross between a yew and a palmetto, produces cones resembling those of a pine. Known locally as coontie, it is a cycad—a species that is not a fern, conifer, or flowering plant but a member of a primitive group transitional between ferns and palms. Today only a few cycad species exist in the world, although the fossil record shows that they were the dominant plants 200 million years ago. Today the coontie is considered rare in most parts of Florida, but Indians and early European settlers used its starchy underground stems as an important source of food.

Big Scrub

Fifty years ago, while hiking in central Florida, biologist Maurice Mulvania recorded that the "nearly pure white sand of the ground surface, when viewed from a short distance, gives the impression of a thin rift of wind-driven snow. The vegetation is mostly dwarfed, gnarled and crooked. Here the sun shed its glare and takes its toll of the unfit." Mulvania was in the Big Scrub section of the Ocala National Forest, a habitat that extends for 35 miles north to south in the forest and nearly 15 miles east to west, stopping a short distance from the community of Ocala. South of the forest, the scrub extends for another 110 miles to Lake Okeechobee.

A notable feature of the terrain here is the way areas of white and cream-colored sand alternate in a continuing but irregular pattern. The white sand apparently derives from ancient dunes of windblown sand, while marine currents deposited the cream-colored sand. The usually abrupt change in sand color is accompanied by a contrast in vegetation. In 1895, botanist George Nash noted, "these two floras are natural enemies and appear to be constantly fighting each other. A bare space of pure white sand that usually separates the two is neutral ground."

The vegetation on the white sand is mostly tangled and scraggly. It is dominated by stands of sand pine with a thick understory of short Chapman's and evergreen scrub oaks and densely branched shrubs that include rosemary, staggerbushes, tough buckthorn, and saw palmettos. This is the true scrub. In the cream-colored sand, which extends as slightly elevated tongues into the scrub or sometimes is completely surrounded by it, grows an open forest of tall longleaf pines, which tower over deciduous turkey oaks, blackjack oaks, and Margaretta's post oaks. These plant communities enclosed by a "sea of white sand" are sometimes called *high pine islands*. An understory of narrow-leaved grasses gives them a parklike appearance.

Some 19th-century botanists thought that the distribution of plants in the two sand types was accidental and depended on which plants got there first. However, when a region of the scrub is burned, the revegetation is always only by plants that are tolerant of the white sand conditions, not by a mixture of scrub and pine island species. The white sand has very little potash, clay, and humus, while the cream sand has ample amounts of these ingredients. This composition accounts not only for the difference in vegetation but also for a contrast in animal life. The lack of cohesive material prevents many animals from burrowing into the white sand, since they are unable to keep their tunnels open. In the cream sand, however, salamanders, gophers, ants, and other burrowing organisms are common.

Rainfall is limited in this part of Florida, and what rain does fall percolates rapidly through the white sand to a depth beyond the roots of the plants. Scrub plants are well adapted to this moisture-deficient habitat. Sand pines have very short needles and scaly branchlets that prevent excessive water loss, while the leaves of the rusty staggerbush are covered with a dense felt of hairs. Rosemary has short leaves that are rolled up like little needles, lessening exposure to the sun's rays.

Where there are not enough nutrients and moisture in the white sand to support leafy green plants, the substrate is often covered by rounded, crunchy, gray-green tufts of deer moss. Deer moss is a lichen whose fungal component gives the plant its light gray color and whose algal part contributes the green hue. Other lichens cling tenaciously to the bark of the sand pine, including one that imparts a pinkish red color.

In the porous white sand, everything dries out rapidly. Dead organic matter is easily blown away, resulting in little accumulation of humus, which would protect potential seedlings. Bare sandy patches, beneath which the decayed roots of long-dead plants may be found, indicate the difficulty of seedling establishment in the sand. Most scrub plants produce great numbers of seeds, improving the odds that a few will develop into mature plants.

The Big Scrub harbors some species and subspecies that are found nowhere else in the world. These organisms developed under austere living conditions and have adapted to deficiencies in nutrients and moisture. Among the animals is the foot-long, bluish Florida scrub jay, which nests in the sand pines, and the Florida sand skink, which lives most of its life buried in the white sand, feeding on termites and beetle larvae. This skink—with its wedge-shaped head, partially recessed lower jaw, tiny front legs that fold into grooves in its body, and highly reduced hind legs that have only two digits each—is streamlined to "swim" in the loose sand. Among the most beautiful plants is the Florida bonamia, a prostrate type of morning glory with a pale blue, funnel-shaped flower that measures up to 2 inches across. This creeping plant grows beneath the sand pines in sunny areas where the understory has been opened up by occasional fires.

With the accelerated development of much of Florida during the past several decades, much of the scrub habitat has been altered or destroyed by housing developments, conversion to citrus groves, and industrial complexes. Even large tracts in the Ocala National Forest have been clear-cut to harvest the pines for the production of paper pulp. As a result of the diminished habitat, many of the sand scrub endemics—from the Ocala National Forest to Lake Okeechobee—are now threatened with extinction.

Juniper Springs

The occasional bald eagle that soars above the Ocala National Forest of central Florida generally sees vast stands of pale, scrubby pine trees, broken by hundreds of bright blue ponds, by strands of heavy growth along the meandering creeks, and by some unusual—even mysterious—islands of thick, dark green vegetation. These patches of broad-leaved evergreens are commonly called *hammocks* in the southeastern United States. Their origins, even the origin of their name, are poorly understood. Some etymologists trace the name to an Indian word meaning "a shady place." Others equate *hammock* with *swamp*. And still others trace the word to the Spanish *hamaca*. My big Webster's dictionary cautiously says the origin of the name is "unknown."

Botanists and physical geographers have long been interested in these islands of broad-leaved plants, but after years of reading about them, I realized that not only their origins but also the reasons for their persistence are not clearly understood. So, as I drove along Florida Route 40 from Ocala and turned into the popular Juniper Springs Recreation Area in the Ocala National Forest, I looked forward to visiting these intriguing patches of vegetation. The campground fortuitously is set in Fern Hammock, one of many hammocks scattered throughout this entirely subtropical forest in the United States.

As you enter a hammock, you immediately sense and feel the abundance of diverse wildlife (don't forget mosquito repellent). In hammocks, the dominant hardwood trees provide myriad cavities in which wildlife can nest, roost, den, or simply hide from predators. In addition, decaying hardwood tissue creates rich microhabitats for insects and fungi. In its diversity of wildlife, the hammock contrasts sharply with the surrounding piney scrubland. There, resin quickly heals diseased or damaged tissues in softwood pines, inhibiting small organisms from establishing themselves. And unlike the surrounding scrubland, where one or two tree species dominate, the hammock contains a great variety of plants, thus ensuring a wide range of seeds, nuts, berries, buds, and other delectables for wildlife.

Botanists classify hammocks into three types: dry, moist, and wet. As your feet sink into the boggy floor, you quickly realize that Fern Hammock is a wet one. Waist-high royal, cinnamon, and chain ferns give a junglelike appearance. In many places, the dense canopy blocks the sunlight from the forest floor and restricts air movement, making the humidity high. In these wet hammocks, the dominant broad-leaved evergreen is the swamp red bay, a medium-sized, cylindrical tree of the laurel family whose leaves are sometimes used to flavor foods. Strikingly different is the cabbage palmetto, a straight-trunked, unbranched palm with huge fan-shaped leaves. A few deciduous trees—water oak, Florida elm, sweet gum, and red maple—interrupt the evergreen canopy. Brush palmetto, wax myrtle, and a shrubby holly known as gallberry form a dense undergrowth in places that makes hiking difficult.

Only a slight elevation changes the hammock and its vegetation into the moist type. About 80 percent of the trees in the moist hammock are broad-leaved evergreens, and the forest floor is covered with their slowly decaying, leathery leaves. Here the dominant trees, bull-bay magnolia, red bay, American holly, laurel oak, and live oak, grow farther apart than the trees in the wet hammock. Some sunlight pierces the canopy and reaches shrubby saw palmettos and fragrant-flowered wild olives.

Dry hammocks are quite different. The trees are irregularly spaced and the canopy has many large openings. This open habitat allows access to sun and wind, so a great loss of moisture occurs through evaporation. In addition, wispy Spanish moss drapes many of the trees, intercepting falling moisture and preventing much of it from reaching the hammock floor. This dry habitat is ideal for picturesque live oaks, cabbage palmettos, bluejack oaks, laurel oaks, American hollies, and red bays. The beautiful yellow-flowered jessamine vine climbs over the shrubby myrtle oaks and Chapman's oaks.

Because of the fertility of their soils, Florida's hammocks are being rapidly cleared and converted into citrus groves and other croplands. Fortunately, the Ocala National Forest preserves good representatives of these three kinds of tree islands.

After a day in the hammocks, a visitor can enjoy a swim in Juniper Springs pool or a 7.5-mile canoe trip down Juniper Creek from the canoe launch at Fern Hammock to the Sweetwater terminus at Florida 19. The route passes beneath huge bald cypresses, oaks draped with Spanish moss, and tall cabbage palmettos that lean over the water. You may sight deer, otter, alligators, herons, egrets, and, if you are lucky, an awkward-looking bird known as the limpkin. The limpkin, now becoming rare, depends almost entirely on apple snails for food.

About 0.25-mile east of Route 19, a small spring-fed brook, known as Morman Branch, enters Juniper Creek. You can hike along the brook to a 10-acre stand of Atlantic white cedar. This is the southernmost location for these evergreens, with their flattened branchlets. Some of the trees are nearly 3 feet in diameter.

Wandering through this stand, you can discover the rare and little-known Florida willow and the attractive large-leaved grass-of-Parnassus—an herb found nowhere else in Florida. You may also see what is, for a botanist, a rather striking phenomenon: small, flowered branchlets growing from the trunk of the Atlantic white cedar, a tree that is not a flowering plant. This mystery, however, can be explained. The flowers and the branchlets belong to a member of the climbing heath family, *Pieris phillyreifolia*, called climbing fetterbush. This vine roots next to the cedar, and its stem actually penetrates the tree's trunk, grows upward beneath the bark for as much as 20 feet, and then bursts out to put forth its white, bell-shaped flowers.

Morman Branch

In the Big Scrub of central Florida, ancient accumulations of windblown sand provide a foothold for scraggly pines, turkey-foot oaks, and palmettos, as well as a number of unique plant species. Several threatened species of animals live there, too—among them, the Florida scrub jay, the sand skink, and the blue-tail mole skink. Within this generally dry region, a 620-acre swamp and a 2-mile-long streamlet known as Morman Branch provide an oasis of moist habitat. Steve Orzell, a former student of mine who is now an ecologist in Florida, invited me to visit this area in the Ocala National Forest.

We trudged in on a Forest Service "road" consisting of deep sand not negotiable by ordinary vehicles. For half a mile, our route wound through scrubby trees; except for lichens, vegetation on the forest floor was sparse, exposing large patches of bare sand. Soon we reached Morman Branch, whose cool, flowing water was so clear that we could easily see the rippled, sandy bottom. The narrow stream, a tributary of Juniper Creek, passes through a swamp that occupies a barely perceptible depression. The swamp is fed by small springs and seepages created by rainwater as it percolates through the surrounding sand hills.

The soil is always moist in the swamp and along the stretch of Morman Branch that leads to Juniper Creek. Botanists Dan Ward and Andre Clewell credit this to the even flow of spring-fed waters throughout the year. Among the plants growing in Morman Branch is sago pondweed, a flowering plant that cannot tolerate any degree of desiccation. Other aquatic species are pickerel weed, eelgrass, and coontail.

While most of the swampy habitat is heavily wooded with black gum, red maple, loblolly bay, and sweet bay magnolia, the occasional openings are dense with grassy vegetation—and poisonous cottonmouths. Each opening has a shallow channel of flowing water supplied by numerous springs. Many plant species rare in this part of Florida turn up here, including Chapman's sedge and the large-leaved grass-of-Parnassus. In the muck grows flaccid quillwort. A distant relative of ferns, this uncommon quillwort produces reproductive spores at the bases of its spreading, grasslike leaves.

Scattered along the stream and particularly through the swamp forest are Atlantic white cedars, cone-bearing evergreens more common farther north on the Atlantic Coastal Plain. A rare species that grows with the white cedars is the shrubby star anise, a relative of magnolias. Its inch-wide pale flowers give rise to ribbed, several-pointed, anise-scented fruits. Another unusual plant is climbing fetterbush, a vine that climbs almost exclusively on Atlantic white cedars or pond cypresses (most vines are less particular, growing on whatever host is nearby).

Climbing fetterbush is one of the few vines in the heath family, which con-sists mostly of shrubs or small trees, including rhododendrons, azaleas, mountain laurels, blueberries, and cranberries. It grows just inside the bark of the white cedar or other host tree, periodically breaking through with short branches that bear clusters of leathery, evergreen leaves and elongated sprays of small white flowers. With age, these branches die and break off, to be replaced by new branches higher up on the host. Ward and Clewell have observed one branch growing 40 feet above the ground. As the bark on the lower parts of the white cedar trunk ages and begins to flake off, the reddish stems of the vine are often revealed.

The stand of Atlantic white cedars that grows along Morman Branch and in the swamp is the world's southernmost, one of two very isolated colonies in peninsular Florida. The species' main range extends from New Jersey to South Carolina, with isolated stands in southwestern Georgia, the Florida Panhandle, southern Alabama, and southeastern Mississippi.

The more northerly stands of white cedar are often nearly pure groves of uniform age. They probably arose when seedlings germinated en masse after a previous forest was logged or destroyed by fire. Their growth rings usually are all about the same width, indicating a fairly uniform rate of growth. Ward and Clewell have found that the Ocala stand, in contrast, is not of uniform age and that the cedars are mixed with several other woody species. In addi-tion, the annual rings show that the growth of these trees was suppressed when they were young, generally for several decades.

These telltale features suggest that the white cedars we see today have filled in gaps created haphazardly in the forest. When a large tree dies or a large branch falls, more light reaches the forest floor, stimulating the germination of numerous white cedar seeds that lie dormant in the soil. Before the seedlings are very old, however, the canopy may close over, causing them to perish. But a new gap may allow some to survive and grow into small trees hardy enough to tolerate periods of low light. Eventually, a few trees reach the canopy, where their growth rate accelerates. Finally, they produce seeds that perpetuate the cycle.

White cedars are able to outdistance all other trees in the canopy because of their resistance to disease and decay. With time, they tower well above other species, becoming a target for lightning strikes, the only natural danger that threatens them. Fire has not played a major role in the Ocala stand, as is evident from the staggered ages of the white cedars. When they occur in this habitat, fires are dampened by the wet soil and by the foliage of mixed broad-leaved hardwood trees and herbaceous undergrowth.

Ward and Clewell have studied environmental conditions along Morman Branch to try to discover why white cedars grow here and not in other similar-looking wetlands in nearby parts of peninsular Florida. One reason may be that even though they grow in a wet environment, white cedars cannot tolerate flooding; consequently, they do not live in the floodplains of Florida's large rivers, where water levels fluctuate drastically with the seasons. Moreover, because of their shallow roots, the trees may be harmed by the deposition of sediments.

Botanists are not the only forest visitors to pay attention to the white cedars. Male Florida black bears mark tree trunks to warn off other bears. The largest white cedar in the stand has been ripped and shredded to a height of 6 feet by bears' claws, and their coarse black hairs are embedded in resin droplets that exude from the scrapes. The layers of scar tissue on the trunk show that this tree has been a territorial marker for several generations of bears.

Osceola National Forest

SIZE AND LOCATION: Approximately 157,000 acres in northeastern Florida, northeast of Lake City. Major access routes are Interstate Highway 10, U.S. Highway 90, and State Routes 2 and 250. District Ranger Station: Olustee. Forest Supervisor's Office: 325 John Knox Road, Suite F-100, Tallahassee, FL 32303, www.southernregion.fs.fed.us/florida.

SPECIAL FACILITIES: Boat ramps; swimming beaches.

SPECIAL ATTRACTIONS: Ocean Pond Recreation Area; Olustee Battlefield Site; Pinhook Swamp.

WILDERNESS AREA: Big Gum Swamp (13,660 acres).

The Osceola National Forest consists of three noncontiguous parcels of swampland and pine flats. The largest part of the national forest is northeast

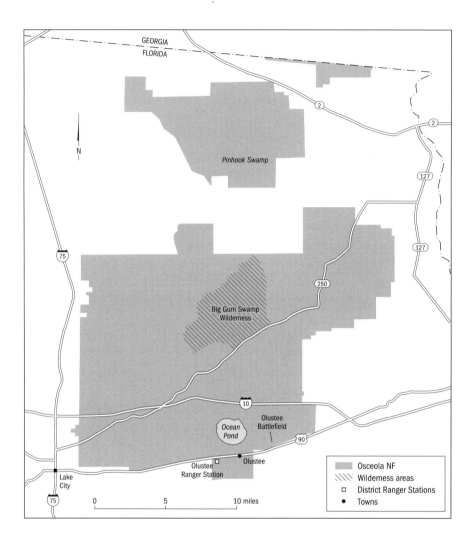

of Lake City, lying between U.S. Highway 90 and Sand Hill Road. North of Sand Hill Road is a significant area known as the Pinhook Swamp Unit that has been purchased during the last few years. This unit lies from 1.5 to 3 miles north of the main part of the Osceola National Forest. One-half mile north of the Pinhook Swamp Unit is a narrow strip of swampland recently incorporated into the national forest. A part of the Okefenokee Swamp, it extends to within a few hundred feet of the Georgia state line. The purchase of these two segments of the Osceola National Forest is part of a plan to form a continuous corridor of swampland from the Okefenokee Swamp south to U.S. Highway 90, a corridor that will be 60 miles long and 20 miles wide when completed.

The only developed recreation area in the Osceola National Forest is around Ocean Pond, a nearly spherical body of water just north of the community of Olustee. Ocean Pond is a shallow, natural, 1,700-acre lake around which grow large bald cypress trees heavily draped with Spanish moss. A fog-like shroud often hangs over the lake in early morning. Ocean Pond Campground at the north end of the lake is nestled beneath the cypresses. Nearby are a swimming beach, boat ramp, and picnic facilities. At the southern end of Ocean Pond is Olustee Beach, which has a swimming area and a boat ramp. A 0.1-mile historic trail at Olustee Beach shows the history of a sawmill and community that thrived here near the end of the 1800s. Also around Ocean Pond is The Landing, which is a group camping area available by reservation only, and Hog Pen Landing, which is a primitive hunt camp. Hog Pen Landing is particularly scenic with a large bald cypress at the end of a narrow spit of land. Four other primitive hunt camps are scattered in the Osceola National Forest.

Four miles east of Lake City, along the north side of U.S. Highway 90, is Mount Carrie Wayside, where there are a picnic area and a mile-long interpretive hiking trail through a forest of longleaf pines. Here you may see the federally endangered red-cockaded woodpecker and the equally rare gopher tortoise.

Immediately south of the town of Olustee is the 3,315-acre Olustee Experimental Forest where research in forest genetics and forest insects is being conducted.

The visitor information center is in the historical Olustee Train Depot that was built in 1888. Nearby is the Olustee Battlefield Historic Site in the national forest where the largest Civil War battle in Florida took place on February 20, 1864; it includes a small museum. Each year during the President's Day weekend, the battle is reenacted featuring artillery and cavalry units.

In the center of the Osceola National Forest is the Big Gum Swamp Wilderness. The interior of this very wild area consists of a large swamp dominated by bald cypresses and tupelo gums. Surrounding the swamp on elevations 3 to 4 feet higher than the swamp are pine flatwoods composed of slash and longleaf pine. Beneath the bald cypresses and tupelo gums is a rather extensive shrub layer of titi, fetterbush, Virginia sweetspire, dahoon, yaupon, myrtleleaf holly, and winterberry. Palmettos and gallberries are the common shrubs beneath the pines. Forest roads almost completely encircle the wilderness area. Two short-loop hiking trails into the wilderness, one along the western side off Forest Road 232, and one on the southeastern side off County Road 250.

Just outside the northeastern corner of the Big Gum Swamp Wilderness is the Osceola Research Natural Area. Similar to the characteristics of the wilderness area, this 7,400-acre region represents typical flatwoods with little changes in elevation. Low ridges in the natural area support pure stands of either longleaf or slash pines, beneath which is a dense cover of wiregrass. Saw palmettos and gallberries are interspersed on the low ridges. A small pond in the natural area supports pond cypress, slash pine, sweet bay, sweet gum, and red maple, with a shrub zone of fetterbush, gallberry, and greenbriers. Sphagnum moss covers much of the ground. Most of the land in the Osceola Research Natural Area is swamp that is often inundated. Trees in the swamp are bald cypress, sweet bay, sweet gum, black gum, and red maple. Some of the bald cypresses are estimated to be more than 500 years old, with diameters above the swollen bases attaining 70 inches.

The Florida National Scenic Trail winds through the Osceola National Forest. You may join the trail at the Olustee Battlefield Site and then circle around the northeastern side of Ocean Pond. The trail passes under Interstate Highway 10. Between the interstate and County Road 250 is a trail shelter. One may also join the Florida National Scenic Trail where the trail crosses County Road 250. The trail then passes the West Tower before leaving the western side of the national forest. The Deep Creek Trailhead is at the western edge of the national forest, a short distance east of U.S. Highway 441.

In addition to the West Tower, the East Tower is east of Big Gum Swamp Wilderness, and the Olustee Lookout is at the Olustee Battlefield Site.

Four interconnected loops traverse over 50 miles of trails through the scenic Osceola National Forest. These trails originate at West Tower where there is a camping area with horse stalls, drinking water, and a flush toilet.

The Pinhook Swamp Unit of the Osceola National Forest is very isolated. Only a few forest roads are in the eastern half of this very swampy area.

NATIONAL FORESTS IN GEORGIA

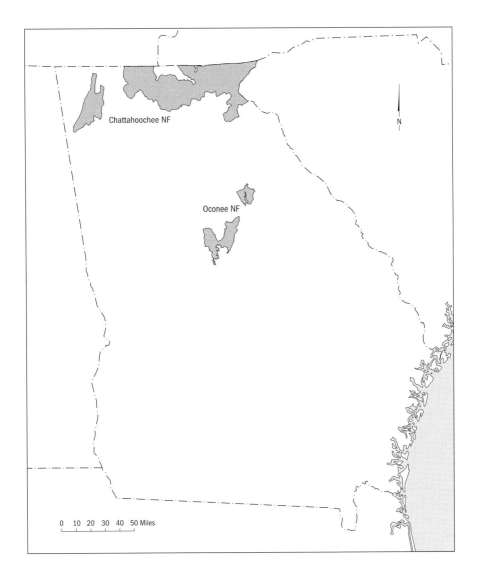

Georgia has two national forests, both under the administration of a single forest supervisor. The supervisor's headquarters for the Chattahoochee and Oconee national forests is at 1755 Cleveland Highway, Gainesville, GA 30501. These forests are in Region 8 of the United States Forest Service.

Chattahoochee National Forest

SIZE AND LOCATION: Approximately 750,000 acres across northern Georgia, from Lafayette and Summerville on the west to the South Carolina border on the east. Major access routes are Interstate Highways 75 and 985; U.S. Highways 19, 23, 27, 76, 123, 129, and 441; and State Routes 17, 25, 52, 60, 105, 136, 151, 180, 184, 197, 201, 348, 356, and 715. District Ranger Stations: Blairsville, Blue Ridge, Chatsworth, Clarkesville, and Clayton. Forest Supervisor's Office: 1755 Cleveland Highway, Gainesville, GA 30501, www.fs.fed.us/conf.

SPECIAL FACILITIES: Boat ramps; swimming beaches; off-road vehicle (ORV) areas.

SPECIAL ATTRACTIONS: Ridge and Valley National Scenic Byway; Brasstown Bald; Anna Ruby Falls; Sosebee Cove Scenic Area; Springer Mountain National Recreation Area; Chattooga National Wild and Scenic River; Russell-Brasstown National Scenic Byway.

WILDERNESS AREAS: Blood Mountain (7,800 acres); Brasstown (12,896 acres); Cohutta (35,268 acres); Ellicott Rock (8,274 acres, partly in the Nantahala and Sumter national forests); Mark Trail (16,400 acres); Raven Cliffs (9,115 acres); Rich Mountain (9,476 acres); Southern Nantahala (23,473 acres, partly in the Nantahala National Forest); Tray Mountain (9,702 acres).

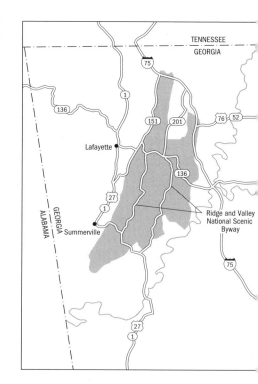

The Chattahoochee is the southernmost of the five national forests in the southern Appalachian mountain system. These southern mountains have played a great part in the vegetational history of the eastern United States. While other areas have been glaciated, submerged, and exposed to great climatic change, the southern Appalachians have been more stable, offering a sanctuary for many species of plants and animals. The Blue Ridge Mountains have been occupied continuously by plants and animals for perhaps 200 million years.

In the early Tertiary, the higher elevations of the Blue Ridge Mountains probably supported temperate forests ancestral to those now in the area, while subtropical floras prevailed at sea level. During the climatic changes to which eastern vegetation was subjected, the topography of the southern Appalachians offered varied conditions of moisture and elevations in which species of diverse climatic adaptations might survive while sometimes destroyed elsewhere.

The vast area covered by the Chattahoochee National Forest can be attested to by the five ranger districts that are in the forest. The forest is more than 200 miles east to west and nearly 70 miles north to south.

The mountains between Interstate Highway 75 on the west and the South Carolina border are the southernmost of the Appalachian Mountains, rising to a height of 4,784 feet on Brasstown Bald. West of Interstate Highway 75 is the small Armuchee Ranger District where the low mountains and ridges are more reminiscent of the Cumberland Plateau than the Blue Ridge Mountains.

The Chattooga National Wild and Scenic River, which separates the northeast corner of Georgia from the northwest corner of South Carolina, is also the dividing line between the Chattahoochee National Forest of Georgia and the Sumter National Forest of South Carolina. Rugged and remote, this river is a challenge to even the most experienced whitewater enthusiast.

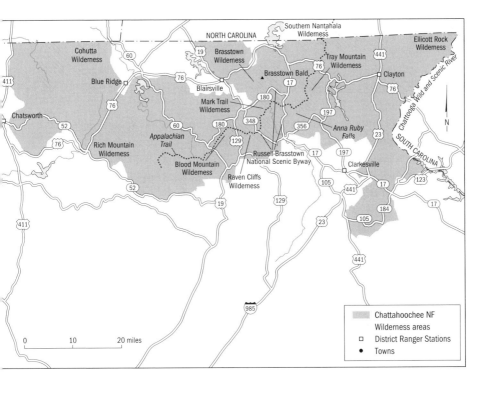

(See Sumter National Forest.) The upper reaches of the Chattooga River are within the Ellicott Rock Wilderness, a wild forested land that includes Ellicott Rock, a boulder in the river that was designated to mark the area where North Carolina, South Carolina, and Georgia converge. From the settlement of Pine Mountain, Georgia, one can follow State Route 28 south along the West Fork of the Chattooga River for approximately 4 miles to the Sumter National Forest in South Carolina. Almost 1 mile before reaching the Sumter, narrow and gravelly Forest Road 646 heads northeast on its way to Burrell Ford, where there is a place to put in your whitewater craft. The road crosses Hedden Creek, and if you were to hike or drive north along the creek, keeping your eyes on the densely wooded slopes above the creek, you might see the very rare and pretty Oconee bells, a plant endemic to the region. It has nearly circular, heart-shaped leaves and bell-shaped flowers on a stalk that arises from the cluster of basal leaves. The Hedden Creek Road soon becomes suitable only for four-wheel-drive vehicles as it makes its way between Cedar Cliff to the west and Glade Mountain to the east.

A hike up to Cedar Cliff will bring you to the habitat of three uncommon plants: twisted spikemoss, grass-pink orchid, and sand myrtle (*Leiophyllum buxifolium*). On Glade Mountain, sourwood, pale hickory, and rock chestnut oak dominate the rocky summit, with deerberry, a type of wild blueberry, the most common shrub. Common wildflowers in this dry habitat include pencilflower, spiderwort, rock saxifrage, alumroot, large coreopsis, blue bellflower, mountain mint, early meadow rue, and a couple of St. John's worts.

Just north of the community of Pine Mountain, along Forest Road 86, is West Fork Campground, picturesquely nestled beneath hemlock and hardwood trees along the West Fork of the Chattooga River.

The paved highway west from Pine Mountain, known as Warwoman Dell Road, goes to the town of Clayton in about 20 miles. On your way you will come to the road to Holcomb Creek Falls, the road to Rabun Bald Trail, and Warwoman Dell Campground, the latter just before reaching Clayton. By driving Forest Road 7 north for 4 miles, you will reach Holcomb Creek. A hike of about 1 mile down the creek brings you to these lovely falls. Where Holcomb Creek, Overflow Creek, and Big Creek come together, each after going through scenic gorges, they form the West Fork of the Chattooga River.

Forest Road 156 winds north and west through dramatic scenery north of Warwoman Dell Road. Just before this forest road peters out after about 12 miles, there is an entrance to the Bartram Trail. Forty miles in length, the Bartram Trail closely follows the route of William Bartram, a naturalist and explorer during the latter part of the 18th century. By taking this trail to the north, you will pass the twin knobs of Double Knob, then Wilson Knob and

Flat Top, before climbing to Rabun Bald. The 4,696-foot summit of the bald provides a commanding view of the surrounding area. The bald itself is home to the incomparable Catawba rhododendron as well as *Rhododendron maximum*. The woods surrounding the bald contain yellow birch, mockernut hickory, sweet bay magnolia, mountain ash, mountain maple, and moosewood (another kind of maple), with a remarkable number of shrubs that includes bush honeysuckle, witch hobble, bear honeysuckle, honey bells, mountain camellia, mountain holly, silky willow, and dwarf pussy willow. In moister areas are spicebush, ninebark, and black chokeberry. A staggering number of wildflowers grow here as well. Some of these are fly poison, featherbells, crow poison, turk's-cap lily, bead lily, trout lily, rosy twisted stalk, bunchflower, green hellebore; plus 9 species of trillium, all members of the lily family; 11 species of wild orchids, firepink, starry campion, goldenseal, doll's eyes, black cohosh, hepatica; several species of buttercups; two wild gentians; and many others.

Warwoman Dell Campground is located along Warwoman Creek and is adjacent to Warwoman Dell Road. The Bartram Trail also crosses this lovely area. An abandoned concrete basin here was used in the past as a holding facility for trout.

Several interesting areas in the Chattahoochee National Forest are south of Clayton and east of U.S. Highway 441. A few miles south is the spectacular Tallulah Gorge, made famous when the "Great Wallenda," Karl, walked a tightrope across the 1,200-foot-deep gorge on July 18, 1970. Although the gorge itself is in a state park, smaller gorges with similar vegetation are adjacent. Along Bad Creek just north of the state park is a site for the federally endangered persistent trillium, which is also in Tallulah Gorge.

U.S. Highway 441 crosses Panther Creek less than 2 miles south of the Tallulah Gorge State Park. From a parking area along the creek and U.S. Highway 441, a hiking trail leads through a pretty part of the Chattahoochee National Forest. The Panther Creek Trail passes through forests of hemlock and white pine and a series of cascades that cut through solid rock. Panther Creek, which forms several steep-sided gorges, is popular for trout fishing. The very uncommon bay starvine (*Schisandra glabra*), a woody vine with small, reddish-tinted flowers related to magnolias, is in the area.

West of the community of Toccoa is lovely Toccoa Falls (not in the national forest) and the Locust Stake ORV area (which is in the national forest). The southeastern-most part of the Chattahoochee National Forest, located between Cornelia and Toccoa, consists of the Lake Russell Wildlife Management Area and 100-acre Lake Russell. Lake Russell has many facilities, including a campground, picnic area, fishing piers, boat ramps, swimming beaches, and hiking trails.

U.S. Highway 76, the Lookout Mountain Scenic Highway, from Clayton west to Ellijay, provides access to all points of interest in the northern part of the Chattahoochee National Forest. Seven miles west of Clayton, the 3-mile-long Forest Road 162 will bring you to Timpson Falls. Four miles farther west on U.S. Highway 76, take Persimmon Road north to Forest Road 70. This forest road follows the turbulent Tallulah River, forming the western border of the Coleman River Scenic Area. There is a scenic hiking trail along the Coleman River in the scenic area; its trailhead is between Tallulah River and Tate Branch campgrounds. Rhododendrons line the Coleman River beneath towering hemlock and tulip poplar trees. A virgin stand of hemlocks occurs along the upper reaches of Tate Branch. Forest Road 70 continues north to Sandy Bottom Campground, then past the old town site of Tate City, and enters the Nantahala National Forest in North Carolina. The road from Tate City northward bisects the two halves of the Southern Nantahala Wilderness. The western half of the wilderness in Georgia contains several balds and rounded mountain tops called *knobs*. By hiking toward 4,568-foot Hightower Bald, you will pass through several lovely forested coves containing record-sized specimens of sweet buckeye and yellowwood trees. The deep shade caused by the large trees keeps the coves very moist, providing optimum conditions for the growth of large Goldie's ferns, intermediate ferns, and glade ferns. Green hellebore of the lily family, with broad, strongly veined leaves, grows in most of the coves, as do mountain elderberry, two types of *Viburnum* known as moosewood and nannyberry, and the rare Biltmore sedge. The Appalachian Trail crosses this part of the wilderness, passing Wheeler Knob, Rocky Knob, and Rich Knob. If you take the road along Little Persimmon Creek to the southeast corner of the Coleman River Scenic Area, you will come to Keener Creek. Along the creek is a high-elevation bog containing such species as the ginger-leaved grass-of-Parnassus and the undulate trillium. A little farther north on Wolf Knob is a nice population of golden corydalis.

U.S. Highway 76 climbs to a pull-out at Popcorn Overlook, which has picnic tables. From here is a vast panoramic view to the north over the valley that has been carved by Popcorn Creek. If you cross the highway carefully on foot, the ridge on the south side of the highway consists of an unusual formation known as *greenstone*. The vegetation that grows on this dry ridge includes pitch pine, wintergreen, nodding onion, Alabama grapefern, Shuttleworth's wild ginger, and the shrubby ninebark.

A few miles west of Popcorn Overlook, U.S. Highway 76 climbs to Dicks Creek Gap where the Appalachian Trail crosses the highway. By following the Appalachian Trail south through rugged terrain, you will hike through the center of Tray Mountain Wilderness. The trail actually crosses 4,430-foot

Tray Mountain near the southwestern corner of the wilderness where Catawba rhododendron grows on the summit of the mountain. A beautiful cove forest is on the north side of the mountain where you may see yellow birch, sweet buckeye, cucumber magnolia, rosy twisted stalk, Canada mayflower, black cohosh, and green hellebore. Other uncommon plants in the wilderness are Fraser's broad-leaved sedge, mountain lily-of-the-valley, purple fringed orchid, whorled pogonia orchid, pink lady's-slipper orchid, yellow lady's-slipper orchid (pl. 6), and obscure clubmoss.

State Route 197 branches off U.S. Highway 76 to the south, passing along the west side of Lake Burton (not in the national forest). From Cherokee Cove at the southern tip of the lake, a trail enters the Chattahoochee National Forest and climbs Long Ridge to Oakey Mountain, where a colony of gorgeous deep purple lesser rhododendron grows. A boggy area along Bad Branch to the east of Oakey Mountain has such interesting plants as sheep laurel, purple pitcher plant, and raisin tree.

East of Oakey Mountain is Lake Rabun where the Forest Service operates Rabun Beach Campground, a swimming beach, and a hiking trail. The 1.3-mile trail follows Joe Branch, fording it occasionally to Angel Falls.

By taking scenic but curvy State Route 17/75 south from U.S. Highway 76 for 3 miles before reaching the town of Hiawassee, you will be able to access High Shoal Falls, Andrews Cove, Anna Ruby Falls, Mark Trail Wilderness, and the Russell-Brasstown National Scenic Byway. High Shoal Falls is the major attraction of a 170-acre scenic area of the same name, reached by a short road just south of the intersection of State Routes 17/75 and 180. It is a series of five waterfalls along High Shoal Creek, together dropping a total of 100 feet. A 1.5-mile hiking trail lined by a dense growth of rhododendrons and mountain laurels goes to an observation deck overlooking one of the falls. Slippery and often dangerous side trails lead to the various cascades. The High Shoal Falls Scenic Area is at the western edge of the extensive Swallow Creek Wildlife Management Area, a rugged and fairly inaccessible area of mountain knobs and creeks. Of significance is Ramp Cove that contains some of the largest buckeye trees in the country and one of the largest concentrations of wild ramp, or leek.

Andrews Cove is a scenic forested cove with a campground nestled along Andrews Creek. The campground is adjacent to State Route 17/75 about 8 miles south of High Shoal Falls Scenic Area after a very crooked segment of the highway.

Anna Ruby Falls is one of the most spectacular highlights of the Chattahoochee National Forest. The 1,600-acre scenic area containing the falls may be reached off State Route 356 a few miles northeast of Helen and after traveling through a part of Unicoi State Park. A paved 0.5-mile trail from the modern visitor center leads to the base of the double falls. Located high on the slopes of Tray Mountain, the two falls are formed at the junction of Curtis and York creeks. Fed by underground springs and frequent rains, Curtis Creek on the left drops 153 feet, and York Creek drops 50 feet; they both form Smith Creek, which rushes rapidly downhill to Unicoi Lake. There is also a rugged 5-mile trail along Smith Creek, and a trail for the blind begins behind the visitor center. Civil War colonel John H. Nichols, who owned the land, named the falls after his only daughter, Anna Ruby.

Across from High Shoal Falls Scenic Area, State Route 180 branches off State Route 17/75. State Route 180 has been designated the Russell-Brasstown National Scenic Byway. If you drive State Route 180 to its junction with State Route 348, and then follow State Route 348, which loops back to Helen, you will have access to Brasstown Bald, Mark Trail Wilderness, Raven Cliffs Wilderness, Raven Cliffs Scenic Area, and Dukes Creek Falls. About 10 miles after you begin the Russell-Brasstrown National Scenic Byway, you will come to Jacks Gap where a spur road climbs steeply for 4 miles to a parking lot at the base of Brasstown Bald. The summit of the bald has the highest elevation in all of Georgia, topping out at 4,784 feet. A well-stocked gift shop is adjacent to the parking area, which is about 0.5 mile below the summit of Brasstown Bald. On the summit is a visitor center with several exhibits and a theater that shows a short film about the Chattahoochee National Forest. You can climb to the top of the tower at the visitor center for an unparalleled view in all directions. To reach the summit and visitor center from the parking area below, you may either hike the 0.5-mile trail that seems to go straight up or take a shuttle van for a small round-trip fee.

The visitor center and parking area are surrounded by the 12,975-acre Brasstown Wilderness. The bald is a spur off of the Blue Ridge Mountains. On the Brasstown Bald summit, Catawba rhododendrons flower in mid- to late June; if you look around, you may also find pink lady's-slipper orchids, kidney-leaved twayblade orchid, and three-toothed cinquefoil—all uncommon in northern Georgia—in flower. Big Bald Cove immediately to the north is known for its very large yellow birch trees towering over myriad common and rare wildflowers. Among the rarest ones are yellow lady's-slipper orchid, starflower, rosy twisted stalk, fringed gentian, squirrel corn, and green hellebore.

If you are up to rugged, long-distance hiking, you may wish to try the two trails that emanate from Brasstown Bald. The Arkaquah Trail goes to the west for 4.5 miles to Track Rock, descending very steeply in places. The trail skirts the edge of Plott Cove Research Natural Area, where there is a majestic stand of yellowwood trees. Jacks Knob Trail heads south from the Brasstown Bald parking area for 4.5 miles, eventually connecting with the Appalachian Trail at Jacks Knob.

Forest Road 180, Forest Road 348, and State Route 17/75 completely encircle the Mark Trail Wilderness. The Appalachian Trail traverses much of the wilderness area and then crosses over Spaniards Knob and Jacks Knob before passing near Horsetrough Mountain, Low Gap (which has a shelter), Sheep Rock Top, and Poor Mountain before coming to State Route 348. State Route 348 is all that separates Mark Trail Wilderness to the north from Raven Cliffs Wilderness to the south. Raven Cliffs Scenic Area occupies the northeast corner of the wilderness, while to the west are such landmarks as Cowrock Mountain and Levelland Mountain. Raven Cliffs Wilderness extends as far west as Neels Gap on U.S. Highway 19/129 and contains numerous waterfalls, gorges, and exceptionally steep slopes. The Appalachian Trail runs west-east through this wilderness area. Within the scenic area are Raven Cliffs Falls, which splits a solid rock face that is perpendicular to the ground and drops 150 feet. The scenic area has several other waterfalls.

As State Route 348 continues eastward between the two wilderness areas, it eventually comes to spectacular Dukes Creek Falls, which drops about 300 feet. An extremely steep trail down into Dukes Creek Gorge leads to the base of the falls after about 0.8 mile.

Lake Chatuge spreads out around Hiawassee on either side of U.S. Highway 76 and extends into North Carolina. The Forest Service maintains the Lake Chatuge Campground just east of State Route 288 west of Hiawassee. A boat ramp is located here as well. Farther to the west is Nottely Lake, surrounded by both private and Forest Service land. The only Forest Service development is the Davenport all-terrain vehicle area on land near the southwest corner of the lake.

South of the town of Young Harris, the Track Rock Road comes to Track Rock Archeological Site in about 4 miles. Here are six sandstone boulders with Indian petroglyphs carved into the surface of the rocks. The 4.5-mile Arkaqua Trail connects Track Rock with Brasstown Bald. Just west of Track Rock is the Beasley Knob ORV area.

U.S. Highway 19/129 south from Blairsville follows the Nottely River for several miles until it reaches the Chattahoochee National Forest. By taking State Route 180 to the west just prior to U.S. Highway 19/129 reaching Vogel State Park, you will climb to a parking area at the upper edge of Sosebee Cove (pl. 7), one of the most marvelous Appalachian forest coves in the area. By hiking the short trail through a small part of Sosebee Cove Scenic Area, you will encounter huge trees more than 4 feet in diameter. Some of these are buckeyes; others are tulip poplars. Particularly during the spring, the forest floor is carpeted by a great variety of wildflowers. Less than 1 mile west of the

Sosebee Cove parking area, the Duncan Ridge National Recreation Trail crosses State Route 180. By taking the south leg of this trail, you will hike along the eastern side of Slaughter Mountain, then drop into Slaughter Gap before climbing onto Blood Mountain in the heart of the Blood Mountain Wilderness. Two miles beyond Blood Mountain, the trail enters Neel Gap and crosses U.S. Highway 19/129. The tops of the mountains are rocky and covered by great amounts of rhododendrons. If you explore the bald knob that covers Bald Mountain, you will likely see three-toothed cinquefoil and silvery whitlowwort in the rocky terrain.

South of Blood Mountain and Neel Gap is the DeSoto Falls Scenic Area accessed by a short spur road off U.S. Highway 19/129. In the heart of gorgeous mountain scenery, this 650-acre area contains at least six beautiful waterfalls as Frogtown Creek drops 600 feet in less than a third of a mile. A trail leads to several of these falls. The streams in the scenic area are lined by dense thickets of rhododendron and mountain laurel. The uppermost of the falls, which drops more than 150 feet, can no longer be reached by a maintained trail.

If you hike the Duncan Ridge National Recreation Trail north of its junction with State Route 180, you will stay on the ridge for a long time, passing such interesting places as Coosa Bald, Buckeye Knob, Bryant Gap, Wildcat Knob, Mulky Gap, Clements Mountain, Fish Gap, Sarvis Gap, and Rhodes Mountain before you turn abruptly southward and eventually connect with the Appalachian Trail on Springer Mountain. If you decide to explore the wonderful coves on the north face of Coosa Bald, you will be treated to nice specimens of yellowwood trees and such interesting understory plants as running strawberry bush, wild ramp, starflower, monkshood, glade fern, Goldie's fern, and intermediate fern. Forest Road 39, more or less paralleling the Duncan Ridge Trail, is a scenic driving route to reach the Cooper Creek Scenic Area. Nice stands of yellow birch grow here. Cooper Creek, which flows through the scenic area, is heavily stocked with trout. The Eyes on Wildlife Trail is a 1.2-mile trail that originates at the western side of the scenic area along Forest Road 236. There is also a mile-long trail along Cooper Creek. The Cooper Creek and Mulky Gap campgrounds are at the western edge of the scenic area.

From Sosebee Cove, State Route 180 turns southward and arrives at 18-acre Lake Winfield Scott and the surrounding recreation area after about five tortuous miles. The recreation area has a campground, picnic area, swimming beach, boat ramp, and a few hiking trails. The lake is stocked with rainbow trout, and fishermen report good luck with largemouth bass and bluegill. For hikers, an easy 0.4-mile trail encircles the lake, and two trails access the Appalachian Trail to the east. Adjacent to the parking area behind the beach, look for a short loop trail that passes through a good stand of a rare shrub known as buffalo nut.

Dockery Lake, a small lake southwest of the DeSoto Falls Scenic Area, may be reached via State Route 60 and then a short spur road, Forest Road 654. Nestled in a nice cove, this 8-acre lake supports a good population of trout. A small campground and the 0.5-mile Lakeshore Trail are found here. If you continue north on State Route 60 past the lake, you will come to the Chestatee Overlook and, just beyond, the Woody Gap Campground located along the Appalachian Trail.

To the west of Dockery Lake is the Springer Mountain National Recreation Area with 3,782-foot Springer Mountain centrally located. The mountain is significant in that it is the southern terminus of the Appalachian Trail.

East of the community of Blue Ridge and south of U.S. Highway 76 is Blue Ridge Lake, featuring Forest Service campgrounds and several boat ramps. Southwest of Blue Ridge Lake is the Rich Mountain Wilderness, with several interesting areas including Wolfpen Mountain and Tickanetley Bald. In the cove on Wolfpen Mountain is a huge silverbell tree as well as uncommon plants in Georgia such as ginger-leaved grass-of-Parnassus and starflower. North of the wilderness area in the Rich Mountain Wildlife Management Area is Horse Cove—worth a visit because of its assemblage of rare plants and one of the largest wild black cherry trees anywhere. Forest Road 295 passes through the cove.

The Cohutta Wilderness, the largest in the Chattahoochee National Forest, lies west of the town of Blue Ridge. The dominant features here are the Cohutta Mountains, part of the Blue Ridge chain, which rise to 4,200 feet at the summit of Big Frog Mountain. There are many trails in the wilderness, almost all of them steep and arduous. Hikers should be aware of flooding after heavy rains because several of the rivers and streams in the wilderness area are often out of their banks. Points of interest in this region include Beach Bottom along Jacks Creek to the north, Tumbling Creek creating the northeastern border, and Cowpen and Bald mountains near the southern end. Several Forest Service roads form boundaries of the wilderness area. One may get a panoramic view of the area from the Cohutta Overlook along State Route 52 between Chatsworth and Ellijay. Just outside the southwest corner of the Cohutta Wilderness and west of Bald Mountain and Little Bald Mountain is Lake Conasauga, where the Forest Service has a campground and a swimming beach. Grassy Mountain Tower Trail, approximately 1.5 miles long, runs from the Lake Conasauga Dam to the old fire tower on the summit of Grassy Mountain.

The westernmost ranger district in the Chattahoochee National Forest is the Armuchee, located between Interstate 75 on its eastern side and State

Route 151 on its western side. The district's northern end is south of Ringgold and west of Dalton, with Summerville at its southwest corner. The area consists of low mountains and long, narrow ridges. Nonetheless, much of the area is rugged, with steep-sided gorges and occasional waterfalls.

Taylor Ridge is a long, very narrow ridge that extends from a few miles east of Lafayette to a few miles southeast of Summerville. The Pinhoiti Trail is on the crest of the southern part of the ridge, as are several Forest Service roads. One of the more popular areas along the ridge is known as The Narrows. The Chattahoochee National Forest maintains the Houston Valley ORV Area at the northern end of the Forest Service land.

To the east of Taylor Ridge and separated from it by a valley are the narrow Johns Mountain and Horn Mountain. The Pinhoiti Trail crosses the northern part of these two mountains. Two popular Forest Service areas in this region are Keown Falls and, 4 miles to the southeast, The Pocket. Keown Falls consists of two waterfalls that drop a total of about 75 feet. A trail leads to an observation deck for a great view of the larger of the two falls. The Pocket, a U-shaped valley in the Horn Mountain area, is of interest because many spring-fed streams arise from it. State Route 156, the Ridge and Valley National Scenic Byway, runs the entire length of this section of the national forest and provides easy access to Keown Falls and The Pocket.

Panther Creek Cove

Mixed mesophytic forests—moist, shaded forests with a mixture of trees, mostly hardwoods—are the dominant forests in the southeastern United States. Some of the best are found in the southern Appalachian Mountains, located in steep, shaded ravines known as *coves*. The ample moisture and rich, fertile soil of the coves support a lush growth of trees, shrubs, wildflowers, and ferns.

A few years ago I ventured into one of these coves on a visit to the mountains of northern Georgia, near the eastern edge of Chattahoochee National Forest. Here a picturesque mountain stream known as Panther Creek snakes its way between alternating ridges and ravines. By hiking the ridge from the summit of Black Mountain toward Panther Creek, I was able to sample the magnificent forest of Panther Creek Cove, one of several coves in the immediate area. The ravines drop off abruptly on either side of the ridge, but by clinging to the standing vegetation, you can reach the cove, which is not easily accessible from other directions.

Oaks, hickories, and pines cover the dry ridge, a solid mass of green from spring until fall, when the brilliant red of the scarlet oak and the golden tones of the black oak contrast with the green needles of the yellow, table mountain, and pitch pines. In some places, a dense undergrowth of shrubby pink rhododendrons and mountain laurel makes hiking difficult, but a few feet down from the ridgetop, these plants disappear as the increasing shade favors the growth of carpets of maidenhair and broad beech fern. Licorice-scented sweet cicely, blue-berried cohosh, and doll's eyes (named for its white, glasslike fruits) poke through the fern layer in great numbers. Above the ferns and wildflowers rises an abundance of flowering dogwoods and pawpaws. High overhead, beech, tulip poplar, white basswood, sweet buckeye, red maple, and some two dozen other tree species form a canopy so dense that little sunlight penetrates to the forest floor, even at midday. Some of the trees are of exceptional size, soaring well above 100 feet.

Here and there the forest floor is littered with the half-decayed, moss-covered trunks of trees struck down by lightning or old age. Occasional boulders provide crevices for wildflowers such as the leathery-leaved hepatica and the dainty bishop's cap, with its snowflake-shaped petals. Walking fern sends its slender-tipped leaves out across the rock surfaces. Deeper into Panther Creek Cove, the vegetation becomes even lusher. The maplelike leaves of the Canada waterleaf seem to be everywhere, and at one place, an acre is solid with the arching, verdant fronds of glade fern.

On my visit to Panther Creek Cove, as I neared the foot of the ravine, just where the cove bottoms out into the flat floodplain adjacent to the creek, I came upon a vine with red-tinted flowers that I couldn't identify at first glance. It was the bay starvine. An uncommon vine related to magnolias, the bay starvine's nearest relatives live in China, Japan, and the Himalayas. It is one of many plants living in the Appalachian coves that have close relatives in eastern Asia. The red maple, for example, has a counterpart in Japan, and the sweet buckeye is closely related to a Himalayan buckeye. Only two kinds of tulip poplars exist in the world: the tall, stately species of the eastern United States and a similar one in central China. Similarly, the white-flowered silverbell, a plant generally restricted to Appalachia, resembles the Macgregor's silverbell of central China.

The list of comparable species in the Appalachians and eastern Asia extends over many genera and includes magnolias, sweet gums, dogwoods, wild hydrangeas, blue cohoshes, horse sugars, May apples, and ginsengs. Ginseng, whose root is highly prized in the Orient, is a story in itself: Two hundred years ago, the first American ship to trade with China sailed to the East with 30 tons of ginseng collected in the wilds of Virginia and Pennsylvania. The parallels between the two regions can also be described in broader terms: the trees that make up the forests are almost all deciduous, the wildflowers usually have stout rootstocks, and the forests develop on mountainsides that have dense shade and rich soils.

Ever since 1790, when Italian botanist Luigi Castiglioni first noted the similarity between the flora of the eastern United States and that of eastern Asia, many diverse theories have been offered to explain the origin of these separated floras. Hui-Lin Li, a botanist at the University of Pennsylvania's Morris Arboretum and one familiar with the floras of both continents, proposes that both are remnants of a great, moist forest that extended over all of the Northern Hemisphere and arctic region during the Tertiary, some 50 million years ago. Geological changes, including mountain uplifting, climatic variations, and glaciation, have altered the flora of many lands in the Northern Hemisphere. But the Appalachians and central China, which occupy the same latitudes, are both geologically very old and have remained relatively stable for the past several hundred million years.

The species of the two areas are not identical, as we would expect given their long isolation. But the visitor exploring Panther Creek Cove may pause to reflect that a similar cove, with magnolias, silverbells, tulip poplars, bay starvines, and May apples, exists somewhere halfway around the world.

Oconee National Forest

SIZE AND LOCATION: Approximately 110,000 acres in the Piedmont of north-central Georgia, between Athens and Macon. Major access routes are Interstate Highway 20; U.S. Highways 23, 129, 278, and 441; and State Routes 11, 12, 15, 16, 18, 24, 44, 83, 212, 213, and 300. District Ranger Station: Eatonton. Forest Supervisor's Office: 1755 Cleveland Highway, Gainesville, GA 30501, www.fs.fed.us/conf.

SPECIAL FACILITIES: Boat ramps; swimming beach.

SPECIAL ATTRACTIONS: Scull Shoals Historical and Archeological Area; Hitichi Experimental Forest; Lake Oconee; Lake Sinclair.

Located on either side of Interstate Highway 20, the Oconee National Forest consists of numerous parcels of land that are mixed with private land in the Piedmont province of Georgia. The Piedmont is the plateaulike region between the mountains to the north and the coastal plain to the southeast. All of the woodlands in the national forest have been cut over in the past, but several of them have redeveloped into attractive that woods are popular with hunters. Man-made reservoirs such as Lake Oconee and Lake Sinclair, along with the Oconee River, also attract fishermen, boaters, and swimmers.

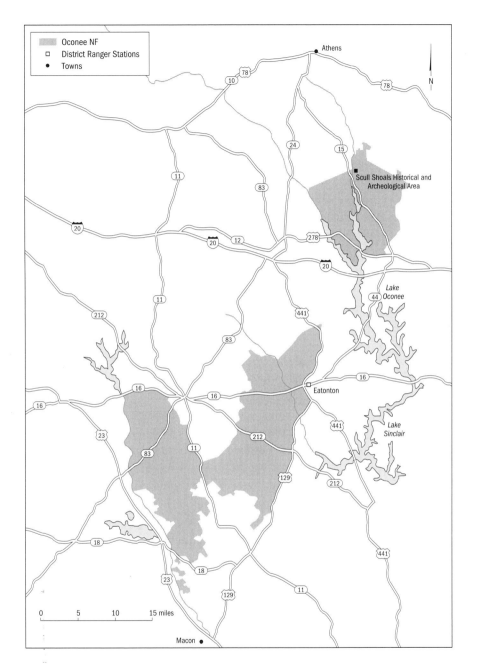

Oconee NF
□ District Ranger Stations
• Towns

Athens

Scull Shoals Historical and
Archeological Area

Lake Oconee

Eatonton

Lake Sinclair

0   5   10   15 miles

Macon

A significant historical and archeological area at Scull Shoals is at the northern tip of the Oconee National Forest.

Eatonton, at the eastern edge of the national forest, is where author Joel Chandler Harris lived when he wrote about Uncle Remus. As you hike near here, you can envision Brer Rabbit and his cohorts scurrying through the thickets. Lake Sinclair is in the vicinity, offering everything from boat ramps to a swimming beach, fishing piers, campsites, and hiking trails. The Oconee River is good for largemouth bass, catfish, bream, and crappie fishing. The Oconee is also a fine canoeing river. The smaller Hillsboro Lake to the west has facilities for camping and boating, too.

The Ocmulgee River on the southwestern side of the Oconee National Forest is a great canoeing river with few obstacles. Largemouth bass and catfish are plentiful in the river. Three hiking trails along the river are routed through hardwood forests. The Wise Creek Trail begins at the Wise Creek Hunt Camp and extends for 4.5 miles. The Ocmulgee River Trail, a 2.5-miler, is where State Route 83 crosses the river. In the Hitichi Experimental Forest is the 2.5-mile Hitichi Nature Trail along Falling Creek; an interpretive brochure is available for this trail.

The Hitichi Experimental Forest Station is an isolated tract south of the Piedmont National Wildlife Refuge. This 4,735-acre forest is on the east bank of the Ocmulgee River in a setting of loblolly pine, shortleaf pine, and hard-wood trees such as tulip poplar and black walnut. Research at the station is geared to loblolly pines for the benefit of nonindustrial forest landowners.

North of Interstate Highway 20 and northwest of the town of Greensboro is Lake Oconee, which extends into the Oconee National Forest. The Forest Service has boat ramps at Redlands and Swords. North of the lake and east of the Oconee River is Scull Shoals, where you may wander around the old foundations and other structures around this old mill town. Also near here are the remains of Fort Clark. Settled in 1794, Scull Shoals was the site of Georgia's first paper mill in 1811 and the first cotton gin and cotton factory in 1834. Fort Clark was built in 1794 to protect the early settlers. Peter Early, one of Georgia's first governors, spent his youth in Scull Shoals. At one time the town had grist mills, sawmills, and a four-story textile mill as well as stores and homes. Five hundred people worked at 4,000 spindles in the mill. During the 1880s, a huge flood started the mill's demise, and by 1920, the last resident of Scull Shoals moved away. A mile-long hike from the Oconee River Campground will bring you to this historical site.

Three miles north of the Oconee River Campground are two Indian mounds dating back to 1250–1500 A.D. The Scull Shoals Experimental Forest is nearby.

NATIONAL FOREST IN ILLINOIS

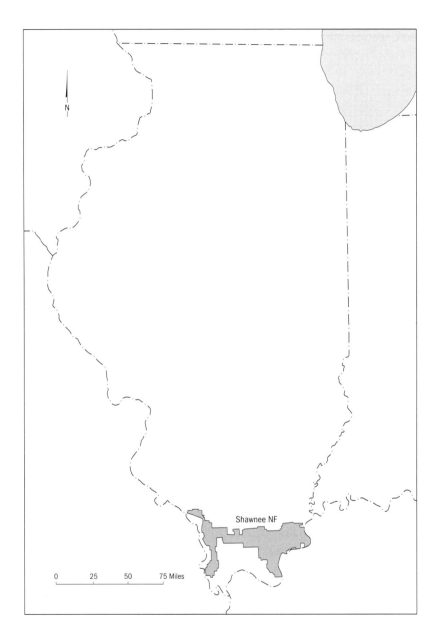

The Shawnee is Illinois's only national forest, beginning a short distance
north of the confluence of the Ohio and Mississippi rivers and extending east
to west from river to river. The Shawnee National Forest is in Region 9 of the
United States Forest Service.

Shawnee National Forest

SIZE AND LOCATION: Approximately 270,000 acres across the southern end of Illinois, extending from the Mississippi River to the Ohio River. Major access routes are Interstate Highways 24 and 57; U.S. Highways 45 and 51; and State Routes 1, 3, 34, 127, 145, 146, and 151. District Ranger Stations: Jonesboro and Vienna. Forest Supervisor's Office: 50 Highway 145 South, Harrisburg, IL 62946, www.fs.fed.us/r9/shawnee.

SPECIAL FACILITIES: Boat ramps; swimming beaches; horseback trails; all-terrain vehicle (ATV) trails.

SPECIAL ATTRACTIONS: Ohio River National Scenic Byway; Millstone Bluff National Register Site; LaRue–Pine Hills National Natural Landmark; Belle Smith Springs National Natural Landmark; Saline Springs National Register Site; Illinois Iron Furnace National Register Site.

WILDERNESS AREAS: Bald Knob (5,863 acres); Bay Creek (2,866 acres); Burden Falls (3,671 acres); Clear Springs (4,730 acres); Garden of the Gods (3,990 acres); Lusk Creek (6,838 acres); Panther Den (774 acres).

The Shawnee National Forest, in the southern tip of Illinois, is very unlike most of the rest of the state, which is flat and, in many spots, covered by corn and soybean fields. Instead, the national forest consists of precipitous scenic cliffs, rocky streams, and heavily wooded forest land that is recovering from intensive lumbering during the first half of the 20th century. The southern edge of the Shawnee National Forest lies just above the confluence of the Mississippi and Ohio rivers and extends northward for nearly 75 miles. It spans the distance west to east between the Mississippi and Ohio rivers. Only one small tract of the forest, along scenic Lusk Creek Canyon, appears to be virgin, the remainder having been harvested time and again by settlers who began replacing the Shawnee Indians in the early 1800s.

The major landform across the Shawnee National Forest is a picturesque range of sandstone cliffs known as the Shawnee Hills. Sometimes referred to as the Ozark Hills, the Shawnee Hills is the preferred name because the landform does not have the same origin as the much older Ozark Mountains of adjacent Missouri. Sheer limestone cliffs occur adjacent to the Mississippi and Ohio rivers on either side of the national forest and at a few isolated areas in its interior. Near the southern edge of the forest, on land once covered by the Atlantic Ocean when the sea extended into extreme southern Illinois, are remnant bald cypress swamps. These are the northernmost cypress swamps in the mid–Mississippi Valley.

Historical features of the past exist in the Shawnee National Forest in the form of an old salt mine along the Saline River, the remains of an iron furnace, and occasional Indian pictographs.

Surprisingly, the Shawnee National Forest is a botanical treasure, with plants from the Gulf Coast migrating northward along the Mississippi River to find a home in the bald cypress swamps, plants from the Appalachian Mountains moving westward to secure their places in deep, forested coves, plants from the north pushing southward through the grinding action of past glaciers, and plants from the Great Plains being blown by strong westerly winds to prairie openings on the tops of some of the limestone cliffs.

The Shawnee National Forest consists of a western district and an eastern district. A part of the western district is near the Mississippi River, with sheer limestone cliffs sometimes rising up to 350 feet above the floodplain below. The remote northwestern corner of the national forest consists of the Kinkaid Hills, which are low, continuous sandstone cliffs along the north side of State Route 3 as far north as the settlement of Cora. At the edge of this region is wild Degognia Canyon where a now-small creek has carved a picturesque ravine filled in the spring with wildflowers.

The focal point of the Kinkaid Hills region is the man-made Kinkaid Lake. The Shawnee National Forest manages the popular Johnson Creek Recreation Area at the northern tip of the lake. There is a boat ramp and swimming beach, and the Kinkaid Hills Hiking Trail begins here. Nearby is the small but scenic Sharp Rock Falls. Toward the eastern side of the lake is the Buttermilk Hill Campground, reached only by boat.

From its junction with State Route 149, State Route 3 takes a southward course through the western edge of the district and provides access to the fine areas along it. The highway stays in the floodplain of the Mississippi River, through bottomland forests and private farms that were once part of a continuous floodplain forest. Although most of the distant bluffs seen from State Route 3 are east of the highway, a large monolith looms up alongside the west side of the highway near the community of Gorham. This oblong cliff, known as Fountain Bluff, is about 4 miles long and up to 1 mile wide; it is named for natural springs, or fountains, on the eastern slopes of the bluff. Most of the exposed rock is sandstone, with small limestone outcrops at the southern end. Botanists have recorded more than 900 ferns and flowering plants from this area. Groups of Indian pictographs appear on some of the sheer west-facing rock exposures. One very rough and narrow rocky road winds through Happy Hollow before climbing steeply to the summit of Fountain Bluff, which provides an outstanding view of the Mississippi River below.

Opposite Fountain Bluff and leading east from State Route 3 is Forest

Road 186. This gravel road in the bottomland forest provides access to the Oakwood Bottoms Greentree Reservoir before ending at Turkey Bayou Campground. The Greentree Reservoir occupies a part of the extensive "bottoms" that possess dense stands of pin oak, swamp chestnut oak, kingnut hickory, and many swamp-loving plants. The Greentree Reservoir is an area where the Forest Service periodically floods and drains the area in an effort to increase the production of acorns by the oaks. With an increase in acorns, the area attracts great quantities of waterfowl. Near the forest road, the Forest Service has constructed a sturdy boardwalk across a part of the bottomland forest that will give the hiker a firsthand look at the plants and animals in this type of habitat. Turkey Bayou is a deep, slow-flowing body of water that empties into the Big Muddy River. On either side of the bayou is an often-inundated swamp forest where you may see water locust, swamp cottonwood, and several rare plants more common to the southeastern United States such as copper iris, parsley hawthorn, the very strange inflated and jointed-stemmed featherfoil (*Hottonia inflata*) of the primrose family, and a

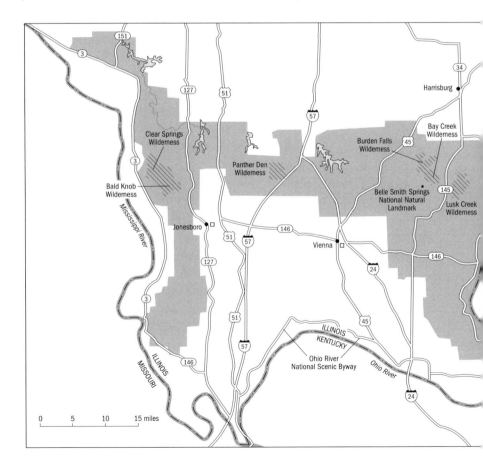

thread-leaved member of the carrot family known only as *Trepocarpus aethusae*. At the end of Forest Road 186, along a part of the bayou, is Turkey Bayou Campground. Since this campground is located between the bayou and the Big Muddy River, it is home to hordes of mosquitoes during the summer. In addition, cottonmouths and timber rattlesnakes (pl. 8) sometimes meander out of the adjacent forests and wetlands.

From Turkey Bayou Campground you may look across the Big Muddy River into the mouth of Little Grand Canyon. The mouth is surrounded by huge rocky cliffs. To reach Little Grand Canyon, however, one has to approach it from the east via forest roads out of Murphysboro or off State Route 127. Little Grand Canyon is one of the most scenic canyons in southern Illinois. It is a box canyon, surrounded on three sides by sandstone cliffs, with the fourth side opening into the Big Muddy River. In addition to its remarkable scenery, Little Grand Canyon is home to countless wildflowers, ferns, and animals, including venomous and other snakes that use the cavities in the cliffs for dens during the winter. Immediately south of Little Grand Canyon is impressive Horseshoe Bluff, and the two areas make up the Little Grand Canyon National Natural Landmark.

From the Turkey Bayou Campground, the Big Muddy River makes five huge arcs, called *bends*, as it literally winds its way through dense bottomland forests before emptying into the Mississippi River south of the historic village of Grand Tower. Just after State Route 3 crosses the Big Muddy River, a levee road along the Big Muddy River branches off to the east. After driving this levee road for about 3 miles, you round a curve and see massive white limestone cliffs looming ahead. These are the spectacular Dolomite limestone cliffs of the Pine Hills, some of them towering 350 feet above the valley. Consisting of ancient Devonian limestone, these sheer west-facing cliffs seem barren of vegetation, but closer observation will reveal colorful lichens plastered against the rocks, and purple cliffbrake fern, baby lip fern, cleft phlox, and Drummond's goldenrod clinging tenaciously in some of the crevices.

At the base of the cliffs, the levee road ends at Forest Road 345. Forest Road 345 to the south follows along the base of the cliffs on the east side of the road, with a standing-water swamp known as LaRue Swamp (pl. 9) on the west side of the road. The swamp is formed from numerous freshwater springs at the base of the cliffs. The summit of the Pine Hills above the road contains shortleaf pine and two species of pink azaleas, among other species, while the LaRue Swamp is filled with wetland plants, many of which have migrated up the Mississippi Embayment from the south. The area is so unique that it was the first federal land to be designated as a national ecological area. It is known today as the LaRue–Pine Hills Research Natural Area. When designated in a ceremony presided over by then–Forest Service chief Dale Robertson, LaRue–Pine Hills became the 250th Research Natural Area in the national forest system. Instead of turning south on Forest Road 345, if you turn north, you soon come to Forest Road 236, which twists its way to the top of the Pine Hills and then back again on the other side.

Adjacent to the north end of the LaRue–Pine Hills area is Clear Springs Wilderness. The clear natural springs that give the wilderness its name flow into clear and rocky Hutchins Creek, one of the few streams in southern Illinois that is not muddy. The wilderness is heavily forested, with steep slopes above narrow ravines. A narrow dirt road follows and sometimes crosses Hutchins Creek, and the road is all that separates Clear Springs Wilderness from Bald Knob Wilderness.

Bald Knob Wilderness has the same kind of terrain and vegetation as Clear Springs Wilderness. Excluded from but surrounded by the wilderness is a treeless summit, Bald Knob, that has a huge cross that is well-known in the area and can be seen for miles. At the south end of the wilderness is a narrow ravine known as Rambarger Hollow. This is the only place in Illinois where the American holly has been found.

State Route 146 crosses the Shawnee National Forest south of the LaRue–Pine Hills Research Natural Area. A series of back roads south of the highway make their way to Atwood Ridge, a sandstone escarpment where one of the few stands of rock chestnut oak in Illinois occurs. Southwest of Atwood Ridge are extensive wetlands along Clear Creek. A natural impoundment known as Upper Bluff Lake is here, and several rare plants live in this wetland complex. The lake may be reached by a paved road off of State Route 3 at Reynoldsville.

The Shawnee National Forest extends south from Upper Bluff Lake to the village of Olive Branch. Although this region has no developed recreation areas, it is a good area to explore on your own in such places as North Ripple Hollow, South Ripple Hollow, along Lingle Creek, and, north of Olive Branch, along Wolf Creek. Exposures along Lingle Creek of a purplish mottled rock that resembles marble are known as calico rock. Along Wolf Creek is the only stand in Illinois of the gorgeous yellowwood tree and the only location in the state of the bigleaf snowbell bush.

South of Murphysboro on the west side of State Route 127, near the community of Pomona, is a symmetrical natural sandstone bridge in the midst of an oak–hickory forest. The bridge spans 90 feet and ranges in width from 6 to 9 feet. The center of the bridge stands 25 feet above the intermittent stream below. A short hike from the picnic area to the bridge is rewarding. Across State Route 127 from Pomona is Clear Lake. The Shawnee National Forest maintains a hiking trail, horseback trail, and boat ramp at the southwest side of the lake. In one of the coves near the lake is the westernmost station for the wildflower black cohosh.

Nearly 150 years ago, kaolin was discovered on a wooded slope about 1 mile south and 2 miles west of the village of Cobden, a short distance east of State Route 127. The kaolin proved profitable, and local potters used it to make many beautiful pieces. When kaolin was found to be of strategic value during World War I, a kaolin mine was established that employed hundreds of men, and the village of Kaolin came into existence. At the height of the excavation, there was a pit about 100 feet deep, but the pit has gradually filled in. The old mine site is now a part of the Shawnee National Forest, and the town has disappeared.

The Ohio River forms part of the boundary of the eastern district of the Shawnee National Forest. The Forest Service maintains two recreation areas on the banks of the river. At the north edge of Golconda is the Ohio River Recreation Area, which includes Rauchfuss Hill and Steamboat Hill, both providing great overlooks of the Ohio River. Between Elizabethtown and Cave-in-Rock is Tower Rock Campground, adjacent to a limestone cliff. There is a boat launch area for entering the Ohio River.

South of the Ohio River Recreation Area along either side of the Massac–Pope county line are a number of seep springs that flow into the adjacent forest, creating boglike wetlands. Many plants unexpected for southern Illinois are in the seep springs, including whorled pogonia, white nut sedge, meadow beauty, little wood orchid, and tubercled orchid. One of the best of these wetlands is in the vicinity of the site of the old Massac Lookout Tower.

North of Elizabethtown is the Illinois Iron Furnace, which was in operation between 1839 and 1883. At its peak, particularly during the Civil War, the furnace produced 9 tons of pig iron each day. The ruins of the blast furnace along Hogthief Creek may be reached by a short, easy trail. Only a well marks the location of the village of Iron Furnace that sprang up around the furnace.

North of the Illinois Iron Furnace National Register Site is the Kaskaskia Experimental Forest. The main thrust of the research there is documenting long-term changes in an old-growth natural area. Several forest types surround the research facility, and visitors to the area will have a chance to see northern red oak, white oak, black oak, black gum, American beech, tulip poplar, sugar maple, post oak, scarlet oak, southern red oak, blackjack oak, several kinds of hickories, and a small stand of rock chestnut oak. The rare deerberry shrub has also been found there.

About 10 miles northeast of the Kaskaskia Experimental Forest is Pounds Hollow Recreation Area, centered around scenic sandstone bluffs with numerous overhangs that served as shelters for Indians and early settlers. If you were to draw a line from Little Grand Canyon on the western side of the national forest to Pounds Hollow, similar bluffs of Pennsylvanian sandstone occur regularly within a few miles on either side of that line. Each of these areas is very similar in geological and biological makeup, having exposed sandstone blufftops that support a vegetation of scrubby post and blackjack oaks, red cedar, winged elm, and farkleberry. At the base of the bluffs are ravines, often called canyons, that are densely shaded by sugar maples, tulip poplars, American beeches, and American elms. On the slopes between the blufftops and the ravine bottoms are forests dominated by northern red oak, black oak, white oak, mockernut hickory, pignut hickory, and shagbark hickory. Under the sandstone overhangs, which are not exposed to sunlight and therefore moist and shady, grows the rare French's shooting star. Although all of these areas have much in common, almost each one has one or more species of plants not found in the others, for some strange reason. From Pounds Hollow westward, the Shawnee National Forest includes Garden of the Gods, Stone Face–Denison Hollow, One Horse Gap, War Bluff, Lusk Creek Canyon, Hayes Creek Canyon, Belle Smith Springs, Trigg Bluff, and Jackson Hollow. All of these highly scenic areas are destinations for hikers, campers, and picnickers.

Some of these, such as Pounds Hollow, have low stone walls near the rim of a bluff with a small opening. It is believed that Indians would usher native bison into these stone wall enclosures, called *pounds*, and then frighten the animals, which would then panic and jump off the cliff to the valley below to their deaths. At Pounds Hollow, the Rim Rock National Recreation Trail makes a circle around the main sandstone bluff. A man-made lake with a swimming beach is also here.

Garden of the Gods is unique in its rock formations that have taken on bizarre shapes. One rock is known as Camel Rock, another as Monkey Face, and many others have fanciful names. A paved hiking trail connects most of these rocks. The Pharaoh Campground is nearby. The wild area surrounding the unusual rocks is the Garden of the Gods Wilderness. Between Pounds Hollow and Garden of the Gods is a high isolated hill known as High Knob that affords a 360-degree view of the surrounding countryside. A picnic area is on the summit.

Stone Face is named for the outline of a bulbous-nosed face of a man at the edge of a sandstone cliff. A trail to the bottom of the face will provide a good view of the formation. On the dry, sandstone blufftop just behind the Stone Face's "forehead' are the federally threatened Mead's milkweed and the usually bog-inhabiting black chokeberry. Stone Face is a research natural area. Denison Hollow Research Natural Area to the north has a fine stand of rock chestnut oak, and in one of the rocky streams in the hollow, there is a colony of the rare heartleaf plantain. Mead's milkweed is also in Denison Hollow, as well as in the nearby Cave Hill Research Natural Area.

One Horse Gap and War Bluff are two of the wildest and most remote areas in the Shawnee National Forest. The trail between the cliffs is wide enough for only one horse to pass through at a time. There is a small lake at One Horse Gap.

Lusk Creek Canyon is arguably the most scenic of the sandstone bluff areas in the Shawnee National Forest. Lusk Creek makes a nearly complete horseshoe turn, with the sides of the creek lined by sheer sandstone cliffs. On rocky ledges above the creek are fan-leaved clubmoss, cinnamon fern, and royal fern. The area around the horseshoe bend has been designated as a National Natural Landmark, and the vast area surrounding the canyon is in the Lusk Creek Wilderness.

Just outside the small village of Eddyville is Hayes Creek Canyon, another typical area with dry, barren sandstone blufftops, oak–hickory slopes, and a shaded canyon below. A small waterfall drops into Double Branch Hole following a rain. On a ledge above Hayes Creek is the only Illinois location for the barren strawberry.

Not far away from Hayes Creek is beautiful Belle Smith Springs National Natural Landmark. Picturesque Bay Creek flows through this area, and a loop trail circles part of the creek. The Forest Service has built a long stone staircase to enable hikers to go more easily from the clifftop to the ravine or vice versa. The ravine has nice specimens of American beech and tulip poplar. Natural springs provide Bay Creek and its tributaries with cool, clear water. The Redbud Campground and Hunting Branch Picnic Area are popular regions in the forest.

North of Belle Smith Springs are Bay Creek Wilderness and Burden Falls Wilderness, separated only by a forest road. Just outside the southwest corner of Bay Creek Wilderness is Teal Pond Campground. The focal point of the Burden Falls Wilderness is the falls, the highest in southern Illinois. After a rain, the water drops over a sandstone ledge for about 40 feet.

In the vicinity but not in the wilderness area is wild Jackson Hollow. Only a primitive trail enters this area, which is the location for the first discovery of filmy fern in Illinois. Adjacent to Jackson Hollow is one of the largest expanses of a near-barren sandstone blufftop known as Trigg Tower Sandstone Barrens Ecological Area. A historic lookout tower stands on the bluff at Trigg.

Not far from the Trigg Lookout Tower along Forest Road 424 is Simpson Township Barrens. Thirty years ago the road passed through a typical-looking oak–hickory forest. It was noticed, however, that big bluestem, purple coneflower, and a few other prairie plants were growing at the periphery of the forest. A local naturalist thought that the prairie plants may have indicated that a more extensive prairie occupied the area in prior years. After the Forest Service cut trees in the forest and followed with a prescribed burn, the oak–hickory forest became transformed into prairielike barrens.

Areas of interest in the Shawnee National Forest south of the sandstone escarpments include Lake Glendale, Millstone Bluff National Register Site, and Grantsburg Swamp. Lake Glendale, a man-made lake, has a swimming beach and campground. Nearby is Millstone Bluff where archaeologists have found relics from earlier Indian habitation. Grantsburg Swamp is a northern remnant of a bald cypress swamp that covered much of the southern tip of Illinois long ago. The swamp is reminiscent of swamps much farther south. Bald cypress, tupelo gum, pumpkin ash, water hickory, and water elm are in the swamp or its vicinity.

At the northern edge of the eastern district of the Shawnee National Forest is the Saline Springs National Register Site. As early as 1778, salt was being mined by European settlers in a small area along the Saline River. Salt-loving plants such as *Atriplex patula*, or spear scale, live in the salty waters near the old mine site.

Although the Shawnee National Forest is divided into a western and an eastern district, one area in the national forest lies more or less in between. This is another of the sandstone bluff–ravine complexes known as Panther Den. The small area around the den is the designated Panther Den Wilderness. On one of the moist sandstone cliffs is the rare American harebell, a plant usually found much farther north. It is believed that this species was pushed southward by the Wisconsinan glacier and remained after the glacier receded.

Belle Smith Springs

At Belle Smith Springs, in the heart of southern Illinois's Shawnee National Forest, one major and several minor springs feed Bay Creek as it flows placidly through a wooded ravine. Huge beech trees, tulip poplars, and sugar maples form a dense, closed canopy throughout the growing season, except in April and early May, when sunlight breaks through to bring out carpets of violets, trilliums, larkspurs, gingers, geraniums, and other wildflowers. These woods are framed by exposed sandstone bluffs, which bear the scars of erosion. House-size boulders have broken off and tumbled down, and at one place, a 30-foot-high natural bridge, carved by a now-dry tributary stream, spans about 125 feet. Surrounding all this is an upland forest dominated by oaks and hickories, part of what ecologist E. Lucy Braun called the western mesophytic (medium moist) forest, which stretches from southwestern Ohio to northern Mississippi.

Lying 20 miles southwest of Harrisburg and 45 miles southeast of my own home in Carbondale, Belle Smith Springs can be reached by several routes. I favor a dirt and gravel road that approaches from the northwest and winds around the property lines of several struggling farms before ending abruptly at a parking lot carved into a densely wooded slope. From the parking lot, a trail meanders slightly downhill through the upland forest of scarlet oak, black oak, northern red oak, shagbark hickory, and pignut hickory. These nut trees provide shelter and ample food for many of the mammals and birds that live there. Then, as the trail approaches the cliff overlooking the ravine, the thin mantle of soil that supports the upland forest disappears, revealing extensive openings of smooth, flat sandstone, known locally as *sandstone pavements*. During the summer, when the daytime air temperature hovers around 100 degrees Fahrenheit, the sandstone is often 20 degrees hotter. Nevertheless, some plants brave the heat and consequent dryness, taking root in the pockets of soil that accumulate in shallow depressions and cracks in the "pavement."

Where the back edge of a sandstone pavement borders the upland woods, stunted blackjack and post oak trees eke out a meager existence, sending an extensive root system through narrow crevices to reach moisture that has seeped under the rock layers. These trees grow very little each year, and the rigorous climate gives them a gnarled appearance. Trees only 20 feet tall may be more than 100 years old. To lessen the loss of water, the thick, leathery leaves have a shiny, heavy layer of wax on their upper surface and, in the case of the post oaks, a dense mat of minute, branched hairs on their undersides.

Shallow pockets of soil provide a habitat for a surprisingly rich variety of flowering plants. Prickly pear cactus, whose 2-inch-wide, waxy yellow flowers provide endless color in early June, is abundant. The thick, succulent stems of this cactus store water for the inevitable droughty conditions of summer. The plant has no leaves; photosynthesis is carried out in the stems. Growing with the cactus is the Illinois agave, a species related to the century plants of the western states. Its wax-coated leaves, which form a rosette at the surface of the sandstone pavement, are thick and leathery and are able to store some reserve water. Flower of an hour is another succulent that survives here. As its name suggests, its bright pink flowers bloom for only 1 hour and then wither.

Wild petunia and goat's rue, whose showy flowers bloom throughout the summer, have a dense covering of hair on their leaves and stems. The hairy leaves of the rushfoil, a 6-inch-tall member of the spurge family, also have flat, brownish scales on their lower surface, which protect the plant from sun reflecting off the sandstone pavement. Several species have minimal leaf surfaces: St. John's wort has very short scalelike leaves along its wiry green stem; while threadleaf sundrops, a delicate-looking member of the evening primrose family, has leaves as slender as threads below its attractive yellow flowers. Pinweed makes it through the heat of summer with very narrow, short leaves but forms larger leaves that persist throughout the winter.

A small plant that sprawls on the sandstone is St. Andrew's cross. The roots of this low-growing species penetrate deep into every available crack in the sandstone pavement. The purple oxalis stores food reserves in underground bulbs. Often protruding downward from the bulb of the purple oxalis is a long, icicle-shaped root that serves as a water storage organ.

Other plants beat the heat and drought by germinating in March, flowering in April, setting seeds in May, and withering by the first of June as the summer approaches. These annuals include tiny bluets, or quaker-ladies; a dwarf species of plantain; and a slender grass known as sixweeks fescue because it completes its life cycle in about 40 days.

Garden of the Gods

On hot, sultry summer days, a blue haze often hangs above the densely wooded hills of the Shawnee National Forest, providing an appropriately unearthly backdrop to a fairyland of rock formations known as the Garden of the Gods (fig. 4). Camel Rock, Monkey Face Rock, Anvil Rock, and Devil's Smokestack are some of the fanciful names visitors have given the shapes that make up this sparsely forested area. The rocks are of sandstone, laid down more than 300 million years ago when the area was washed by an inland sea. A mantle of windblown soil subsequently covered the sandstone, but much of it has worn away, exposing the bedrock to erosion. One rock has been ground down to a perfectly flat surface; it is called Table Rock.

The exposed sandstone erodes very slowly as the iron oxide that cements the sand particles together gradually dissolves. Where water has moved along joints and bedding planes, tiny crevices have expanded into fissures, eventually creating intricately carved boulders, pinnacles, overhanging rock shelters, and even tunnel-like caves. In places, giant blocks of sandstone have separated from 50-foot-high cliffs and slid downslope.

Figure 4. Garden of the Gods, Shawnee National Forest (Illinois).

As a result of a poorly understood weathering process, targetlike patterns up to 3 feet across have appeared on some cliff faces. These patterns are made up of concentric ribbons of rock, known as *liesegang rings*, which project up to 2 inches above the intervening surface. The liesegang rings are saturated with iron oxide from the adjacent sandstone, rendering them relatively resistant to erosion.

Even where the sandstone at Garden of the Gods is not exposed, the soil is usually very shallow. Moisture and nutrients required for plant growth are at a premium, and summer temperatures often soar to more than 100 degrees Fahrenheit. The trees that are able to live on the ridgetops under these conditions grow exceedingly slowly, often taking on a gnarled appearance.

Blackjack oak and post oak are the principal deciduous trees, and the abundant red cedar is the only evergreen. A blueberry-like plant called farkleberry dominates the shrub layer. Beneath the woody plants, vegetation is sparse. Prickly pear, southern Illinois's only cactus, adds a desertlike touch, along with succulent-leaved sedums. A striking thick-leaved species is the

Illinois agave, whose 4-foot-tall flowering stalk bears 2-inch-long green-yellow blossoms in early summer. Ridgetops with this assortment of plants dot southern Illinois between the Mississippi and the Ohio rivers, contrasting with the surrounding forest of oak and hickory. Because of their rocky nature and the general paucity of vegetation, these areas are referred to as *sandstone glades*.

In spite of its generally hostile environment, Garden of the Gods contains another, lusher kind of glade, characterized by a continuous carpet of plants—mostly grasses—and few trees. Some ecologists call this opening in the forest a *sandstone prairie*, because its chief grasses typify some of the prairies in America's heartland. Big bluestem is the most conspicuous of these, growing more than 10 feet high by autumn. Nearly as tall are Indian grass and little bluestem; somewhat smaller are Canadian wild rye and tall dropseed. Several prairie wildflowers that bloom during the summer and autumn are also found among the grasses: Virginia bush clover, bristly sunflower, pineweed (a St. John's wort), goat's rue, and the flowering spurge.

Differing soil conditions apparently account for the contrasts between the glades. Soils beneath the scrub oaks on the sandstone glades are rarely more than 1 inch deep. In comparison, sandstone prairies arise where the soil is several inches deep and gets deeper each year as it incorporates more dead plant material. The sandstone prairie at Garden of the Gods slopes from north to south, dropping 200 feet over a quarter of a mile. A rocky cliff at the south end seems to provide enough protection from the wind to prevent rapid erosion of the soil.

Grantsburg Swamp

Four hundred million years ago, a great sea covered what is now the southeastern United States, extending inland as far as southern Missouri, Illinois, and Indiana. As this ocean retreated, it left behind a vast area of low terrain pocked by depressions, some of which developed into bald cypress swamps. Many such swamps remain in the southern United States to this day, but those in the north, less extensive to begin with, have now mostly disappeared as a result of pioneers' attempts to create agricultural land. Southern Illinois has several associated with the Cache River, an old channel of the Ohio River. One contains some exceptionally massive trees and is the focal point of the new Cypress Creek National Wildlife Refuge. Another is Grantsburg Swamp (formerly known as Bell Pond) in the Shawnee National Forest.

Bisected by Illinois Route 146, Grantsburg Swamp and its surroundings are managed as a 751-acre ecological area. Approaching from the parking area, you pass through a forest of oaks and hickories to the rim of a basin.

From there a wooded slope strewn with small sandstone boulders drops quickly to the swamp. Spring wildflowers that bloom on this forest floor include the white trout lily, dutchman's breeches, Jack in the pulpit, and purple wakerobin. Low sandstone cliffs that border the northern side of the swamp are a favorite roosting and nesting place for black and turkey vultures.

Within the swamp grow bald cypress, tupelo gum, pumpkin ash, water hickory, and swamp cottonwood, woody species restricted to wetland habitats and more commonly found in the southern United States. The lower-growing plants, including blue hydrolea, climbing buckwheat, pink St. John's wort, and swamp milkweed, are also typical southern species. At the inner depths of the swamp is a small heronry. Venomous water moccasins, or cottonmouths, and copperheads are common and may be found on the rocky slopes. Although not aggressive if unprovoked, these snakes command caution.

From mid-April to the first frost in October, mosquitoes are an ever-present nuisance, particularly at dawn, dusk, and on cloudy days. Nevertheless, the aquatic mosquito larvae and airborne adults play a useful role in the food chain, providing a meal for a number of species, including birds and frogs.

Twelve of the 50 species of mosquitoes recorded in Illinois have been found in Grantsburg Swamp. Three species of *Aedes* and three of *Psorophora* overwinter in the egg stage, the first adults usually emerging in April. Nearly all of these species produce more than one generation during the summer. The eggs are laid on moist soil or humus and hatch when the breeding habitat is flooded by torrential rains or rising waters. The adult females (which require a blood meal for the development of their eggs) are generally fierce biters. Most prefer the shade, but one species of *Psorophora* bites voraciously in bright sunlight.

The other six species of mosquitoes overwinter in the adult female stage, usually taking shelter in protected places, including homes. Two species of the genus *Culex* and one of *Culiseta* appear late in April and live in dense clumps of vegetation, particularly grasses and sedges. They seldom can be found during the hottest days of summer but reappear during September and October, until the first signs of cold weather are detected.

Unlike the other mosquitoes that live in Grantsburg Swamp, the tree hole mosquito (*Orthopodomyia signifera*) feeds on birds and does not bite humans. The adults usually sit in and around the cavities of trees, which fill up with rainwater or sap from wounded tissue. They breed in the tree holes and then lay their eggs at the swamp's edge.

Anopheles, the mosquito genus that includes species responsible for transmitting malaria, has a single, innocuous species in Grantsburg Swamp. A difference between this genus and other mosquitoes is that the larval form does not breathe through an air tube extending from its hind end (in most genera, the end of the air tube clings to the water's surface, and the larva hangs head downward). Instead, *Anopheles* larvae lie horizontally just below the water's surface and breathe through an opening on the back of their eighth segment. If disturbed, they may drop to the bottom until danger passes.

The last mosquito in Grantsburg Swamp is *Coquillettidia perturbans*, usually found in patches of cattails. Although the larvae of this species have air tubes, they are modified into sharp, piercing structures that penetrate the submerged roots and stems of cattails and other vascular plants, entering the air chambers of the cells. As a result, the larvae do not have to come to the surface of the water where they would be exposed to many common predators.

LaRue–Pine Hills

Not long ago I guided a group of naturalists from Chicago through an area in southwestern Illinois just 40 miles from where I taught. As we stood before a spring-fed swamp that stretches to the base of 350-foot bluffs, one member of the group exclaimed appreciatively, "It sure doesn't look like Illinois!" No, LaRue–Pine Hills, a 5-mile by 2-mile strip running north-south in the Shawnee National Forest, does not evoke the flat fields of corn and soybeans that dominate the central and northern portions of the state. Instead, one finds limestone cliffs like those in the Missouri Ozarks, swampland that mimics parts of Louisiana, and densely wooded coves reminiscent of the Appalachians. The plant and animal life also reflects a relationship to these peripheral areas.

Several events shaped the character of the region and made it a biological crossroads. The uplifting of the Ozark Mountains, the inward spread of shallow Paleozoic seas, the crushing and crunching of continental glaciation, the strong westerly winds blowing across the prairies and plains, and the development of the Appalachian Range have all left their mark. As a result, LaRue–Pine Hills is the most diverse natural area of its size in the Midwest and possibly in the entire nation. Biologists who have used it as an outdoor laboratory for more than a century have found that it supports a wide selection of Illinois's plant and animal species, including some that are otherwise rare in the state. Unfortunately, certain of them have been nearly extirpated by overzealous collectors. Prodded by local naturalists, the U.S. Forest Service now manages LaRue–Pine Hills as an inviolate research natural area, where plants, animals, and all things natural are not to be disturbed.

The Pine Hills part of the name refers to a pine-laden ridge of chert-bearing limestone that extends along the eastern edge of the area. The sheer, west-facing limestone cliffs, with steep, gravelly talus slopes, result from the same uplifting that some 400 million years ago created the Ozark Mountains to the west. As in adjacent southern Missouri, some of the limestone has dissolved, exposing acidic, orange-red pebbles of chert. Plants adapted to this habitat in Missouri have also found a home in this corner of Illinois. One is the shortleaf pine, whose presence was cause for comment as early as 1860, when Frederick Brendel, an immigrant German physician who lived in Peoria, wrote of it in *Prairie Farmer* magazine. Aside from a few ridges in the Pine Hills, this tree inhabits only one other Illinois location, although it is abundant in the Missouri Ozarks. Growing in the chert beneath the pines are pink azaleas, also confined in Illinois to this extension of the Ozarks. Other species range from colorful wildflowers, such as Boott's goldenrod and bird's-foot violet, to inconspicuous grasses and sedges.

The dry, rocky ridges also support numerous post oaks, blackjack oaks, black hickories, and red cedars, which grow stunted and gnarled in response to the harsh environment. Hikers clambering over the ridges are often surprised to emerge abruptly from the woods into grassy, treeless areas. Because these isolated openings harbor some of the same kinds of plants common to the great prairies to the west, they are called *hill prairies*. In the Pine Hills, they range in size from less than 1 acre to about 2 acres, although farther north on the bluffs along the Mississippi River, hill prairies may extend over several hundred acres. They develop mostly on west-facing slopes, where the hot afternoon sun beats down relentlessly on summer days, causing such dryness that trees cannot grow. Westerly winds, which contribute to drying the slopes, pick up seeds from the prairies and plains and propel them eastward to lodge in these openings. They sprout into grasses, such as big and little bluestems, Indian grass, and the curious-looking sideoats grama, and into showy wildflowers, such as pink prairie phlox, yellow prairie evening primrose, butterfly weed, prairie clovers, and rosinweeds.

The rare eastern wood rat builds its nests of twigs and leaves along the narrow ledges of the limestone cliffs. This animal is a true pack rat, usually incorporating anything it can find into its nest. At some time long past, it probably migrated into the Pine Hills from the Missouri Ozarks. Another phenomenon of the erosion-pocked cliff faces is one 200-foot stretch that is a haven for small evergreen ferns known as spleenworts. No fewer than nine different kinds grow here, more than at any other place in the world. Why there are so many, biologists aren't sure, but in the crevices of chert and limestone, side-by-side combinations of acidic and basic soils vary to an extent rarely repeated in nature.

At the foot of the limestone cliffs and talus slopes lies a vast area of standing water known as LaRue Swamp. Named after a now-abandoned whistle stop along the nearby Missouri Pacific railroad, the swamp is partly fed by clear springs that issue from the base of the cliffs. These water sources harbor the Ozark spring cave fish, a small, blind organism whose eyesight became dispensable as it evolved in the darkness of springheads. As its name implies, this fish is associated with similar springs in the Missouri Ozarks.

After the uplifting of the Ozarks, Paleozoic seas covered the low-lying areas of southern Illinois. When these shallow waters subsided, the extensive swampland left behind became inviting habitat for scores of coastal plain species. Organisms migrated up the Mississippi Embayment, the tongue-shaped trough that extends from near the mouth of the Mississippi River north into southern Illinois. Although the swamps now remaining are but vestiges of a former era, migratory birds that land in the wetlands probably continue to carry plant species along this corridor. Among the species that established themselves in LaRue Swamp were the rusty-flowered copper iris and the sponge plant, a floating, flowering plant whose leaf cells inflate with gas bubbles to keep the heart-shaped leaves buoyed up to the water's surface. Other southern plant species include the tupelo gum and pumpkin ash, two trees whose swollen trunk bases help anchor them in their watery home. The southern rice rat also made its way northward into the swamp, as did large numbers of fish, reptiles, and amphibians, including the banded pygmy sunfish and the green water snake.

Floating on the surface of LaRue Swamp, often in sufficient numbers to obscure the water completely, are duckweeds, the smallest flowering plants in the world. Each one consists of a tiny green body, which may or may not have roots. The tiniest of all is *Wolffia*, with a body less than 1 millimeter in diameter. On rare occasions, duckweeds produce flowers so minute that they can only be seen through a microscope. Botanists recognize about 30 different kinds of duckweeds in the world, including 17 in North America; of these, 11 appear in LaRue Swamp. Occasionally interrupting the continuity of the duckweeds are patches of wine-colored, floating mosquito fern and yellow-flowered bladderwort. The latter's underwater structures bear tiny traplike bladders that suck in minute aquatic animals for nutrition.

On slightly elevated areas adjacent to the swamp, in the coves of the limestone cliffs, the rich, moist land supports stately forests of tulip poplar, American beech, cucumber magnolia, and Ohio buckeye, trees with an Appalachian connection. Several hundred miles to the east, the Appalachian Range is 4,000 to 5,000 feet above sea level. The drainage patterns that have developed from the Appalachians have opened up paths of migration through which

Figure 5. Celandine poppy, Shawnee National Forest (Illinois).

these elements may have come. The cool, moist coves are also perfect for spring wildflowers that are suited to the shade. Freed from competition, they grow so thickly that walking without trampling them is difficult. There is 1 acre of uninterrupted bluebells; another cove is filled with Gleason's white trillium and yellow celandine poppy (fig. 5); elsewhere, Miami mist forms a light blue haze across the forest floor.

Four great glaciers have ground their way south into Illinois, push-ing before them anything that stood in the way. The deepest penetration, made by the Illinoisan glacier some 200,000 years ago, stopped a few miles north of LaRue–Pine Hills, but the groundwork was established for new routes of migration for plants and animals. The bulblet hemlock, a yellow loosestrife known as swamp candles, and such fish as the spottail shiner and the northern hog sucker may have reached LaRue–Pine Hills along this pathway.

The area's diversity seems inexhaustible. Kenneth Weik, a botanist at Lake Forest College and a former student of mine, studied LaRue Swamp for *Tra-chelomonas*, a genus of microscopic, euglena-like algae. His identification of 33 different species of these green spheres—including two new to science—tripled the number previously known in the state. Not to be outdone, Jay McPherson of nearby Southern Illinois University has recorded 50 kinds of stinkbugs from the Pine Hills; possibly the region now deserves the unenvi-able title of "Stinkbug Capital of the World."

Little Grand Canyon

Hidden within the forested hills of southwestern Illinois, just 8 miles from where I grew up, is the Little Grand Canyon, renowned for ferns, wildflow-ers, and an abundance of snakes. This roughly east-west box canyon has sandstone cliffs and steep wooded slopes that rise nearly 400 feet on the east, north, and south. The west end, guarded by two enormous, white-faced escarpments, flares out onto the low floodplain of the Big Muddy River. Huge, cavelike overhangs in the shaded, north-facing cliffs make ideal dens for snakes, including rattlesnakes, copperheads, cottonmouths, and nonpoisonous species, which hibernate by the thousands among the rocks.

Earlier last century the snakes would emerge from their dens in huge numbers on a warm, sunny day in early April and slither westward along the canyon floor. Crossing the 100-foot-wide Big Muddy, they would disperse all over southern Illinois. A local resident who witnessed the river crossing described it to me as a continuous rippling brown wave of snakes from one bank to the other, the procession lasting for several hours. Nothing like that has been seen since the 1930s because the snake population has been much reduced, in part because of overcollecting by greedy scientists seeking choice specimens and also because of the area's increased human population. Little Grand Canyon is now a special management zone in the Shawnee National Forest, and the snakes are legally protected.

Although Little Grand Canyon has a reputation for snakes, they pose little danger. In fact, on my latest visit, I did not encounter a single one. Still, hikers should exercise caution, especially in April and October, when the snakes are more likely to be on the move.

A round-trip hiking trail into the canyon begins on the south side at the summit of Hickory Ridge, once the site of a lookout tower. The trail proceeds from the parking lot down a series of gentle slopes and through a pine plantation to the edge of the north-facing cliff, near the eastern end of the canyon and about 150 feet above the canyon floor. The opposite, south-facing cliff is about 100 feet away. The only plants that survive on its exposed rim are such desert species as prickly pear cactus and Illinois agave, beyond which grows a sparse covering of trees.

From the cliff edge, the trail doubles back, descends slowly through woods, and then follows an 8- to 15-foot-wide drainage. (After a heavy rain, this becomes a fast-flowing waterfall.) Between dripping, fern-laden sandstone walls, the trail drops down a series of natural, 4-foot steps to the canyon floor. At the base of each step is a depression containing cool dripwater; above the water the riser is crowded with layers of leathery liverworts, feathery ferns, and miniature mosses.

At the bottom, a right-hand turn leads quickly to the eastern wall of the canyon, which only the most agile climber can ascend. The main trail meanders westward, more or less paralleling a shallow, clear, gravel-bottomed stream. Ranging from 100 to 300 feet wide, the canyon floor is covered with trees, shrubs, and wildflowers, while ferns prevail on the moist, shaded, north-facing cliffs. Near the mouth of the canyon, cracked brown soil nearly devoid of wildflowers is a sign that this area is frequently inundated by the Big Muddy River.

Just inside the mouth of the canyon, the trail turns south into a narrow side canyon, which is bordered by a sheer vertical cliff on the west and a series of higher and higher small cliffs on the east. At first glance, no exit seems possible, but the trail winds its way upward, finally emerging from the canyon between rock walls dripping with moisture. Once above the canyon, the route continues higher through upland forest and circles back to the trailhead.

The upland forest consists mainly of black oak and pignut hickory, with coralberry and lowbush blueberry as the principal shrubs. Wildflowers, which begin blooming in June, include goat's rue and pencilflower, both members of the pea family; Indian physic, a member of the rose family; and American columbo, a tall, robust member of the gentian family. On the tops of south-facing cliffs, the main trees are post oak and winged elm. Because of the aridity and thin, acidic soil, they may grow only 15 to 20 feet tall in 100 years. Beneath the trees is a shrub layer of aromatic sumac and farkleberry. The sumac leaves resemble those of poison ivy, but they emit a pleasant, spicy odor when crushed. In March and early April, tight clusters of small yellow flowers come out either before the leaves appear or as they are unfolding. Fuzzy, edible red berries adorn the sumac in late summer, while the leaves turn a deep purple or red in the autumn. Farkleberry, with small leathery leaves, is a type of blueberry, but its small fruits are acrid and not suitable for human consumption. Wildflowers in this habitat bloom in April and May, before the heat of summer, and include tiny bluet, succulent widowscross, and hairy wild petunia.

The north-facing cliffs, shaded throughout most of the day, are usually moist or dripping with water. Ferns that find a home in the sandstone crevices and ledges include marginal shield fern, maidenhair spleenwort, and Christmas fern. Wide ledges may support small shadbush trees, whose white flowers bloom before the leaves unfold, and wild hydrangea, a shrub with spherical white flower heads. North-facing slopes support a moist forest of red maple, flowering dogwood, and shagbark hickory.

Canyon floor wildflowers run riot in April and May, when acres of bluebells bloom, along with clusters of squirrel corn, dutchman's breeches, blue cohosh, celandine poppy, and giant white trillium. The tree canopy consists of American beech, sugar maple, and tulip poplar. Below are shrubs of spicebush and bladdernut. Near the mouth of the canyon, approaching the Big Muddy River, a floodplain forest takes over, dominated by sweet gum, river birch, and box elder.

Simpson Township Barrens

Nearly 20 summers ago, while driving along a country road through Shawnee National Forest, just north of the tiny village of Simpson, botanist Max Hutchison of the Nature Conservancy noticed an occasional purple coneflower and butterfly milkweed blooming along the roadside at the edge of the woods. These two plants do not usually grow in forests but in prairies and glades, open habitats with few trees. Later, when Hutchison and Forest Service botanist Lawrence R. Stritch, a former student of mine, walked into the dense upland forest adjacent to these wildflowers, they found that it contained primarily post oaks, chinquapin oaks, and eastern red cedar, with a sparse understory of woodland herbs. The purple coneflower and butterfly milkweed did not grow under the trees.

A check of the first government surveyor's report for the region, from the 1830s, turned up the term *barrens* applied to some parts of the Simpson area. Today, *barrens* refers to a forest area where prairie plants grow beneath a canopy of sparse, usually stunted trees. Botanists have not used that term to describe a natural community in Illinois for the last 110 years. We do not know what these early surveyors were looking at, but it was obviously not the closed forest we see there today. To test whether more open conditions would encourage the growth of purple coneflower, butterfly milkweed, and other species that may have been present in the former barrens, Stritch proposed subjecting a patch of the forest to some drastic management.

In the fall of 1987, 6 acres of the forest were thinned of small-diameter, forest-grown post oaks, sparing the large, misshapen post oak trees, called wolf trees, that had been left previously because of their minimal commercial value. The removal of the small-diameter material allowed ample sunlight to reach the ground layer and the prairie flora to restore itself. Only the healthiest post oaks were left. In all, 13 cords of wood were removed. In the early spring of 1988, a fire was set and allowed to consume the leaf litter that had built up during the past decades. Such fires retard the growth of and eventually kill the shrubs that live beneath the forest trees.

Within weeks, as the new growing season got underway, prairie grasses, including big and little bluestems and Indian grass, began to spring up throughout the burned-over area. Purple coneflowers and butterfly milkweeds increased manyfold, flourishing even beneath the remaining trees. Rosettes of leaves came up everywhere; most were perennials, not mature enough to flower and thus be identified in 1988. The seeds for some of these plants may have been lying dormant in the soil for years; most were probably blown in from the limestone glades that dot the upper slopes in the vicinity.

Botanists eagerly awaited the 1989 growing season to see the transformation continue, and many local groups and individuals, caught up in the project, volunteered their help. The local Boy Scout troop constructed a split-rail fence to mark off the area, the county highway department donated and installed a new culvert to divert drainage water from the road away from the project, inmates from a nearby correctional center cleared the debris that had resulted from the tree cutting, and the Nature Conservancy donated funds to purchase needed equipment. For its part, the Forest Service offered the 13 cords of wood removed from the area as free firewood to the local residents.

The spring of 1989 brought one pleasant surprise after another as the barrens community took shape. Mead's sedge, a prairie species with bluish leaves and slender rhizomes, found a niche and spread out over an area of about 10 square feet. Pinkish purple flowers of the hairy phlox contrasted brilliantly in May with the orange-yellow blossoms of hoary puccoon. Low clumps of bird's-foot violets, not recorded from the Simpson area previously, abounded. In the vicinity of exposed limestone pebbles, colonies of marble-seed appeared for the first time. A 3-foot-tall herb whose leaves and stems are covered by short, stiff, silver hairs, marbleseed produces inch-long, tubular, creamy flowers followed by shiny, pearl-white seeds.

By summer, the barrens contained a riot of colorful wildflowers, including purple milkweed, yellow coneflower, showy rudbeckia (a type of black-eyed Susan), and tuberous Indian plantain. Autumn brought forth the yellows of prairie rosinweeds, sunflowers, and goldenrods, and the purples and whites of numerous kinds of asters, as well as several typical prairie grasses. Prairie rose, a common, low-growing shrub, bore 2-inch-wide pink flowers from June through September.

Across the road from the barrens is another post oak woods with barrens species on the fringes. The soil is more acidic there because of the chunks of sandstone that fall from 50-foot cliffs above the woods. Six acres of this woodland were also burned in 1988, but few trees were removed. The burning has resulted in an increase of shade-tolerant, acid-loving species, including Enslen's blackberry (designated an endangered species by the state), the rare shrubby evening primrose, and several other species that had not been recorded before from the area. Beyond the burned area, where the post oak woods slope down to a rocky creek below this second barrens, heartleaf plantain and the superb lily grow. Heartleaf plantain is one of the rarest plants in the country, while the superb lily is another species designated endangered by Illinois.

Stritch and his colleagues burned both barrens again in the spring of 1990. Each year, more plant species are discovered in the barrens. The total already exceeds 200.

Trigg Tower Sandstone Barrens

While most of Illinois is flat—a one-time prairieland now mostly devoted to corn- and soybean fields—the southern eighth of the state is much more varied. Three-hundred-foot limestone bluffs line the Mississippi River to the west and the Ohio River to the east as they roll toward their confluence at the southern tip of the state. In between is a more or less continuous outcropping of sandstone, known as the Shawnee Ridge, much of which is managed by the Shawnee National Forest. Within it are exposed sandstone surfaces up to several acres in size. I sometimes have referred to them as *glades* or *barrens*, but a better name for them is *sandstone pavements*.

In the Shawnee Ridge zone are 200- to 300-foot-deep ravines, locally referred to as *canyons*, filled with large American beeches, sugar maples, shagbark hickories, and tulip poplars. On the midslopes above the shaded canyon floors, a rather dry forest is dominated by several kinds of oaks, while the driest ridgetops support a forest of stunted, gnarly post oaks, blackjack oaks, winged elms, black hickories, and red cedars. The patches of exposed sandstone, often no more than a few feet wide, lie mainly on west- or southwest-facing ridgetops, where the full heat of the afternoon sun bears down on the rock surface. Trees have not been able to encroach on these spots because conditions are so hot and dry during the growing season.

One of the most extensive and pristine bare areas is Trigg Tower Sandstone Barrens Ecological Area, accessible only by a network of one-lane gravel roads. From a distance, it appears bare of vegetation, but a closer look reveals a variety of colorful lichens cemented to the rock surface, as well as mats of black moss. The lichens, and to a lesser extent the mosses, anchor themselves on the pavement by secreting an acid that etches the rock surface just enough for them to become established. After hundreds or even thousands of years, the lichens and mosses, along with the weather, have gradually eroded the sandstone, creating pockets and crevices filled with thin soil. This soil provides tiny footholds for some flowering plants and ferns that can tolerate the harsh habitat.

One summer afternoon, when the air temperature had soared into the mid-90s, I recorded the temperature at the exposed pavement surface at 120 degrees Fahrenheit (in comparison, the soil of the nearest canyon floor was only 76 degrees). Among the plants I found surviving this desertlike heat was a prickly pear with yellow, waxy flowers. This cactus stores lots of water in its flat, thick stems, while its leaves, reduced to hard spines, offer no surface for water loss.

Other succulent plants on the pavement have fleshy, water-storing leaves. Among them are a 6-inch-tall, pink-flowered sedum known locally as widowscross, a small agave related to the century plants of the western United States, and a delicate plant related to portulacas called flower of an hour (this plant is unrelated to the flower of an hour in the hibiscus family). The 0.75-inch-wide, pink flowers of flower of an hour open about 10 o'clock in the morning and begin to wither an hour later. Beginning in mid-July, one or two flowers open on the plant each day until all the flower buds have opened, usually by late August. The timing is almost accurate enough to set your watch by.

A dense covering of hairs or scales on the leaves of other plants insulates them from the direct rays of the sun and prevents moisture from escaping. Lip fern, a 6-inch fern with a purplish black leaf stalk, has fronds so densely covered with hairs that they hide the surface of the leaves. Another 6-inch plant with hairy leaves is small croton, while an equally diminutive one whose protection consists of flat, silvery scales is rushfoil.

Several species on the sandstone pavement have tiny leaves, sometimes so reduced that they scarcely resemble leaves. These provide less surface for evaporation. Among them are a tiny evening primrose known as threadleaf sundrop, a dwarf St. John's wort called pinweed, and rock clubmoss, which is a dwarf evergreen plant closely related to ferns.

Some of the plants send down unusually long taproots through the rock crevices, where they may reach additional moisture. One of the most striking is pencilflower, a yellow-flowering member of the pea family. While its aboveground stem may be only 6 to 8 inches tall, its taproot may penetrate 2 to 4 feet.

Some plants avoid the hottest, driest days on Trigg Tower Sandstone Barrens by simply making their exit before midsummer. These are annual plants, which sprout early in the year, usually in March, and produce their full set of leaves, flowers, and fruits during April, May, and early June. They then wither away by July, relying on their seeds to repeat the cycle the following year.

NATIONAL FOREST IN INDIANA

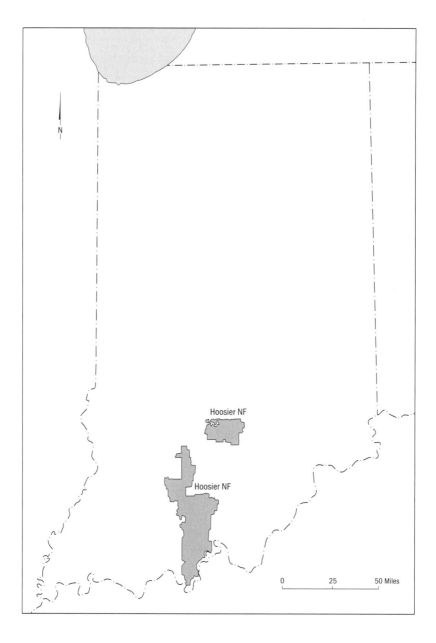

The Hoosier is Indiana's only national forest, located in the central and south-central parts of the state. It is in Region 9 of the United States Forest Service.

Hoosier National Forest

SIZE AND LOCATION: Approximately 200,000 acres in south-central Indiana, extending southward to the Ohio River. Major access routes are Interstate Highway 64; U.S. Highways 50 and 150; and State Routes 37, 46, 60, 62, 64, 66, 70, 135, 145, and 446. District Ranger Stations: Bedford and Tell City. Forest Supervisor's Office: 811 Constitution Avenue, Bedford, IN 47421, www.fs.fed.us/r9/hoosier.

SPECIAL FACILITIES: Boat ramps; swimming beaches; horseback trails; bicycle trails.

SPECIAL ATTRACTIONS: Hemlock Cliffs; Pioneer Mother's Memorial Forest.

WILDERNESS AREA: Charles C. Deam (12,945 acres).

The Hoosier National Forest sprawls across much of south-central Indiana, with many private parcels of land interspersed. The forest is divided into two districts. The smaller northern district lies in two blocks southeast of Bloomington. One is referred to as the Brown County Hills and extends from Bloomington to Nashville, south to Brownstown and west to Bedford. The second block within the Brownstown District lies west of Bedford and south of Shoals in Martin and Orange counties. The southern district stretches from Paoli south to the Ohio River; it contains the upland forests of the Crawford Hills, the limestone region of the Mitchell Karst Plain, and the rocky ravines of the Shawnee Hills.

Occupation of this land by Indians began about 12,000 years ago following the retreat of the last glacier. Bands of hunters and gatherers gave way to Indians who lived in camps and villages and who began to cultivate native plants. Europeans and other adventurers and pioneers began to arrive in the late 1600s, burning and clearing the land, farming, and establishing villages.

Early settlers used the network of rivers for transportation as well as the Buffalo Trace, a wide path made by migrating bison from the Falls of the Ohio near what is now Clarksville, Indiana, to Vincennes. Much of the flat land used for cultivation was cleared of timber in the early 1800s, but the more difficult-to-reach steep hills and ravines weren't logged until after the Civil War. By 1910, most of the forests in what is now the Hoosier National Forest had been harvested.

During the early 1930s, with the land depleted and the settlers abandoning their land to move on, Indiana's governor petitioned the United States Forest Service to establish the national forest and restore the eroding lands in south-central Indiana on February 6, 1935.

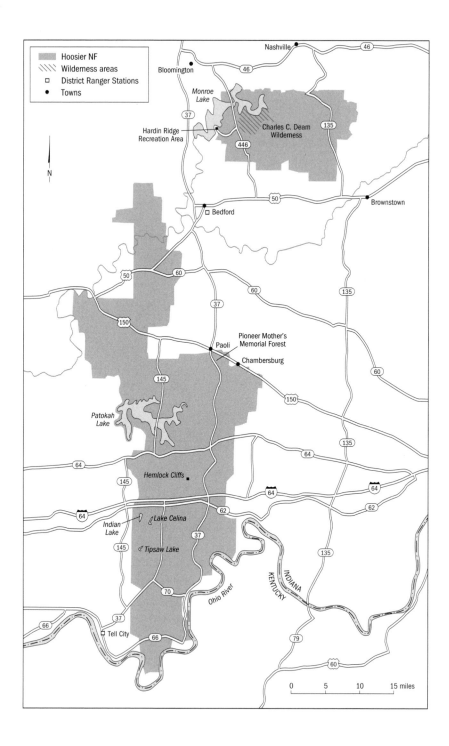

Hoosier NF
Wilderness areas
□ District Ranger Stations
● Towns

Nashville

Bloomington

Monroe Lake

Hardin Ridge Recreation Area

Charles C. Deam Wilderness

Bedford

Brownstown

Paoli

Pioneer Mother's Memorial Forest

Chambersburg

Patokah Lake

Hemlock Cliffs

Indian Lake

Lake Celina

Tipsaw Lake

Ohio River

KENTUCKY

INDIANA

Tell City

0   5   10   15 miles

N

The eastern edge of the northern district of the Hoosier National Forest is occupied by the 10,750-acre Monroe Lake. Although much of the area adjacent to the lake is owned by the state of Indiana, the United States Forest Service does manage the 1,200-acre Hardin Ridge Recreation Area at the southern end of the lake. In addition to a large campground, this area includes a 50-foot-wide concrete boat ramp, a sheltered sand beach with a bathhouse, and an amphitheater. Interpretive programs are given from Memorial Day to Labor Day. The Twin Oaks Interpretive Trail forms a loop behind the amphitheater.

The southeast corner of Monroe Lake is adjacent to the northern edge of the Charles C. Deam Wilderness. Named for Indiana's famous forester and botanist who scoured all of the state for plants during the mid-1900s, the wilderness is divided into two sections, separated only by Tower Ridge Road. This densely forested region of nearly 13,000 acres consists of dry wooded ridges separated by narrow ravines, or hollows, usually with a stream at the bottom of the ravine. Monroe Lake forms the northwest boundary of the wilderness, while the Middle Fork of the Salt River denotes the northern boundary. Several small streams, all tributaries to the Middle Fork, penetrate the wilderness. Brooks Cabin, near the junction of State Route 446 and Tower Ridge Road, is an information center at the edge of the wilderness. The log cabin was erected about 1890 along the Little Blue River in Crawford County. The Forest Service purchased Brooks Cabin in 1992 and reconstructed it at the edge of the Charles C. Deam Wilderness. Brooks Cabin is operated by volunteers, and its hours of operation vary.

The ridges in the wilderness feature several species of upland oaks and hickories, while the ravines contain maples, American beech, tulip poplar, basswood, Ohio buckeye, and other hardwood species. Most of the ravines have a diverse spring wildflower flora of trilliums, wild geranium, wild ginger, violets, true and false Solomon's-seals, and many other species. Wildflowers on the dry ridges bloom during the summer and autumn. Goldenrods and asters are plentiful.

Blackwell Campground, along Tower Ridge Road in the center of the wilderness area, is a popular equestrian campground with limited facilities but easy access for excursions into the wilderness.

Another point of interest in the wilderness along Tower Ridge Road is the historic Hickory Ridge Lookout Tower. It is the last remaining lookout tower in the Hoosier National Forest out of eight that once stood in the national forest. The tower has a cabin at the top that was once manned by a forest ranger during the high-danger fire season. At the tower's base was a two-room house for the ranger and his family, but this structure has been torn down. Trails from the lookout tower wind out through the Charles C. Deam Wilderness, with one of the most popular of the 36.3 miles of trails running along Terrill Ridge.

Southeast of the Charles C. Deam Wilderness is the Hickory Ridge Horse Camp. The camp is at the center of a series of long and short multiple-use (horseback, hiking, and mountain bike) trails that total 46.7 miles.

Two and one-half miles southwest of the village of Houston is the Fork Ridge Hiking Trail. Following the dry ridge, the 3.5-mile trail includes several scenic overlooks across unending forests. At the northern end of the trail are pock-marked rocks known as Hominy Mortar. One mile east of Elkinsville is the 8-mile Nebo Ridge Trail that is suitable for hiking, horseback riding, and mountain biking. The Forest Service warns of an unusually high number of rattlesnakes along this trail.

A small unit of the Hoosier National Forest occurs west of State Route 37 between Bloomington and Paoli. This region contains a mixture of Forest Service and private lands, so you should be aware of this when exploring. Three miles north of the junction of U.S. Highway 50 and State Route 60, on the west side of U.S. Highway 50, is Tincher Lake, which reportedly has good fishing. South of State Route 60, about 1 mile from the settlement of Hindostan, is the Shirley Creek Horse Camp. A 19.4-mile trail, used by horseback riders and mountain bikers, passes through a pleasant forest of hardwood trees.

Another isolated unit of the Hoosier National Forest lies immediately south of Paoli and north of State Route 64. Once again, a lot of private property is within this region. Just south of Paoli is the Pioneer Mother's Memorial Forest, 88 acres of old-growth hardwoods that have been virtually undisturbed since Joseph Cox bought the property in 1816. The 88 acres were designated a research natural area by the United States Forest Service in 1944, along with 165 acres of buffer. Within the old-growth forest are tulip poplars and white ashes up to 150 feet tall, black walnuts as much as 140 feet tall, white oaks up to 130 feet tall, and American beech trees 120 feet tall. Other native trees in the forest include American elm, wild black cherry, black gum, sugar maple, chinquapin oak, northern red oak, black oak, honey locust, Kentucky coffee tree, flowering dogwood, redbud, basswood, sycamore, shortleaf pine, and five kinds of hickories. The area is partially enclosed with a cedar rail fence, and it contains an 0.8-mile hiking trail.

Two and one-half miles southwest of the Pioneer Mother's Memorial Forest is the Youngs Creek Horse Camp and a 10.5-mile hiking, horseback, and mountain biking trail through deeply shaded hardwood forests.

Two miles south of Chambersburg is the Lick Creek Settlement Site where free black families led by Jonathan Lindley, a Quaker, settled sometime prior to 1820. The most visible remains near the site of an African Episcopal church is the Thomas and Roberts family cemetery. The 7.5-mile Lick Creek Multiple Use Trail also winds through the former Lick Creek settlement.

South of Paoli to the Ohio River is the southern district of the Hoosier National Forest. Some of the most spectacular scenery in all of Indiana is in this district.

South of State Route 550 and about 4 miles southwest of the town of Shoals is Plaster Creek, a small creek that flows into the east fork of the White River. The creek is fed partially by seep springs, and the water from the springs has formed a lush wetland in the bottom of a deeply shaded ravine. Numerous ferns and several unusual wildflowers lie in the wetland.

Hemlock Cliffs is a remarkable scenic area located 2.5 miles north of Interstate Highway 64 and 4 miles south of State Route 64. The main geological feature of Hemlock Cliffs is a sandstone box canyon where small waterfalls may be observed after periods of rain. The box canyon contains huge overhangs used as rock shelters, small cavelike openings, and a tumbling stream at the base of the box canyon. Rich vegetation occurs. Because of the position of the box canyon, the climate is cooler than in surrounding areas, making the cliffs conducive for several plants more commonly seen much farther north of here. Hemlocks are the most conspicuous trees, remaining evergreen throughout the year, and mountain laurels are present although unexpected. A lush spring flora features several kinds of toothworts, trilliums, violets, and shooting stars, wild geranium, and several rarities including wintergreen. A 1.2-mile hiking trail, often wet and slippery, leads down into the box canyon; its rock steps are particularly treacherous.

The community of Birdseye lies at the junction of State Routes 64 and 145. Two miles southeast of town is the 11.8-mile Birdseye Trail for hikers, horseback riders, and mountain bikers. The trailhead is just north of small Birdseye Lake.

Two miles south of Interstate Highway 64 and west of State Route 37 are Lake Celina and Indian Lake. These are impoundments of the Middle Fork of the Anderson River, and they are connected by the 16-mile Two Lakes Trail. This trail gives the hiker a good cross section of the geology and plant life of the Hoosier National Forest as it crosses ridgetops, drops down into ravines, and passes rocky outcrops. Shorter loops are available if you don't want to hike the entire length. On the banks of Lake Celina is the Rickenbaugh House that dates back to 1874. The house is built of sandstone blocks that were chiseled from nearby rocky outcrops. The T-shaped house has sandstone chimneys at each of the three ends. Upstairs are three bedrooms, with two parlors downstairs. The kitchen fireplace is 5.5 feet high with a 6-foot-long oven built into the rear. Cupboards are made out of black walnut. The house also served as a post office from 1880 to 1951. The Rickenbaugh Cemetery is about 100 yards west of the house.

Five miles south of Indian and Celina lakes is Tipsaw Lake, and 4 miles farther south is Saddle Lake. All lakes have a boat ramp, Celina and Tipsaw have modern family and group campgrounds, Saddle Lake has a primitive campground, and Tipsaw Lake has a swimming beach. A hiking trail encircles each lake.

About 3 miles south of Sulphur, on either side of State Route 66, are the two Oriole Trails, both suitable for hiking, horseback riding, and mountain biking. Each trail winds through a hardwood forest.

Buzzard Roost Overlook is above a huge bend in the Ohio River, 3 miles east of Mount Pleasant. Farther south along the Ohio River is the town of Derby. One mile north of Derby along U.S. Highway 66 is Mano Point, which has a boat ramp. Northwest of Mano Point are the unique Clover Lick Barrens where sandstone, limestone, and shale are all exposed. Where sandstone predominates, the dominant trees are chestnut oak and mockernut hickory. Where limestone and shale prevail, the dominant trees are usually post oak and blackjack oak with an understory of prairie species. The area is managed to preserve the vegetation and encourage the prairie plants. Derby flint, used by the early Indians to make tools and arrowheads, is also found in the area.

Near the Clover Lick Barrens is Mogan Ridge where the Forest Service has two trails, the 12.3-mile West Trail and the 6.6-mile East Trail. The longer trail is also open to horseback riders and mountain bikers.

At the southern end of the Hoosier National Forest and 1 mile north of State Route 66 is the German Ridge Recreation Area. There is a 3.5-acre German Ridge Lake, a campground, picnic areas, and a 24.1-mile multiple-use trail as well as a 1.5-mile hiking trail. As you hike the trail, winding along cliffs and through rocks, you will see shagbark hickory, sugar maple, American beech, persimmon, hackberry, black walnut, pawpaw, slippery elm, red cedar, butternut, hop hornbeam, flowering dogwood, northern red oak, sweet gum, and white pine.

Clover Lick Barrens

Bordering Kentucky, the southern portion of Indiana's Hoosier National Forest consists of rugged, rocky, hilly terrain with picturesque streams that eventually empty into the Ohio River. An upland forest dominated by white oak, tulip poplar, black gum, and red maple covers the ridges. Sourwood, a medium-size tree common in the Appalachian Mountains, is found here and there; smaller trees and shrubs include ironwood, flowering dogwood, redbud (pl. 10), winged elm, sassafras, sumac, and red cedar. Bittersweet with its yellow fruits, prickly-stemmed greenbriers, and other vines climb the trees, and wildflowers occasionally color the forest floor—phloxes, partridge pea, goldenrod, winged crownbeard, joepyeweed, and shining aster.

While surveying this forest in the spring of 1987, botanist Mike Homoya and biologist Jim Aldrich, both with the Indiana Department of Natural Resources, noticed a number of prairie plants, a good indication that the area may have been more open in earlier times. The prairie plants were growing beneath stunted oaks on very rocky soil near the edge of a ridge, about a 50-foot drop-off above Clover Lick Creek. When Homoya checked the original surveyor's notes for the region, he discovered that on September 25, 1805, the surveyor recorded "a mile of poor barrens and grassy hills, with much flint and a few shrub oaks."

Flint, known as Derby chert, was still scattered over the terrain, but the grassy hills with just a few shrub oaks had apparently given way to a dense growth of stunted, gnarly oaks with a sparse understory of prairie grasses and prairie wildflowers. This conversion from barrens—forest openings with prairie plants and stunted trees—to stunted forest probably came about because fires had been suppressed, particularly after the Hoosier National Forest gained control of the area. Without periodic fires, the oaks became plentiful, providing too much shade for the good growth of prairie species.

Until a few years ago, as the government added worn-out farmland to Hoosier National Forest, open areas were replanted with nonnative pines. Recently, however, efforts have increased to restore some of the habitats present during presettlement times. In 1993, forest personnel decided to bring back the former barrens above Clover Lick Creek. Steve Olson, a botanist and former student of mine, assisted in planning a spring burn of the barrens and adjacent old fields, to suppress the growth of woody plants while encouraging barrens-type species. He kept a detailed diary of how the prescribed burn affected members of the community, including animals as well as plants. Following the 2,200-acre burn, which was set April 6 and took 3 days, Olson observed:

> Before the fire, tiny seed ticks were plentiful as one walked through the area. On the first weekend after the fire, no ticks were found! Other invertebrates, including moths and butterflies such as mourning cloaks and blues, were everywhere. . . . As for vertebrates, frogs and toads seem oblivious to the burn. They were calling all over the place this weekend. Fence lizards and skinks were running around and climbing trees. I noted three casualties over the entire burned area—two box turtles and one earth snake, which got caught in the open. . . .
>
> [April 26] The most striking thing I noticed was the distribution of birds. I saw or heard very few species until I was at the fire line. Then everything seemed to be there: spring migrants, summer residents, and permanent

residents; forest interior species, edge species, and open-land species. Moles seemed to be rather active. Their runways crossed the old roadways and parts of the fire line frequently. Flies were in the air, and moths, skippers, and butterflies were active. . . . The barrens had burned, leaving small grassy refugia. Small encroaching shrubs and trees were top-killed. . . . Wildflowers, grasses, and other green things are coming up all over the place. . . .

[June 28] I spent an entire humid morning wandering through neck-deep grasses and found only one tick and no chiggers. Other more innocuous invertebrates of note include countless grasshoppers and a wide diversity of butterflies and dragonflies. I noticed annual cicadas singing as well. In all cases these invertebrates were most conspicuous in the burned parts of the area. Deer are . . . busy munching. Rabbits are also rather abundant. . . . There are hundreds of plants blooming in the open areas and even some in the burned forest.

The following year, in mid-September, Homoya and Olson introduced me to the area, which they have named the Clover Lick Barrens. Big bluestem, little bluestem, and Indian grass, all grasses that once extended across the vast prairies of central Illinois and Missouri and into Kansas, grew abundantly within a sparse, woody cover of post oak, blackjack oak, and rusty nannyberry. Scattered about was an array of prairie wildflowers—hoary puccoon, yellow gentian, purple coneflower, rosinweed, blazing star, tickseed sunflower, rattlesnakemaster, flowering spurge, narrowleaf aster, and false boneset. Small pieces of limestone, calcareous shale, and flint littered the soil.

As we walked away from the barrens' edge above the stream, we crossed over an abandoned dirt roadway and into a forest of rock chestnut oak. Instead of calcareous shale and limestone, chunks of sandstone were scattered about, a type of rock that weathers to form acidic soil. The plants of this woods were acid loving and completely different from the barrens' plants. Conspicuous among them was lowbush blueberry, which produces tasty fruit during the summer.

The old dirt roadway, which happens to serve as a boundary between the barrens and the sandstone-littered forest, provides a drainage corridor after heavy rains. Its vegetation consists of moisture-loving plants, among them dark green bulrush, inland sea oats, grassleaf goldenrod, and mistflower.

Plaster Creek

Botanist Mike Homoya, a former student of mine, first saw a seep spring—a spring-fed boggy area in the forest that supports the growth of tall royal and cinnamon ferns—in his native southern Illinois. In 1980, when he began work at the Indiana Department of Natural Resources, he was told that seep springs were unknown in that nearby state, but his interest was aroused by a 1982 report that cinnamon ferns were growing 5 feet tall in a "swamp" on private property in southern Indiana. Accompanied by fellow explorer Tom Post, Homoya sought out the spot and found that it was indeed a seep spring, although the area was far from pristine. The two searched further and about 0.5 mile away discovered extensive seep springs along Plaster Creek in Hoosier National Forest.

Recently Homoya invited me to see the Plaster Creek seep springs for myself. We parked at the top of a hill, walked across an abandoned field, and hiked through an upland forest of white oaks, red oaks, and a few hickories. We soon found ourselves on top of a narrow sandstone ridge, which juts westward into the canyon carved by Plaster Creek. For most of its length the ridge is less than 40 feet wide and drops off precipitously on either side for nearly 100 feet to the canyon floor.

The major species of tree on the ridge—in some places, the only one—is the rock chestnut oak, so named because its coarsely toothed leaves are reminiscent of those of the American chestnut. Throughout its range in the eastern and central United States, the rock chestnut oak grows on such rocky exposures, tolerating dry conditions thanks to its deeply penetrating roots. Also growing on the ridge are some blackjack oaks, whose leathery, wax-covered leaves conserve moisture, and scarlet oaks, notable for their brilliant autumn foliage. The shrubs beneath the trees reflect the acidic nature of the underlying sandstone. They include members of the acid-loving heath family—lowbush blueberry, farkleberry, deerberry, and black huckleberry—and clumps of poverty oat grass.

The drier southern edge of the ridge has expanses of sandstone where few trees or other flowering plants have been able to gain a foothold. Such sandstone glades, as they are usually called, are common in adjacent southern Illinois but are unusual in southern Indiana. Several kinds of lichens and mosses etch a home on the exposed sandstone. Slowly but surely these plants break down the rock, which together with their decaying bodies form a thin soil where other plants may take root.

As the peninsula-like ridge narrows to its western end, it drops off gradually enough so that one can scramble down to the valley floor below. Immediately beneath the crest the habitat becomes less arid, and the rock chestnut oaks are replaced by white and red oaks. Perennial herbs help fill out the understory and include such brightly colored wildflowers as firepink, purple oxalis, and false woodland dandelion. Toward the base of the slope, where increasing moisture permits greater growth of trees, American beech, tulip poplar, and sugar maple become dominant. Christmas and marginal shield ferns, two handsome woodland ferns that remain green during the winter, grow in large tufts among the moist rocks that have broken away from the cliff.

Nearby, at the foot of the ridge's northwest-facing cliffs, grow the large ferns that signal the presence of seep springs, which are fed by the rain that falls on the ridge and the upland forest. The water works its way down through narrow fissures in the sandstone to an impenetrable layer; then it seeps horizontally across the sandstone until it emerges near the bottom of the cliffs and drains into the valley. The seep springs community is a long, narrow band, about 5 acres in extent, that includes large patches of sphagnum moss, as well as ferns and seed plants. The area is so wet and boglike that walking into it is dangerous.

The royal ferns and cinnamon ferns are the most striking of the seep spring plants because their fronds are as long as 5 feet. Smaller but conspicuous for their abundance are New York ferns, marsh ferns, and sensitive ferns. Here and there grows hay-scented fern, with its graceful drooping fronds, interrupted fern, and the rare crested shield fern.

Trees such as sweet gum, red maple, green ash, and American elm are common, but none grow to a large size. Windstorms are apt to uproot them, since the soupy substrate provides poor anchorage. Shrubs adapted to soggy, saturated soils are also plentiful: winterberry, Virginia willow, swamp rose, black chokecherry, and buttonbush. Perennial wildflowers add color at different seasons of the year: wild blue iris blooms in the spring, clubspur orchid in early summer, white turtlehead in late summer, and swamp goldenrod in the fall.

The bluff face adjacent to the seep springs is kept very moist by water seepage and the heavy shade provided by the trees. The rocky surface is covered by a dense layer of mosses, through which creep ground pines and partridge berries.

On our way back, Homoya and I decided not to retrace our steps up the slope but to follow the base of the north-facing sandstone cliffs. Along the way I saw still more wildflowers, although I then had to clamber up some fairly slippery rocks to get back to the top of the ridge. After I returned home, I realized the entire Plaster Creek area is so untouched that it lacks nonnative plants—no Japanese honeysuckle, woolly mullein, sweet clover, or meadow fescue. With considerable justification, 72 acres that include the Plaster Creek seep springs have been proposed as a research natural area, a place whose features are to remain undisturbed.

NATIONAL FOREST IN KENTUCKY

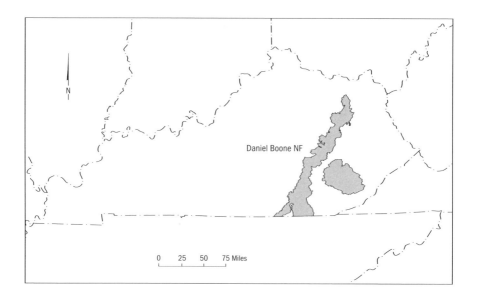

The Daniel Boone National Forest, the only national forest in Kentucky, is mostly in the Cumberland Plateau province of the eastern United States. It is in Region 8 of the United States Forest Service.

Daniel Boone National Forest

SIZE AND LOCATION: 678,000 acres in eastern Kentucky about 50 miles southeast of Lexington. Major access roads are Interstate Highways 64 and 75; U.S. Highways 25, 27, 60, 421, and 460; and State Routes 15, 77, 80, 89, 90, 92, 192, 312, 317, 519, 587, 801, 1045, 1193, 1274, and 1955. District Ranger Stations: London, Morehead, Redbird, Somerset, Stanton, and Stearns. Forest Supervisor's Office: 1700 Bypass Road, Winchester, KY 40391, www .southernregion.fs.fed.us/boone.

SPECIAL FACILITIES: Swimming beaches; boat ramps; horseback trails.

SPECIAL ATTRACTIONS: Red River Gorge Geological Area; Cave Run Lake Recreation Area; Laurel River Lake Recreation Area; Zilpo National Scenic Byway; Pioneer Weapons Hunting Area; Red National Wild and Scenic River; Cumberland National Wild River (state designated); Rockcastle National Wild and Scenic River; numerous natural arches and natural bridges.

WILDERNESS AREAS: Beaver Creek (4,791 acres); Clifty (12,646 acres).

The Cumberland Plateau is a low, mountainous area that extends through eastern Kentucky and central Tennessee into northern Alabama. Much of the Cumberland Plateau in eastern Kentucky is within the Daniel Boone National Forest. In fact, Kentucky's national forest was originally named the Cumberland National Forest. In 1930, the United States Congress officially established what is known today as the Daniel Boone National Forest.

Daniel Boone wandered over much of what is now the national forest that bears his name, and he recorded the region as "Eden . . . a great forest on which stood myriads of trees, some gay with blossoms, others rich with fruits." This extremely beautiful and rugged national forest has low, rounded mountains, steep gorges, miles of sandstone cliffs, and numerous rock formations in the shape of arches, bridges, and overhangs that are known locally as *rock houses*. In addition, some of the wildest, most scenic rivers in the eastern United States are in the Daniel Boone National Forest. Portions of two of them have been designated as National Wild and Scenic Rivers.

The Daniel Boone National Forest trends slightly northeast to southwest. Its northern edge is north of Interstate Highway 64, a few miles north of the city of Morehead. The forest extends all the way to the Tennessee border. This long area consists of four contiguous ranger districts. A fifth district, the Redbird, occurs directly east of the north-south corridor of the national forest.

The most striking features of the Daniel Boone National Forest are the remarkable rock formations in the Red River Gorge Geological Area, located on either side of the Mountain Parkway. The Red River winds through the region, with the eastern part of the river designated as a National Wild and Scenic River. A loop road (State Route 715) from the visitor center on the Mountain Parkway provides access to most of the dramatic features of the

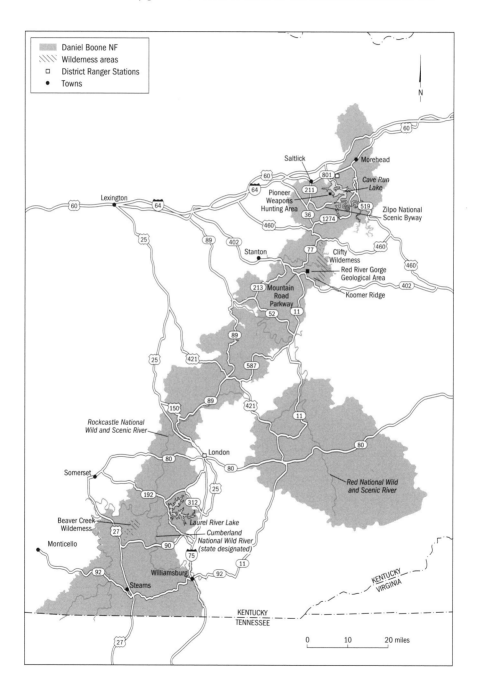

Daniel Boone NF
Wilderness areas
District Ranger Stations
Towns

N

60

Saltlick
Morehead
60
801
Cave Run
Lake
Lexington
64
211
Pioneer
Weapons
Hunting Area
519
Zilpo National
Scenic Byway
60
64
36
1274
460
460
25
89
402
Stanton
77
Clifty
Wilderness
460
Red River Gorge
Geological Area
402
213
Mountain
Road
Parkway
Koomer Ridge
52
11
89
25
421
587
89
421
150
11
Rockcastle National
Wild and Scenic River
80
London
80
Somerset
80
Red National Wild
and Scenic River
192
25
312
Beaver Creek
Wilderness
Laurel River Lake
27
Cumberland
National Wild River
(state designated)
Monticello
90
75
11
92
Williamsburg
92
92
Stearns
KENTUCKY
VIRGINIA
27
KENTUCKY
TENNESSEE
0   10   20 miles

recreation area. The Clifty Wilderness, consisting of 12,646 acres within the Red River Gorge, is the largest designated wilderness area in the Daniel Boone National Forest. Also within the Red River Gorge is the Gladie Historical Site that has a reconstructed log house reminiscent of those that were scattered throughout the forest in the early 1800s.

Before starting the loop road, take Forest Road 24 east from the visitor center to Rock Bridge. From the end of the road is a 1.4-mile hiking trail to this bridge. Picnic facilities are available here. A short distance north of the visitor center, another forest side road leads to Princess Arch. You may hike to this arch on a 9.2-mile trail. From the end of this side road is another short, easy trail to Chimney Top Rock, where you may see a good cross section of the trees of this part of Kentucky.

Returning to the loop road, head north to the Sky Bridge where the imposing arch looms before you. Sky Bridge is the most spectacular of the bridges, spanning 75 feet at a distance 27 feet above the ground. The 0.7-mile trail leads to the top of the bridge. Picnic tables are available as well. The loop road then follows along the north side of the Red River through beautiful scenery. By taking Forest Road 23 from the loop, you will be treated to seeing Double Arch, which has a 1-mile hiking trail. The loop road now becomes State Route 77 and passes through Nada Tunnel. The 900-foot-long tunnel was once a railroad grade for a logging railroad. Turn south onto Forest Road 39 to continue the loop to Gray's Arch, where there is a small picnic area and a 0.3-mile trail to the arch. Before the loop road returns to the visitor center is the Koomer Ridge Recreation Area with a campground, picnic facilities, lookout tower, and a 2.5-mile hiking trail.

Other features within the Red River Gorge Geological Area that may be hiked to are Courthouse Rock, Angel Windows (pl. 11), Rush Ridge, and several attractive streams. West of the Red River Gorge area and reached via Forest Road 227 and 290 is a relic of the past known as the Cottage Iron Furnace.

Between the Red River Gorge Recreation Area and Morehead is the Cave Run Lake Recreation Area, a large forested region surrounding Cave Run Lake. The 8,290-acre Cave Run Lake was built in 1969. From the settlement of Farmers on U.S. Highway 60, a paved road curves around the northern and eastern side of Cave Run Lake. Most of the recreation facilities may be accessed from this highway.

At the northern end of Cave Run Lake, where Scott Creek empties into the lake, is a boat ramp and fishing pier. Nearby is Twin Knobs Campground, with boat ramps, a swimming area, and a hiking trail. From the campground you may drive Forest Road 16 a short distance to Lockegee Rock. There are several boat ramps along the eastern side of Cave Run Lake and along some of its coves that are east and south of the main part of the lake. A campground is at Claylock.

To reach the west side of Cave Run Lake, drive State Route 211 south from Saltlick. When you reach the boundary of the national forest, the 11-mile highway that ends at the Zilpo Campground is the Zilpo National Scenic Byway. After passing Clear Creek Lake, you will come to an old iron blast furnace. The furnace, which produced an average of 3 tons of iron each day, is made of hand-cut limestone stacked 40 feet tall. The furnace has an inside diameter of 10.5 feet. At one time, a small village sprang up in the area around the iron furnace, consisting of a store, school, and church. A nearby picnic area features an experimental 60-foot-long, fiber-reinforced foot bridge that is the first of its kind in the world. By hiking across the bridge, you will be connected to the Sheltowee Trace National Recreation Trail. Farther south along the Zilpo Scenic Byway is a short trail that leads to a rocky area composed of chert particles, a rock used by Indians in making flint.

Between the Clear Creek Picnic Area and the western edge of Cave Run Lake is the extensive Pioneer Weapons Hunting Area. Here hunters using only longbows, crossbows, flintlock rifles, muzzle-loading shotguns, pistols, and percussion cap rifles may hunt for wild turkey, ruffed grouse, deer, rabbit, dove, and quail. Forest Road 918 penetrates part of the area as far as the Tater Knob Fire Tower. This is the last tower in the Daniel Boone National Forest open to the public. Built on solid rock in 1934 by the Civilian Conservation Corps, the tower was reconstructed in 1959 and restored in 1993. The scenic Chestnut Cliffs lie south of the forest road. From the fire tower, the scenic byway descends to the Zilpo Campground on Cave Run Lake.

The area between the Red River Gorge Recreation Area and Interstate Highway 75 to the south is a mixture of Forest Service land and private property. Points of interest in the national forest in this region are a large overhang known as the Great Saltpeter Cave, the S-Tree Campground and Picnic Area (which has a hiking trail), and Turkey Foot Campground.

South of the town of Livingston, reached from Exit 49 on Interstate Highway 75, is the site of the Battle at Camp Wildcat. This area marks the site of the first encounter of regular troops in Kentucky during the Civil War on October 21, 1861. The historic area is reached by a gravel road off Hazel Patch Road.

The part of the Daniel Boone National Forest that lies south and west of Interstate Highway 75 and extends to the Tennessee state line contains some of the wildest and most breathtaking scenery in the eastern United States. Rockcastle River has cut a narrow defile lined by sheer cliffs down the center of this area. The part of the river between State Route 1956 and Bee Rock has been designated as a National Wild and Scenic River. Northwest of Bee Rock and reached via Forest Road 56 is the delightful Daylight Arch.

Bee Rock stands high above an extremely narrow gorge of Rockcastle River, appropriately called the Rockcastle Narrows. The river takes two sharp bends at this point. Bee Rock has a campground, picnic facility, and hiking trail. As you drive along State Routes 192 and 1193 above the east side of Rockcastle River, you will come to the Rockcastle Marina on the river featuring boat ramps, picnic tables, a campground, and a hiking trail.

To the east of Rockcastle River is Laurel River Lake. Between the river and the lake, short but mighty Rock Creek has formed a deep gorge before the creek empties into Rockcastle River. The gorge is so remote that it is very difficult to explore.

State highways completely encircle Laurel River Lake, and the Forest Service maintains campgrounds and boat ramps at several places along the shoreline. West of the dam, where the Laurel River enters the Cumberland River, is the Mouth of Laurel Recreation Area, with a marina, campground, boat ramp, picnic area, and hiking trail.

Across the Cumberland River from the Mouth of Laurel Recreation Area is Savage Campground and Cliffside Hiking Trail that follows the Cumberland River. From this point south to where Jellico Creek enters the Cumberland River is the state-designated Wild and Scenic River portion of the Cumberland River. This stretch of the river below State Route 90 includes the wild Thunderstruck Shoals.

A wild forested area with cliffs and overhangs between the Cumberland River and U.S. Highway 27 is the Beaver Creek Wilderness. The county roads that used to cross the wilderness have been blocked, but signs of human habitation may be seen in the form of a cemetery, a bridge over Beaver Creek, abandoned home sites, coal and saltpeter mine remains, and tunnels. The wilderness is known for its sheer vertical sandstone cliffs. Many of the cliffs form overhangs that Indians and early settlers used as home sites or shelters. These overhangs, known locally as *rock houses*, support some unusual plant species. Lucy Braun's white snakeroot and the whitehair goldenrod are two fall-blooming plants that live only in the protection of the rock houses. Several waterfalls that flow after rains are present in the wilderness. Five trails wind through the wilderness, ranging in length from 0.5 mile to 3.5 miles; most follow Beaver Creek or one of its three main forks. The Alpine Campground along U.S. Highway 27 is a good base for exploring the wilderness. The campground is on the site of the old railroad town of Alpine. A chimney from an old farmhouse still stands in the area. The campground also includes two short trails and a children's playground.

Natural Arch Scenic Area is one of the gems of the Daniel Boone National

Forest. The sandstone arch, 90 feet across with the center standing 50 feet above the ground, was formed by the erosive power of wind, water, and ice. It is surrounded by forested ridges. Several trails are available. The trail to the Natural Arch from the picnic area is 1 mile long. Buffalo Canyon Trail, also accessible from the picnic area, passes through the arch and follows the western side of Buffalo Canyon for 5 miles. The other trails in the Natural Arch Scenic Area begin from State Route 927 west of the Natural Arch Picnic Area. From the Great Gulf Overlook Parking Area, the Panoramic View Trail follows a ridge line for 0.5 mile to a splendid overlook. Gulf Bottom Trail is a 1.7-mile loop that follows the edge of a cliff to a gap and then goes down steep metal stairs, or ladder steps, into the gorge known as the Great Gulf.

A short distance south of Natural Arch is the Yahoo Falls Scenic Area. To reach this area, take U.S. Highway 27 to the Stearns Ranger Station and then a forest road to the edge of the area. Yahoo Falls was at one time in the Daniel Boone National Forest, but when the Big South Fork Recreation Area was created, it fell just within the boundaries of the recreation area and is now managed by the National Park Service. The road to the falls as well as the surrounding land are still a part of the Daniel Boone National Forest.

State Route 92 from Monticello to Williamsburg crosses the southern end of the Daniel Boone National Forest, passing over the Big South Fork of the Cumberland River on the Yacamaw Bridge. Near the bridge are two waterfalls worth seeing: Princess Falls and Lick Creek Falls. South of the Yacamaw Bridge is Koger Arch, reached by a very short but lovely trail off of Forest Road 582.

At the extreme southwestern corner of the Daniel Boone National Forest, just above the Tennessee line, is the Great Meadow Campground. Nearby are Gobblers Arch and Buffalo Arch. Gobblers Arch is nearly 12 feet high and 50 feet across. A good hiking trail starts at Hemlock Grove Picnic Area adjacent to Forest Road 137.

The Sheltowee National Recreation Trail passes through the entire length of the Daniel Boone National Forest from Morehead to the Tennessee state line. From this trail are easy access trails to several of the significant areas of the national forest, including Koger Arch, Gobblers Arch (a few miles south of Koger Arch), and Buffalo Arch.

The Redbird District of the Daniel Boone National Forest is not contiguous with the remainder of the forest. This isolated section is about 10 miles east of the main part of the forest. The surrounding country is characterized by low, steep mountains with narrow stream valleys. In this district is Chestnut Knob, the highest point in the national forest at 2,510 feet. It is located on the Harlan-Leslie county line, approximately 5 miles northwest of Harlan, Kentucky. Just south of the ranger station at the end of Forest Road 1501 is Big Double Creek Picnic Area. Within this district is a rare, beautiful, red-flowering azalea known as the Cumberland azalea. Look for this shrub in bloom in May. The Redbird Crest Trail runs for 64 miles across the district.

Koger Arch

Koger Arch is one of several sandstone arches scattered across the Cumberland Plateau and part of the Daniel Boone National Forest. The arch, caused by millions of years of erosion by water, wind, and ice, is 91 feet across, 54 feet wide, and 18 feet high. A short lovely trail from Forest Road 582 goes through a fine forest before reaching the arch. It is a wonderful way to see some of the typical vegetation of the Cumberland Plateau.

During spring, wildflowers abound along the trail, including trilliums, cucumber-root, golden bellwort, foamflower, Solomon's-seal, false Solomon's-seal, rattlesnake plantain orchid, fire pink, hepatica, dwarf crested iris, wild geranium, rue anemone, wild ginger, and pussy-toes. Flowers that bloom during the summer and autumn along the trail are starry campion, with five white, fringed petals; zigzag goldenrod; agrimony; mountain mint; monarda; skullcap; and white avens. Growing in the shade of the arch are alumroot; a white violet; roundleaf fire pink; columbine; giant chickweed; and Emmons' sedge.

The canopy above the wildflowers has a great diversity of species as well. The more common of the numerous species of trees are northern red oak, white oak, rock chestnut oak, tulip poplar, American beech, mockernut hickory, bitternut hickory, black gum, hemlock, pine, white ash, sassafras, mountain laurel, basswood, red mulberry, sweet gum, sycamore, witch hazel, sugar maple, red maple, sourwood, cucumber magnolia, bigleaf magnolia, flowering dogwood, and serviceberry. Shrubs in the midlayer are farkleberry, lowbush blueberry, white rhododendron, devil's walkingstick, strawberry bush, wild hydrangea, American beautyberry, winged sumac, smooth sumac, and mapleleaf viburnum.

You will need to walk or jump across a very narrow, clear stream at the beginning of the trail. The banks of the stream harbor an incredible number of ferns—Christmas fern, lady fern, common polypody, spinulose wood fern, New York fern, broad beech fern, ebony spleenwort, fragile fern, maidenhair fern, rattlesnake fern, climbing fern, and, in drier areas, bracken fern.

Red River Gorge

Except for the Canyonlands and Arches areas of the Colorado River drainage in Utah, no part of the country boasts as many natural stone arches as Red River Gorge, the focal point of Daniel Boone National Forest. At least

100 arches carved by erosion dot the forest lands. Of modest size, and often obscured by the surrounding foliage, these features look like the abandoned bridges of some overgrown garden. Drama on a greater scale is provided by vertical cliffs that rise as much as 600 feet above the twisting Red River.

The foundation for the rocky gorge was laid 300 million years ago, as shallow seas, swamps, and rivers covered the region a number of times. Sometimes, water deposited a lime-bearing ooze; at other times, a clay-rich mud accumulated or sand washed up along ancient shores and streams. As new layers were added, these sediments were compressed into rock, the lime-rich ooze turning into limestone, the clay-rich mud becoming shale, and the sandy deposits forming sandstone. The earliest strata in most regions consist of limestone, while the sandstone lies above.

After these sedimentary rocks had formed, a twisting of the Earth's crust caused a domelike uplift from southern Ohio through central Kentucky. Following the uplift, the Red River and its tributaries began their erosive action, cutting through the hard sandstone into the softer rocks below. Today the landscape is deeply dissected by valleys rimmed by sandstone bluffs and ridgetops.

According to Kentucky botanist and geologist A. C. McFarland, most of the arches of the Red River Gorge are the remains of sandstone ridgetops. As the softer shale and limestone below the sandstone erode away, large overhangs are left, creating rock houses. Sometimes two rock houses develop back to back under the same ridge. As erosion continues, the two rock houses may meet. At first, only a small opening, called a *lighthouse*, is formed. In time, the lighthouse becomes a window and ultimately the opening of an arch. The most spectacular of the natural arches is Sky Bridge: it measures 75 feet long and 27 feet high.

The vegetation in the Red River Gorge is varied. The ridgetops are dry, supporting a forest dominated by shortleaf, Virginia, and pitch pines; black, scarlet, white, and post oaks; shagbark hickory; and red maple. Downslope, where more moisture is available, beech, tulip poplar, basswood, and sugar maple are the primary trees. Eastern hemlock, a tree more common to the northern states, often grows on north-facing slopes.

The ancestors of most of these trees and other predominant plant species established themselves in the region 50 to 70 million years ago, but some additional species arrived during the last Ice Age (the Pleistocene epoch), which began 1.6 million years ago. When the Ice Age glaciers made their southern advance, stopping only 200 miles north of the Red River Gorge, the climate cooled, apparently permitting northern species to penetrate the area. Even after the glaciers receded and the temperatures moderated, some of these northern elements—such as false lily-of-the-valley, small ginseng, purple fringed orchid, and Canada lily—persisted in sheltered ravines.

Similarly, several species from warmer regions reached the Red River Gorge when the Pleistocene climate was suitable for their spread. Among those usually associated with the southern Appalachian Mountains are pygmy pipes (a plant that lacks chlorophyll), spreading pogonia orchid, and the sweet pepper bush. The Georgia lobelia, a rare southern species, has also been discovered in the gorge's moist woods.

The plant in the Red River Gorge with the most restricted range is the whitehair goldenrod, a species confined to a few of the rock houses. It is found nowhere else in the world, although similar-appearing sandstone overhangs exist in Indiana, Illinois, and Arkansas, as well as in other places in Kentucky. The whitehair goldenrod was discovered in 1942 by E. Lucy Braun, a plant ecologist at the University of Cincinnati who devoted much of her life to analyzing the origin of the eastern deciduous forests. The origin of this particular plant remains a mystery. A less hairy goldenrod called zigzag goldenrod, widespread in the Midwest, sometimes grows in the same rock houses as the whitehair goldenrod. Some botanists have suggested that this plant was the ancestor of the rare species.

The discovery of ancient arrowheads and spearpoints indicates that the Red River Gorge was occupied by humans beginning as early as 10,000 years ago. The deeper rock houses often provided shelter for these early inhabitants. Although the forest now receives thousands of visitors annually, so far most of its natural features have withstood human intrusion. The greatest threat to the region is a plan to build a huge reservoir, authorized by the federal government in 1962 but currently on hold, which may someday be reactivated. If this happens, the area will again be covered by water, as it was millions of years ago.

Rock Creek Gorge

Originating high on a ridge crest of the Cumberland Plateau, about 22 miles southwest of London, Rock Creek drops into a deep, narrow canyon that it began cutting thousands of years ago. Perpendicular cliffs up to 360 feet high line both sides of the creek for 2 miles before the waters empty into Rockcastle River. A virgin stand of eastern hemlock mixed with some deciduous trees is but one of seven habitats that botanist Ralph Thompson, a former student of mine, has identified within the area. This stand survived over the years because the tall cliffs and Rockcastle River made it difficult to get logging equipment into the gorge. The area is now managed as a national natural landmark by the Daniel Boone National Forest.

Access to the floor of the gorge is provided by a series of iron rungs that the Forest Service has embedded in the face of a sandstone cliff. The tall, straight trees that fill the gorge form a cathedral-like canopy that maintains much of its color throughout the year because of the eastern hemlock's short, flat, evergreen needles. Other common large trees are American beech, sugar maple, tulip poplar, and cucumber magnolia. The tree cover is so dense that species that fare poorly in shade never get the opportunity to invade this closed forest community.

The shrub layer is composed of pawpaw, spicebush, mapleleaf viburnum, and wild hydrangea. In places, mountain laurel forms nearly impenetrable thickets. A wide variety of shade-tolerant wildflowers often carpets the canyon floor.

The well-named harbinger of spring, only 4 inches tall, ushers in the flowering season in early March with its small white petals and purple-pollen-forming anthers. Following closely in the procession are bloodroot (fig. 6) and hepatica (the latter identifiable even in winter by its evergreen, three-lobed leaves). By late April, wild geranium, doll's eyes, and dwarf crested iris are in bloom. Wild ginger and little brown jug, both members of the birthwort family, are noteworthy for their dark blossoms that form amid the leaf litter on the forest floor. The showy lily family is well represented by Solomon's-seal, false Solomon's-seal, yellow trout lily, cucumber root, golden bellwort, and two kinds of trilliums.

The sheer cliffs of Rock Creek Gorge are composed mostly of coarse sandstone overlying layers of limestone. Where the softer limestone has worn away, large sandstone overhangs (rock houses) have sometimes developed.

Figure 6. Bloodroot, Daniel Boone National Forest (Kentucky).

Because of the additional shade these provide, only the most shade-tolerant plants survive in them. Most have very thin leaves that enable the chlorophyll-bearing cells to make do with the minimal amount of available light.

Among these plants are various ferns with membrane-thin leaves: mountain spleenwort, maidenhair spleenwort, hay-scented fern, bulblet fern, fragile fern, and filmy fern. Some live in the soil beneath the overhangs; others find a niche in tiny cracks in the face of the wet rock cliff. Filmy fern has the thinnest leaves of all and grows at the very back of the rock houses. Wildflowers that survive and blossom in the rock houses include mountain galax, white wood sorrel, and the scarlet roundleaf catchfly. One plant found only in the rock houses of Kentucky is Lucy Braun's boneset, a white-flowered species named for its discoverer, an authority on the flora of Kentucky and the forests of northeastern North America.

Where Rock Creek enters Rockcastle River at the western end of the gorge, a riparian forest of river birch, sycamore, silver maple, black willow, and sweet gum grows above a shrub layer of buttonbush, silky dogwood, and alder. Understory wildflowers, most of which bloom from late July to October, include cardinal flower, jewelweeds, and beggarticks (pl. 12). Be careful when walking in the forest because of occasional colonies of woodnettle—the acid-bearing hairs on its leaves and stems sting the skin.

Around the rim of Rock Creek Gorge and extending to the crests of higher ridges, more open, dry conditions have favored other kinds of forests. On the driest sites, usually on the north side of the gorge, an oak–pine forest prevails in which the dominant trees are shortleaf pine, Virginia pine, pitch pine, chestnut oak, scarlet oak, white oak, black oak, and red maple. Where a little more moisture is available, the pines are present in smaller numbers, while pignut hickory and mockernut hickory join the oaks. Black gum and sourwood, two trees whose leaves turn brilliant colors in the autumn, are also common in this forest. Unlike the gorge, several areas around the rim of the canyon have been clear-cut in recent years.

NATIONAL FORESTS IN MICHIGAN

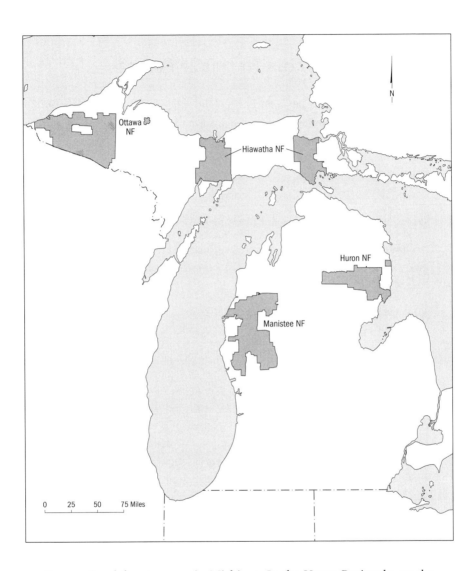

Four national forests occur in Michigan. In the Upper Peninsula are the Ottawa on the western side and the Hiawatha on the eastern side. In the Lower Peninsula are the Huron and Manistee national forests. The Huron and Manistee national forests occupy the central part of the Lower Peninsula of Michigan, extending from Lake Michigan to Lake Huron. For many years, the Huron and Manistee were separate forests, but in order to streamline the forests' administrative offices, they were combined under one administrative unit. The forest supervisor's office of the Huron-Manistee National Forest is at 421 S. Mitchell Street, Cadillac, MI 49601. The national forests in Michigan are in Region 9 of the United States Forest Service.

Hiawatha National Forest

SIZE AND LOCATION: 858,000 acres in the Upper Peninsula of Michigan, divided into western and eastern sections. Major access routes are Interstate Highway 75; U.S. Highway 2 and 41; and State Routes 28, 94, 123, and 134. District Ranger Stations: Manistique, Munising, Rapid River, Sault Ste. Marie, and St. Ignace. Forest Supervisor's Office: 2727 N. Lincoln Road, Escanaba, MI 49289, www.fs.fed.us/r9/hiawatha.

SPECIAL FACILITIES: Boat ramps; swimming beaches.

SPECIAL ATTRACTIONS: Grand Island National Recreation Area; Peninsula Point Lighthouse; Bay Furnace; Point Iroquois Lighthouse; Whitefish Bay National Scenic Byway.

WILDERNESS AREAS: Big Island Lake (5,856 acres); Delirium (11,870 acres); Horseshoe Bay (3,790 acres); Mackinac (12,230 acres); Rock River Canyon (4,640 acres); Round Island (378 acres).

The Hiawatha National Forest has everything you would want in the north woods country of the United States. It has frontage on three Great Lakes— Lake Huron, Lake Michigan, and Lake Superior—and 418 inland lakes. There are 450 miles of rivers and streams. Forests range from dry, jack pine–dominated woods, to majestic cathedral-like woods of white pine and hemlock, to mesic woods of maples and basswood. History is everywhere, showing past cultures of Indian and early European occupation. Lighthouses, iron furnaces, and charcoal kilns remain after decades of use.

The Hiawatha National Forest is divided into two distinct sections. The western section stretches from Escanaba and U.S. Highway 41 at the western edge to State Route 94 and the Lake Superior State Forest at the eastern edge. The eastern section is bordered on its west by the Lake Superior State Forest and on its east by Interstate Highway 75. Both units extend from Lake Superior to Lake Michigan.

Prior to the Civil War, much of the area now occupied by the Hiawatha National Forest was home to the Ojibway Indians, who fished in the streams and hunted game in the forests. This is the area commemorated by Henry

Wadsworth Longfellow in his epic poem "Song of Hiawatha." After the Civil War, European settlers moved in, and the virgin forests of the Upper Peninsula beckoned people from the timber industry. Wood-burning steam engines hauled logs to the sawmills, barges were heavily loaded with timber, and rivers were clogged by logs floating downstream.

When most of the virgin timber had been cut, fires ranged on the debris left behind on the forest floor. By 1928, the land held little commercial value, and the federal government began to purchase the land that President Hoover designated as the Hiawatha National Forest. The forests began to make a comeback; today, forests of white cedars (pl. 13), spruces, balsam firs, jack pines, maples, and birches are found throughout the Hiawatha.

Fishing is still an important activity, with anglers trying their luck to land rainbow trout, brown trout, largemouth bass, smallmouth bass, walleyes, northern pike, bluegill, perch, and crappie. Hunters come to the forest for ruffed grouse, mergansers, Canada geese, mallards, black ducks, buffleheads, goldeneyes, wood ducks, scaup, pintails, blue-winged teal, white-tailed deer, black bear, rabbits, and squirrels. In addition, moose, bobcats, foxes, coyotes, raccoons, weasels, beavers, muskrats, river otter, mink, loons, bald eagles, and sandhill cranes live in the Hiawatha National Forest.

Grand Island is a large island of more than 13,000 acres in Lake Superior less than 1 mile north of Powell Point on the mainland of Michigan's Upper Peninsula, about 2 miles north of Munising. Managed entirely by the Hiawatha National Forest, Grand Island is shaped like South America, but with a large lake at the southeastern corner reached by a narrow peninsula between Trout Bay and Murray Bay. The entire island is a Forest Service recreation area. Echo Lake lies at the center of the island. The Navigation Lighthouse is near North Point at the upper tip of the island. The peninsula connecting the main part of the island with the lobe is primarily marshland. A ferry from Powell Point crosses Grand Harbor Island, transporting passengers to Williams Landing at the southern tip of Grand Island. The island is 8 miles long, 4 miles wide, with 26 miles of shoreline. Sheer rock cliffs rise above sandy beaches. A hiking trail nearly encircles Grand Island. Beginning at Williams Landing, the hiking trail follows the west coast, crossing Echo Lake Creek about halfway up the island. The trail then continues along the western cliffs, eventually curving to the northeast, passing Gull Point, and climbing to North Point. From North Point, the trail crosses North Light Creek to Northeast Point. Four campsites are available along this stretch of the trail from Williams Landing to Northeast Point. At Northeast Point, one trail, suitable for hikers and mountain bikers, follows the cliffs on the eastern side of the island, while a hiking trail stays inland over rolling terrain. These two trails eventually join each other midway down the east side of the

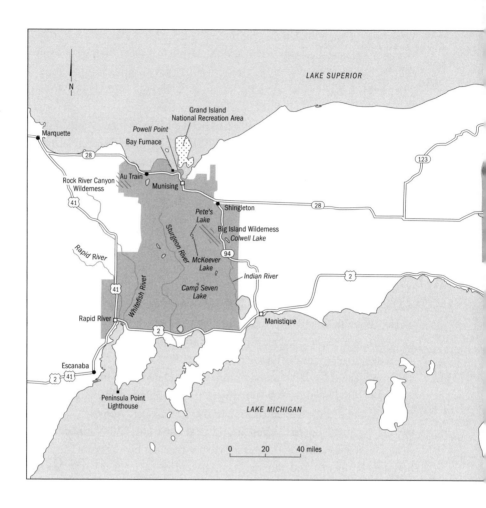

Au Train
Munising
Shingleton
Pete's Lake
Big Island Wilderness
Colwell Lake
McKeever Lake
Indian River
Camp Seven Lake
Manistique
Rapid River
Whitefish River
Sturgeon River
Rapid River
Rock River Canyon Wilderness
Marquette
Powell Point
Bay Furnace
Grand Island National Recreation Area
LAKE SUPERIOR
Escanaba
Peninsula Point Lighthouse
LAKE MICHIGAN
0   20   40 miles
N
28
123
41
28
94
2
41
2
2   41

island. The trail then follows the cliffs, circling around the lower end of Trout Bay and finally making its way into the island's southeast lake. There is also a mountain biking road through the interior of Grand Island, passing to the west of Echo Lake. While on the island, keep a lookout for black bears, bald eagles, and loons. The forests in the interior of the island are dense and dark.

State Route 28 follows the southern end of Lake Superior from Marquette to Powell Point across from Grand Island. This route enters the Hiawatha National Forest at the ruins of historic charcoal kilns just southeast of Spirit Lake. From the ruins you may hike about 2 miles southwest to Silver Falls. One and one-half miles south of Silver Falls is the northern boundary of Rock River Canyon Wilderness. Within this wilderness of 4,640 acres are two canyons separated by a broad, relatively flat ridge. Each canyon is about 150 feet deep, with the Rock River having carved the north canyon and Silver Creek the south canyon. At one point, the Rock River plummets 20 feet

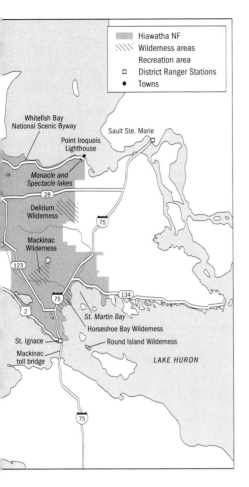

Hiawatha NF
Wilderness areas
Recreation area
□ District Ranger Stations
● Towns

Whitefish Bay
National Scenic Byway
Sault Ste. Marie
Point Iroquois
Lighthouse
Monacle and
Spectacle lakes
28
Delirium
Wilderness
75
Mackinac
Wilderness
123
75
134
2
St. Martin Bay
Horseshoe Bay Wilderness
St. Ignace
Round Island Wilderness
Mackinac
toll bridge
LAKE HURON
75

over a sandstone ledge to a catchpool. Erosion into the sandstone has created caves along the canyon that go back as much as 40 feet. Curtains of ice hang from the caves' ceilings during winter. An old-growth forest of large trees occurs on the northern edge of Rock River Canyon. Oaks, maples, birches, and basswood are the major hardwood trees, with black spruce and pines also present.

State Route 28 continues eastward from the charcoal kiln ruins, staying close to Lake Superior. There is a grand overlook near Coots Pond just before the highway crosses the Au Train River. Attractive Scott Falls is nearby. Au Train Lake is 2 miles south of State Route 28; the Forest Service maintains a campground on the southeast side of the lake. The campground is at the edge of the Au Train Semi-Primitive Motorized Area through which the North Country National Recreation Trail passes. After State Route 28 crosses the Au Train River, it heads east instead of staying next to Lake Superior, coming to the Bay Furnace Recreation Area. This area is centered around the restored Bay Furnace, which dates back to 1870. During the first years of operation, Bay Furnace produced 3,800 tons of pig iron annually, increasing to more than 9,000 tons by 1875. The availability of hardwood trees for charcoal in the vicinity was the primary reason for locating the furnace here. In 1877, a load of charcoal being hauled to the furnace burst into flames, destroying the furnace and most of the small town that had grown up around it. Bay Furnace was never used again after this disaster. The Forest Service maintains a campground near the furnace.

Munising is at the north end of the western unit of the Hiawatha National Forest, while Rapid River and Manistique are near the southwestern and southeastern ends, respectively. In between is a vast area of forests and wetlands. Picturesque streams meander through the region. Forests range from dry, jack pine–dominated woods, to mesic hardwood forests, to wooded swamps dominated by black spruces. Wetlands consist of a huge diversity of types, including open lakes, marshes, bogs, fens, and tamarack and white cedar bogs. The broad range of habitats is reflected by the high diversity of plant species.

Munising is a good place from which to explore the northern end of the western division of the Hiawatha National Forest. Several waterfalls are within 3 miles of Munising, but these are not on national forest land. To explore the western side of this division of the national forest, take State Route 94 that leaves Munising to the southwest. In about 9 miles you will come to Ackerman Lake. From the lake you may pick up the Bay de Noc–Grand Island National Recreation Trail that extends southward for nearly 40 miles, ending 1 mile east of the Whitefish River and 2 miles east of Rapid River. This trail closely follows the old Indian and trappers' foot trail used between Lake Superior and Lake Michigan. It parallels the Whitefish River for most of its length. The trail is fairly easy as it meanders over rolling hills, offering numerous scenic vistas along the way. Near the northern part of the trail in the Ackerman Lake area are mesic hardwood forests dominated by sugar maples and American beeches. As the trail approaches its southern terminus, it plies through sandy plains where jack pines dominate above an extensive shrub zone of lowbush blueberries. You may also make your way through similar habitats by driving Highway HO5 from just east of Ackerman Lake southwest to the Haymeadow Creek Campground, and then by following Forest Road 509 to the southern trailhead of the Bay de Noc–Grand Island Trail. These forest roads travel through several forest types as well as marvelous wetlands. Some of the finest wetlands for species diversity are along the northeastern side of Lake Stella, which has fine bogs, and around the three Eighteenmile Lakes. From the Haymeadow Creek Campground, be sure to take the 0.8-mile loop trail through a hardwood forest to Haymeadow Falls.

South of U.S. Highway 2 and following the east side of Little Bay de Noc is Forest Road 513 that goes all the way to land's end at Peninsula Point. The narrow land mass at the point projects into Lake Michigan. The historic light tower here was built in 1865 to protect wooden ships from the dangerous reefs and shoals that surround this area. Although the adjacent house was destroyed by fire in 1959, the light tower remains. At the point you may picnic, hike along the rocky shoreline, and look for rocks and fossils. Bird-watchers have a field day here; moreover, in September and October, you may watch for monarch butterflies on their southern migration.

About halfway between Rapid River and Peninsula Point, where County Road 513 crosses Squaw Creek, is a unique 64-acre natural area of old-growth trees as much as 250 years old. Here are exceptionally large white pines, hemlocks, red pines, sugar maples, and other hardwood trees.

By driving U.S. Highway 2 eastward from Rapid River, you will have access to several interesting areas both north and south of the highway. Forest Road 2233 to the north goes through marvelous wetlands. Of particular importance is the Ogontz Lake Plains at the end of Forest Road 2233 and the Ramsey Lost Lakes area on the west side of Forest Road 2233 across from Ramsey Lake. Forest Road 503 south from U.S. Highway 2 goes to Ogontz Bay, where the Forest Service has a boat ramp for entering Big Bay de Noc. At the northern tip of Ogontz Bay, between Forest Roads 499 and 503, is another significant natural area. By driving Forest Road 499 south along the Big Bay de Noc shoreline, just past Indian Point and Indian Point Cemetery, you pass through the Nahma Wetlands before crossing the Sturgeon River.

The part of the Hiawatha National Forest south of Munising that lies between the Sturgeon River and State Route 94 contains hundreds of lakes and countless marshes, bogs, and fens. Forest Highway H13 south from Munising goes through the heart of the region, eventually paralleling the Sturgeon River at its southern end. Bruno's Run Hiking Trail is a very interesting loop trail that connects Moccasin Lake, Pete's Lake, Grassy Lake, McKeever Lake, Wedge Lake, Dipper Lake, the Widewater Campground, and Town Lake—all in 7.2 miles. South of Moccasin Lake, the trail crosses an old logging grade and enters the Hemlock Cathedral, a dense stand of very tall, old-growth hemlock. As you hike the Bruno's Run Hiking Trail, you will see many lakes, hike over rolling hills, and observe bogs and streams. A good field guide will help you identify the myriad plants along the trail.

At the Widewater Campground, you may put your canoe in the Indian River for an interesting run. From the campground to 8-Mile Bridge, it is 6.25 miles. In places, the river banks are 30 feet high, and where the river traverses marshland, a feeling of being in a wilderness overcomes you. Fishing for brown trout is excellent in this river. Also at the Widewater Campground is the 1-mile Porcupine Nature Trail that follows the shoreline of Fish Lake. At Pete's Lake Campground, a 2-mile nature trail is worth taking.

Three miles east of Pete's Lake is Big Island Lake Wilderness. Forest Road 2254 follows the west side of the wilderness. Many lakes dot the area, including the larger Big Lake at the northern end and Byers Lake near the southern end. Fishing is at its finest in this wilderness area, with ample opportunities to catch muskellunge, northern pike, bass, trout, yellow perch, bluegill, pumpkinseed, and sunfish. If you look skyward, you will probably observe bald eagles, hawks, and ospreys. Grouse, loons, herons, and sandhill cranes are also here. Mammals in the wilderness include black bear, red fox, porcupine, coyote, white-tailed deer, beaver, river otter, raccoon, mink, red squirrel, gray squirrel, bobcat, striped skunk, woodchuck, and snowshoe hare.

North of Big Island Lake Wilderness is a vast area of wetlands, including Scotts Marsh and Shingleton Bog. The latter is actually one of the best examples of a fen in the Upper Peninsula.

Driving south on State Route 94 from Shingleton, you will pass Colwell Lake and Indian River, both with campgrounds. The area around the lake includes a 1-mile nature trail.

Camp Seven Lake and Campground is several miles south of Big Island Lake Wilderness, reached by a short spur road north of Forest Highway 442. Camp Seven Lake is named for the seventh logging camp established by the VanWinkle Lumber Company. A 1-mile nature trail encircles nearby VanWinkle Lake. Two miles east of Camp Seven Lake is Wolf Lake, a small, shallow lake with a distinct zonation of vegetation around it.

The eastern section of the Hiawatha National Forest includes frontage on Lake Superior, Lake Michigan, and Lake Huron; Round Island and Government Island in Lake Huron; four wilderness areas; some large inland lakes; beautiful stretches of the fast-moving Tahquamenon River and Carp River; and many significant tamarack bogs and cedar swamps.

To experience the Hiawatha National Forest along Lake Superior, drive the Whitefish Bay National Scenic Byway, a 27-mile length of highway along Whitefish Bay. The scenic byway is also Forest Highway 42. The eastern end of the byway is near Monocle Lake and the Point Iroquois Lighthouse. Monocle Lake has a nice campground and a swimming beach. A short hike takes you from Monocle Lake to Spectacle Lake. At the tip of land north of Monocle Lake is the Point Iroquois Lighthouse with a light tower and a two-story brick residence in use since 1870. A trail leads down to the lakeshore from the light tower. All along the Whitefish National Scenic Byway, which never strays far from Lake Superior, are splendid views of Whitefish Bay. Where the scenic byway dips south around Pendills Bay, the Forest Service has frontage on the bay as far as Salt Point and then again from Naomikons Point to the western terminus of the scenic byway at State Route 123. Many places along Whitefish Bay feature broad, sandy beaches. There are a picnic area at Big Pine and a campground at Bay View, in addition to several pullouts for scenic vistas.

The Hiawatha National Forest's eastern section with frontage on Lake Michigan extends from large Brevoort Lake to the St. Ignace Ranger Station. U.S. Highway 2 follows the lake's shoreline, with Lake Michigan Campground along the highway and Brevoort Lake Campground a mile inland. The Ridge Nature Trail is a 0.5-mile loop interpretive trail at the Brevoort Lake Campground across a high, wooded sand dune area. Along the trail you will see pines, hemlocks, quaking aspens, paper birches, and northern red oaks. As the trail climbs onto the wooded dunes, you will encounter jack pines. One point offers a scintillating view of Lake Michigan. As you return to your starting point of the trail, you will pass red maples and American beeches.

Continuing south along U.S. Highway 2 toward the St. Ignace Ranger Station, you will pass through the vast marsh along Pointe aux Chenes Bay. The 0.6-mile Juniper Nature Trail, located on a low sand ridge, leads into the marsh.

U.S. Highway 2 comes to the town of St. Ignace at the northern end of the Mackinac Toll Bridge. By following highway H-63 that parallels Interstate Highway 75 northward, you will be along the shore of Lake Huron. The road curves around the edge of the Horseshoe Bay Wilderness. Beginning only 4 miles north of St. Ignace, the wilderness area contains 7 miles of undeveloped lake frontage on Horseshoe Bay and St. Martin Bay. The land here was exposed when the water in Lake Huron receded. The area contains low, forested ridges separated by narrow, shallow swamps. On the ridges are dense growths of balsam fir and white cedar. Among the plants on the shoreline is the rare Houghton's goldenrod.

Outside the southern end of the wilderness is Foley Creek Recreation Area with a campground and a 2.5-mile loop trail to a sand beach on Horseshoe Bay. Two miles north of the upper end of the wilderness, the Carp River enters St. Martin Bay. The Carp River flows across the Hiawatha National Forest from Rock Rapids on the west to St. Martin Bay on the east. Although all of the forests along the Carp River were logged during the early 1900s, the secondary forests that have developed during the last century contain many nice trees. The Carp River passes through the center of the Mackinac Wilderness. Near the northeastern quarter of the wilderness are good stands of northern red oak, basswood, and sugar maple, with occasional patches of quaking aspen and birches. Low ridges south of the river are covered with quaking aspen, birches, balsam fir, and spruces. In low areas between the ridges are wooded swamps with white cedar and black spruce dominant. Bogs and sedge meadows are scattered throughout the region.

Four miles southeast of the Mackinac Wilderness is one of the finest wooded swamps in the Hiawatha National Forest: Summerby Swamp. State Route 123 passes through the heart of the swamp.

State Route 134 follows rather closely to the shoreline of St. Martin Bay. At the tip of St. Martin Peninsula is an interesting low area surrounding Paquin Lake. County Road 224, which passes through a fine wetland, goes to the lake.

Lake Huron has several islands a few miles offshore, two of which are managed by the United States Forest Service. Round Island, between Mackinac and Bois Blanc islands, was inhabited by Indians until 1762. Most of its 378 acres make up the designated Round Island Wilderness. A lighthouse built in 1873 is situated on a spit of land on the northwest corner of the island, just outside the wilderness boundary. A 75-foot-tall steep cliff of exposed dolomite is on the northeast side of Round Island. Two sandy beaches are on the island, one on the northeastern side and one on the southwestern side. Maples, birches, and northern red oak occur on the higher ground near the center of the island, while the lower elevations support white cedar, balsam fir, black spruce, white pine, and white birch. Two rare plants known from Round Island are the dwarf Great Lakes iris and the fairy slipper orchid. Squirrels, red foxes, raccoons, and white-tailed deer roam the region. Ferries from St. Ignace and Mackinaw City go to the island.

Government Island is one of the beautiful Les Cheneaux Islands located just south of Cedarville from the mainland. The island is uninhabited by humans and has been designated a Semi-Primitive Non-Motorized Area. Private boats may dock at one of two primitive picnic areas on the sheltered side of the island. From 1874 to 1939, the 214-acre island served as a U.S. Coast Guard Station. Woods in the interior of the island are carpeted by bunchberry and bearberry, with an abundance of twinflowers, trilliums, pyrolas, clubmosses, and many other species. The forest is dominated by balsam fir and white cedar. The northern and eastern shores feature rocky beaches.

North of Highway H-40 and south of State Route 28 is the very swampy and boggy Delirium Wilderness. The small Delirium Pond is near the western edge of the wilderness, and Sylvester Creek crosses much of the area. A few hiking trails lead to the interior. Eight miles west of the Delirium Wilderness are the Betchler Lakes and a significant tamarack bog. The East Branch of the Tahquamenon River, north of the Betchler tamarack bog, is a beautiful stream. State Route 123 crosses the river. The Diamond Hill Lookout is immediately south of State Route 28, a short distance before the community of Raco.

Lake Lusters

Thousands of lakes of all sizes dot the Upper Peninsula of Michigan. Some are quite muddy; others are fairly clear, with an organic bottom built up from plant and animal remains. The clearest lakes, however, have a sandy bottom. Among these is Wolf Lake, which lies in the heart of Hiawatha National Forest. An important feature of this Wolf Lake (there are other Wolf Lakes in Michigan) is the presence of various plant species typical of the Atlantic Coastal Plain of the United States.

I visited Wolf Lake in early August, when I knew most of the plants would be flowering or fruiting (grasses, sedges, and rushes are otherwise hard to identify). Forest Road 2696, the closest access to the lake, is a narrow cut through a forest of jack pines, scattered red pines, and such broad-leaved trees as red maple, quaking aspen, and white birch. Through the trees, I could see water gleaming in the middle of a 30-acre opening.

The forest was bordered with shrubs, including two species of *Spiraea*—one with pink clusters of flowers called steeplebush or hardhack, and a white one known as meadowsweet. Beyond this transitional zone, the terrain sloped imperceptibly toward the lake. In the early spring, as a result of snowmelt and heavy precipitation, water nearly covers the open area. As the weeks pass, the water gradually recedes, and the lake is ringed with vegetation. At the outer margins, where the soil is first exposed, tall grasses, sedges, and wildflowers predominate. Smaller plants that flower and fruit more rapidly lay claim to the inner circles.

I first passed through a zone of twig rush, 2-foot-tall plants with reddish brown clusters of flowers. I then entered a green band, perhaps 20 feet wide, containing a mixture of plants 6 to 24 inches high. Ringing the lake next was a bright orange-red swath about 15 feet wide, with plants no more than 6 inches tall. The soil was very soggy, covered here and there with a film of water. The orange-red color was due to an abundance of fruit-bearing bulblet rush. At the edge of the lake, where perhaps an inch of water covered the sand, grew a carpet of hatpins, 2-inch-tall plants aptly named for their tiny spherical heads of white flowers. Farther in, the lake was still shallow enough for me to wade across. Peering through the clear water, I could see submerged plants growing in the sand.

Some biologists consider Wolf Lake a Coastal Plain lake because the plants growing in and around it include species mainly found in the Atlantic Coastal Plain east of the Appalachians. Examples are twig rush, water hyssop, Tuckerman's panic grass, and hatpins. Naturalist-writer Donald Culross Peattie called attention to such coastal plants in the Great Lakes region in 1922. Subsequent studies found coastal plants most prominent in the southeastern Georgian Bay region of Ontario, in the sandy plains near Lake Michigan in southwestern Michigan and northern Indiana, and in sand deposits in Wisconsin. Lesser concentrations appear near Lake Erie, in north-central Illinois, and in the Upper Peninsula of Michigan, where Wolf Lake is located.

Botanist Anton Reznicek has found that the Coastal Plain plants live primarily in sandy, soft-water ponds and small lakes with fluctuating water levels. The plants are common during years when the water level is low. When the level remains high, they survive primarily as seeds in the soil underwater.

Peattie suggested that the Coastal Plain plants gradually spread to the Great Lakes along the shores of lakes and streams that formed when the last Ice Age glaciers melted, about 10,000 years ago. These would not have provided an uninterrupted path for migration, however. Another suggestion has been that the plants spread by random, long-distance dispersal, probably by birds. All the plants have succeeded in migrating to the same far-inland areas, however—a coincidence that is hard to explain if they were dispersed randomly. Reznicek takes an intermediate view, suggesting that the coastal plants migrated into the Great Lakes regions along postglacial drainages, taking short to medium jumps to suitable habitats along the way, probably with the aid of birds.

The outer zone around the lake consists mostly of a green swath dominated by *Euthamia remota*, a grassleaf goldenrod. Named for their very narrow leaves, most grassleaf goldenrods grow in moister areas than other goldenrods, and their clusters of flowers are usually flat topped rather than elongated. Other plants include a pinkish-flowered mint known as water hyssop; a tiny white-flowered mint called water horehound; and a small, round-fruited grass called Tuckerman's panic grass. The twig rush that rings this green swath is actually not a rush but a sedge, *Cladium mariscoides*.

The inner zone receives its orange-red color from the fruit crowding the stems of bulblet rush. This is one way the rush propagates. But many of this plant's flowers, instead of forming pollen grains and ovules (immature seeds), develop into tufts of tightly rolled plant tissue, called *bulblets*, which fall to the ground and sprout directly into new plants. Among the small plants growing inconspicuously alongside the bulblet rush is the delicate creeping buttercup and slender water milfoil, a plant that bears little resemblance to the water milfoils commonly used in aquariums. Most water milfoils have leaves divided into many threadlike segments and rather obscure flowers and fruits. Slender water milfoil consists of erect stems nearly devoid of leaves; near the top of each stem a few purple, four-petaled flowers appear, which eventually give rise to small nutlets. Slender water milfoil may grow submerged or stranded on land.

Lake-bottom plants include spiny quillwort, which reproduces by microscopic spores that form in swollen pockets at the base of the plant. Another is shoregrass, a dwarf member of the plantain family. This family includes several familiar coarse lawn weeds, but the shoregrass is a diminutive succulent with short stems and tiny, urn-shaped flowers. Hatpins grow at the edge of the lake, where the water is only 1 inch deep.

Shingleton Bog

A 5-mile-square wetland in Hiawatha National Forest is known locally as Shingleton Bog. Because most of the area is not very acidic, however, the term *bog* is inappropriate under the definitions developed by Michigan botanist Howard Crum. Since it contains ample sphagnum, or peat moss, it is a peatland. Its various open areas, which are best termed *fens*, are interspersed with tree-studded patches known as white cedar swamps and black spruce muskegs.

Among its habitats, Shingleton Bog has a "poor" fen and a "patterned" fen. To see them, I followed Hiawatha National Forest ecologist Jan Schultz, then–regional forest botanist Lawrence Stritch, and research natural area coordinator Lucy Tyrrell through a rather impenetrable white cedar swamp adjacent to Forest Highway 2251. The white cedar swamp is a natural community that gradually arose following the retreat of the great glaciers that covered the region some 12,000 years ago. At that time, heavy, waterlogged soil began to build over the limestone bedrock. Sphagnum mosses eventually covered much of the soil, and their decomposed remains began to accumulate as peat.

The considerable calcium in the underlying limestone kept the peatland from becoming acidic, so that the fen maintained itself until white cedar seedlings began to invade. As more and more trees became established and grew to maturity, their dense cover promoted the growth of shade-tolerant plants.

The white cedar swamp was difficult to walk through because of low-hanging branches that often reach the ground. In addition, the mat of sphagnum beneath the trees contained weak areas where a visitor could easily step through and twist an ankle. Filling the understory were shoulder-high clumps of royal fern and cinnamon fern. Here and there, occasional pink lady's-slipper orchids and bluebead lilies grew among thick patches of low-growing, evergreen club mosses.

The ground sloped down imperceptibly as we made our way through the cedar swamp. Even though I could not detect the difference, the plants responded to the slight change in soil and moisture. Almost abruptly, the crowded, large white cedars gave way to open habitat containing few woody plants, all of them dwarfed and gnarled. Apart from cedars, there was an occasional tamarack, a few red maples, and a scattering of shrubs—red chokeberry, mountain holly, and raisin tree. As we proceeded, the ground became wetter, and water rose above the toes of our boots with every step.

Crum describes this type of community as a poor fen because of its greater degree of acidity, not because it lacks a diversity of plants. Dozens of low-growing wildflowers grow on the sphagnum-dominated soil, all species having adapted to saturated soils, cool summers, and frigid winters with long durations of snow cover. They include bushy-branched horsetail, wintergreen, starflower, and bunchberry (a 4-inch-tall, nonwoody type of dogwood). Carnivorous sundews and pitcher plants, as well as a wide variety of slender, delicate sedges, are also common.

After making our way for a few hundred feet through this fragile, watery terrain—being careful not to step on the flowering plants—we left behind most of the scraggly trees and faced a meadowlike area with small rivulets of water running between ridges covered by sphagnum moss and other vegetation. This region was the patterned fen, although the pattern was not immediately visible. If we could have looked down from above, however, we would have seen that the ridges and rivulets were all more or less parallel to one another, oriented east-west at right angles to the slight slope of the terrain.

Peatlands all across the more northerly regions may contain patterned fens. Scientists in Europe recognized them many years ago, calling them *aapamires*. The rivulets are referred to as *flarks*; the adjacent ridges of soil and vegetation are called *strings*. Biologists have come up with several hypotheses concerning the origin of patterned fens. One suggestion is that the alternate freezing and thawing of the soil over a long period of time eventually gave rise to the alternating flarks and strings.

While freezing and thawing may play a role in creating patterned fens, a more important factor may be at work. Patterned fens usually arise where the terrain has a gradual, nearly imperceptible grade of about 2 percent. Through time, soil slides down this small gradient. When one edge of the slipping soil hooks onto something, such as a small tree or even a rock, the soil tears, forming a flark along the tear line. After many years of constant sliding and tearing, a distinct pattern of alternating flarks and strings becomes evident.

At Shingleton Bog, the strings and flarks may be as narrow as 1 foot or as wide as 30 feet, and they are usually from 10 to 100 or more feet long. The strings may stand as much as 3 feet higher than the flarks, but usually the contrast is more subtle. The amount of water in the flarks varies with rainfall, ranging from inconspicuous amounts up to pools 6 inches deep. The water is nearly neutral, with a pH of about 6.

Several plants seem confined to the flarks: a tufted little sedge known as *Carex exilis*, the intermediate sundew, one kind of bladderwort, and the white beaked rush. The strings, on the other hand, provide habitat for Kalm's lobelia, bog rosemary, shrubby cinquefoil, a wild lily, and several flowering plants exceptionally rare for the region. Most of the rarities—including a sedge, an orchid, a sundew, a tiny raspberry, and a willow herb—are arctic species that were left behind when the great glaciers of the Ice Age receded northward.

Summerby Swamp

Summerby Swamp is among the countless wetlands that dot northern Michigan, northern Minnesota, and adjoining parts of Canada. Bisected by Michigan Highway 123, the swamp is about 3 miles square. On one side of the road, the swamp is rather soupy looking, with hummocks of vegetation forming hundreds of tiny islands in shallow, standing water. On the other side, it is forested with northern white cedar. The contrast in vegetation is related to differences in soil chemistry and drainage. This type of variation in wetlands is also a clue to how these habitats gradually change from one type to another, or even into a dry habitat, as a result of plant growth.

Terms such as *bog*, *fen*, *marsh*, and *swamp* are often used interchangeably, even by professional botanists. But biologist Howard Crum, in his book *A Focus on Peatlands and Peat Mosses* (1988), proposes a more precise terminology. One of the differences he emphasizes is between peatlands, where sphagnum (peat moss) grows and accumulates, and nonpeatlands. Peatlands develop where the ground is water soaked throughout the growing season, causing the sphagnum to grow faster than its dead remains can decompose. The built-up deposit is known as *peat*.

Peatlands vary depending on the degree of acidity. *Fens*, according to Crum, are peatlands that are rich in minerals and low in acidity or even slightly alkaline. They develop where water near the surface of the wetland is well aerated and supplied with minerals such as calcium. Northern Michigan has "rich fens" with abundant calcium and a pH value between 6.0 and 7.5. (On the pH scale, 7 is neutral, values from 7 to 14 indicate increasing alkalinity, and values from 7 to 0 indicate increasing acidity.) Where the calcium is low, a sedge-dominated "intermediate fen" will develop, with a tendency to become increasingly acidic. Crum designates a wetland as a "poor fen" when the pH is between 4 and 6 and the vegetation, dominated by sphagnum, is still in contact with groundwater. If the pH falls to 3 or less, it is a bog.

Crum notes that peatlands form in lowlands that have a constant water supply and may even encroach on open water. In a fen, where the water is well aerated and not too acidic, the habitat will support a diversity of plants, often dominated by sedges. But sphagnum mosses are the key to the peatland ecosystem: usually several species are present, and they may come to dominate, depending on conditions.

In some calcium-rich fens in Michigan, spring flooding or other changes in water level may restrict the growth of sphagnum, which is a perennial. Such locales may be invaded by white cedars to become cedar swamps. But in fens where peat accumulates rapidly, the water flow is restricted, trapping nutrients so that they are no longer recycled. Such fens end up as bogs, as the waterlogged peat slows down oxygen movement and reduces the rate of decomposition. Fewer and fewer plant species other than sphagnum are able to survive in the habitat. Some, perched on the peat, must obtain their water and nutrients strictly from rain, absorbing these necessities mostly through aboveground tissues rather than through roots.

As a bog matures, more and more shrubs invade it, most of them members of the heath family. In northern Michigan, bogs eventually become dominated by black spruces, forming a type of swamp referred to as a *muskeg*. This process may take several thousand years.

Unlike peatlands, marshes and swamps are flooded at least part of the year, so sphagnum has little chance to become established and to accumulate. Their soils are well aerated and rich in minerals. Marshes are dominated by grasses, with few woody plants. When similar habitats are dominated by sedges, they are called *sedge meadows*; when forested, *swamps*.

In Crum's terms, Summerby Swamp consists of both rich fen and cedar swamp zones. I toured the area one July with botanist Donald Henson, a former student of mine. The fen, on the north side of Michigan Highway 123, was dotted with sphagnum hummocks. Although the fen's surface water and groundwater are charged with magnesium and calcium, these sphagnum hummocks are acidic enough to accommodate the growth of acid-loving plants, including wintergreen, leatherleaf, cranberry, and Labrador tea, all members of the heath family. Scattered throughout were thickets of stunted tamarack, white cedar, and black spruce.

The fen was colorful with the orange flowers of wood lily, the yellow and orange blossoms of Indian paintbrush, and the purplish pitchers of pitcher plants. Closer observation revealed the much smaller flowers of arrowgrass (not a true grass) and a diversity of sedges and rushes.

After surveying the fen, we crossed to the south side of the road. Here we observed a mature white cedar swamp with occasional stands of black spruce. Beneath the trees grew royal fern and many species of flowering plants that had bloomed earlier in the year, including starflower, goldthread, and bunchberry (a dwarf type of dogwood). Henson speculates that the construction of the road has restricted the draining of water from the north to the south side, speeding the establishment of the swamp zone.

While most of the plants in the fen and cedar swamp are common throughout northern Michigan, several are rare for the region. Black crowberry, bird's-eye primrose, butterwort, and the hyssopleaf fleabane (which looks like a small daisy), all more common much farther to the north, have found the right conditions to thrive in Summerby Swamp.

Worldwide, peatlands are often found in cool temperate zones near oceans. This is because mild winters and long growing seasons with cool, humid, foggy conditions favor the growth of sphagnum moss. Peatlands also arise in poorly drained topography sculpted by glacial action. This is true of the Great Lakes area, where the poor drainage of the shallow soil, combined with an even distribution of rainfall throughout most of the year, allows peatlands to form despite short growing seasons, low humidity, and long, cold winters.

Huron National Forest

SIZE AND LOCATION: 415,000 acres in east-central Michigan, from Au Sable and Oscoda on the banks of Lake Huron to Grayling on the west. Major access routes are Interstate Highway 75; U.S. Highway 23; and State Routes 18, 33, 65, and 72. District Ranger Stations: Mio and Oscoda. Forest Supervisor's Office: 421 S. Mitchell Street, Cadillac, MI 49601, www.fs.fed.us/r9/hrmnf.

SPECIAL FACILITIES: Boat ramps; swimming beaches; bathhouses.

SPECIAL ATTRACTIONS: River Road National Scenic Byway; Au Sable National Wild and Scenic River; Lumberman's Monument; Kirtland Warbler Breeding Area.

An abrupt change in the vegetation is evident as you just enter the Huron National Forest from the south. Gone are the numerous hardwood forests dominated by oaks and hickories and by maples and basswood. Present now are forests dominated by coniferous species—pines, spruces, white cedars, and tamaracks. The Huron National Forest marks the southern end of the great boreal forests that extend northward toward the Arctic Circle. The entire landscape is now pockmarked by numerous lakes of all sizes, and boglike areas in swampy forests become more prevalent.

Crossing the upper part of this beautiful region of the Huron National Forest is the clear Au Sable River, so pristine that 23 miles of it was designated a National Wild and Scenic River in 1984. During the summer, canoeists and kayakers crowd the waters of the Au Sable, and a famous canoe race from Grayling to Au Sable on the shore of Lake Huron is an annual highlight for river enthusiasts. From Grayling to Mio, the Au Sable River forms the northern boundary of the Huron National Forest. Along this stretch of river are the White Pine and Parmalee Bridge campgrounds, as well as several places where you may put your canoe or kayak in the river. That part of the Au Sable River that extends east from Mio for 23 miles to Forest Road 4001 that crosses the river has been designated a National Wild and Scenic River, and both sides of the placid river in this section are in the Huron National Forest.

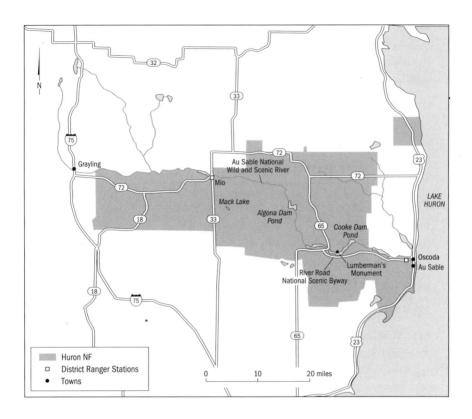

East of Mio, canoeists and kayakers may stop at the Louds Rest Stop to take advantage of restroom facilities. The river then loops south around the Au Sable Loop Campground, which provides walking access to the river. By canoeing eastward, you will find several rest stops where rustic camping facilities are usually available. Where Forest Road 4004 crosses the Au Sable River is the McKinley Trail Camp, which is also on the route of the Michigan Shore-to-Shore Hiking and Biking Trail. At the end of the designated National Wild and Scenic section, where Forest Road 4001 crosses the river, is the Gabion Campground, and there is a scenic overlook of the river from the forest road.

From Forest Road 4001, the Au Sable River continues for many miles, bearing southward and then eastward, finally emptying into Lake Huron in the community of Au Sable. The last stretch passes through several lakes that have been formed by the Au Sable River, including Algona Dam Pond, Loud Dam Pond, Cooke Dam Pond, and Foote Dam Pond. At each of these lakes, the Forest Service provides campgrounds, picnic areas, and access to the water.

The eastern section of the river is paralleled by the River Road National Scenic Byway for approximately 22 miles. The eastern end of the road is less than 2 miles from Oscoda and Au Sable on the shore of Lake Huron. Most of it follows State Route 65. The western end of the byway begins at the junction of River Road and State Route 65 at the southwest corner of Lower Dam Pond. Most of the road here follows the southern edge of the Au Sable River on a trail used historically by Indians. Known as the Saginaw to Mackinaw Indian Trail, this route passes through beautiful forest land dominated by red and jack pines. At the scenic byway's West Gate is a viewing platform above Loud Dam Pond, an interpretive trail, and the Rollway campground and picnic area.

Four miles east of the scenic byway's West Gate is a side road to Loud Dam, constructed in 1912–1913. The dam is 2,000 feet long, and the lake generates hydroelectric power. The stretch of the Au Sable River from Loud Dam Pond to Cooke Dam Pond is called Five Channels Dam Pond. State Route 65 crosses the Au Sable River at the end of Five Channels Dam Pond, and the scenic byway is now on River Road as it stays along the south shore of the river.

Shortly after River Road branches off State Route 65 is Iargo Springs, which includes a low but picturesque waterfall. A trail with wooden stairs allows for nice views of the surrounding area. Interpretive displays explain the geology and plant and animal life of the area. Iargo Springs is also the trailhead for the Highbanks Trail, which stays near the edge of some high bluffs above the south side of Cooke Dam Pond. At the junction of River Road and Forest Road 4408 is a monument for canoeists who participate in the Au Sable Canoe Race. Less than 2 miles to the east is the Eagle Nest Overlook where you get a fine view of not only Cooke Dam Pond but also a bald eagle's nest in use since 1985.

A short distance to the east is the Lumberman's Monument and Visitor Center. The monument, a tribute to lumbermen both past and present, depicts these men, one with a cross-cut saw, one with a compass, and one with a peavey. (A *peavey* is a hook with a sharply pointed end.) The monument was completed in 1932 at a cost of $50,000. A log visitor center was added in 1982. From the visitor center is a long wooden stairway to Cooke Dam Pond. Be sure you have the energy to climb back up the 263 steps to the visitor center. Across the road to the south is the Kiwanis Monument, a tribute to Michigan Kiwanians who sponsored the replanting of more than 10,000 acres of forest land in the Huron National Forest during the 1920s and 1930s.

A side trip off the scenic byway passes the community of Sid Town and reaches Cooke Dam in about 4 miles. This 864-foot-long dam is on the site of the first power plant on the Au Sable River.

Returning to the River Road National Scenic Byway, the highway eventually comes alongside the Foote Dam Pond, at the eastern end of which is the 4,000-foot-long Foote Dam, completed in 1918. Two and one-half miles east of the dam is a short side road to the north to Whirlpool Access on the Au Sable River with a concrete boat ramp and a fishing pier accessible to wheelchair users.

South of the River Road National Scenic Byway is a Kirtland's Warbler Management Area. The rare Kirtland's warbler, reduced a few years ago to no more than 300 pairs, lives only in jack pine forests with a history of frequent fires. This bird breeds only in this part of the world. A larger Kirtland's Warbler Management Area exists around Mack Lake.

At the scenic byway's East Gate are interpretive signs as well as the Eagle River Trailhead where hikers and cross-country skiers may venture on several miles of trails.

Fishing is good in the Au Sable River for muskellunge, walleye, bluegill, northern pike, and bass.

Although there are no designated wilderness areas in the Huron National Forest, several semiprimitive nonmotorized areas are scattered throughout the forest. From west to east, these semiprimitive areas are Wakely Lake, South Branch, Hoist Lakes, Au Sable River, Reid Lake, Cooke Dam Pond, and Silver Creek.

Wakely Lake is 11 miles east of Grayling off State Road 72. A short walk from the parking area along the highway will bring you to the lake, where you may see loons that nest here. South Branch Creek occupies about seven square miles west of State Route 18 and south of State Route 72. Two abandoned Forest Service roads allow for easy access to this wild area that is crossed by Douglas and Sauger creeks. The South Branch of the Au Sable River is just outside the western border of this semiprimitive area. Hoist Lakes Semi-Primitive Area occupies 10,600 acres and contains 20 miles of foot and cross-country ski trails. Passing through forests of pine, hardwoods, and aspens, these trails provide access to numerous small lakes and marshes. A parking area along Forest Road 4126 and another along State Route 65 are good starting points for the west and east sides of the area, respectively. Forest Road 4119 forms the southern boundary of the Hoist Lakes area. In addition to Hoist Lakes, the area also contains Penoyer Lake, No Name Lake, and Byron Lake.

Two miles east of Hoist Lakes is Reid Lake Semi-Primitive Area consisting of 3,000 acres and 6 miles of trails over gently rolling terrain. Parking for a hike into this area is at the north end along State Route 72. Within this area are Reid Lake, Little Trout Lake, Beaver Lake, Mossy Bog, Big Marsh, and Fanny Marsh. The northern half of the area is very boglike.

The Au Sable River Semi-Primitive Area lies between Algona Dam Pond and Loud Dam Pond. Forest Road 4128 follows the eastern side of the area for its entire length. The Au Sable River winds through the upper part. The

South Branch Trail Camp off River Road provides good access for entering the area.

Cooke Dam Pond Semi-Primitive Area on the north side of Cooke Dam Pond is marshy and boggy throughout. Silver Creek Semi-Primitive Area, 3 miles south of Cooke Dam Pond, includes the Silver Valley Campground.

Six miles east of Grayling and 1 mile south of State Route 72 are the Kneff Lakes, where there are a campground, picnic area, and swimming beach. Near the southeast end of the Huron National Forest are Sand Lake and Round Lake, each with swimming facilities.

Two areas of particular interest in the Huron National Forest are around Mack Lake southeast of Mio and the Tuttle Marsh Wildlife Area southwest of Oscoda. Mack Lake is in the heart of the major breeding grounds for the federally endangered Kirtland's warbler. This area of jack pine forests is carefully managed to provide optimal habitat for this tiny bird.

Tuttle Marsh Wildlife Management Area consists of seven square miles of wetlands and not-wetlands managed to improve wildlife habitat. The south end of the area is a shallow marsh with occasional areas of aspen, hardwoods, and tamarack. Willows and alders are continuously invading the marsh. With increasing elevations at the northeast corner of the area, conifers, aspens, and red maples dominate. Over half of the Tuttle Marsh Wildlife Management Area is dominated by shrub swamps and wildlife openings. Tuttle Marsh Road crosses the area diagonally. Along its northern end are some old-growth red pines, a 17-acre stand of white pines estimated to be more than 150 years old, and a 6-acre stand of old-growth balsam fir.

One of the fascinating habitats in the Huron National Forest is the bog-like condition in forests dominated by white cedar and tamarack. The very hummocky soil is often carpeted by mosses from which grow many interesting flowering plants, including two-seeded sedge, three-seeded sedge, goldthread, and wild cranberry.

Where State Route 72 crosses the South Branch of the Au Sable River at Smith Bridge is a very scenic site with typical streamside vegetation. Here grows a forest of American elm, black ash, and red maples. Two species of *Viburnum* and elderberry are common shrubs in the understory. The herbaceous layer includes marsh marigold, mad dog skullcap, swamp betony, and several species of orchids.

Mack Lake

Many of my excursions take me through pristine forests, unplowed prairies, or scenic rock formations. The terrain around finger-shaped Mack Lake does not offer these features, however. It is mostly flat, with scraggly jack pines and numerous areas of burned and cut trees. Yet this is one of the few places where the Kirtland's warbler nests and rears its young. Accordingly, signs attached to trees along the back roads that crisscross this part of the Huron National Forest notify the visitor that the interior is off-limits between May 1 and September 10. During this time, only ranger-led guided tours into the Mack Lake area are available.

Kirtland's warbler first came to scientists' attention in the spring of 1851, when Charles Pease shot a migrating bird near Cleveland, Ohio. He gave it to his father-in-law, the noted naturalist Jared Kirtland, who in turn presented it to ornithologist Spencer F. Blair. The following year, Blair described the bird as being new to science, naming it in honor of Kirtland. After that, individual Kirtland's warblers were occasionally spotted during spring and fall migrations, but the species' breeding and wintering grounds long remained a mystery. The first wintering bird was found in January 1879, on Andros Island in the Bahamas. For the next 20 years, more than 70 specimens of Kirtland's warblers were collected in the Bahamas, but the place where they nested and spent their summers was still unknown.

Finally, more than a half century after the species' original description, a nesting bird was found in June 1903 near the Au Sable River in Michigan's Lower Peninsula. It was taken to ornithologist Norman A. Wood, who identified it as a Kirtland's warbler and, upon investigating, observed a nest of the birds for himself the next month. From that time until now, all nests of this species have been found in the 13 counties of jack pine habitat in the Lower Peninsula, with 90 percent of the nests located in the Au Sable River basin.

Lawrence Walkinshaw, an amateur ornithologist, has spent much of his free time from his occupation of dentistry studying Kirtland's warbler. He has noted that when the birds first arrive in Michigan in May, the males immediately select and defend their territories, even battling intruding males in midair by fluttering their breasts and feet against their opponents. Usually by June 1, the female bird has selected a mate, built a nest on the ground, and laid an average of five eggs. The young birds hatch out after an incubation period of 14 days; 9 or 10 days later, now called *fledglings*, they leave the nest. Walkinshaw notes that the coloration of the fledglings is similar to that of an unopened jack pine cone, and the young birds resemble cones when they sit on the lower pine branches with their bills pointed upward.

Called the *jack pine plains* by the locals, the habitat favored by Kirtland's warbler has a poor, sandy soil that supports, in addition to jack pines, a variety of deciduous trees and a sparse collection of low-growing shrubs, grasses, sedges, and occasional wildflowers. The sand and the fallen pine needles create acidic soil especially suitable for members of the heath family:

wild blueberries, bearberry, wintergreen, and trailing arbutus. Bracken fern and sweet fern are also plentiful (the latter has fernlike leaves but is actually a flowering plant). All the plant species that live here have to be able to withstand fire, drought, and temperature extremes.

Because the bird's population appeared small and showed no sign of increasing, concern for the survival of Kirtland's warbler arose as early as the 1940s. In 1951, in commemoration of the 100th anniversary of the bird's discovery, an effort was made to find out how many of them actually summered in Michigan. Since male warblers arriving on the breeding grounds sing loudly and clearly, the census was taken by counting the number of singing males and doubling the total, on the assumption that each male attracted a mate. The 1951 census recorded 432 singing males, or an estimated breeding population of fewer than 900.

Apparently contributing to the species' survival problems are the birds' narrow nesting requirements. They nest only where there are stands of jack pine, building their nests on the ground under the trees, in association with a clump of grass or other vegetation. The nest consists of dead grasses, leaves, and small pieces of bark, lined with deer hairs, slender stalks of mosses, and fine porcupine quills. Almost always, the birds choose sites shaded and protected by low-hanging branches. As a result, they prefer stands of jack pine containing trees that are 5 to 7 feet tall, or about 8 years old. As jack pines mature beyond 13 years and grow taller, the lower branches lose their needles, depriving the ground nests of shade and cover. By the time jack pines are 21 years old, the Kirtland's warblers will have nothing to do with them.

Based on the number of birds observed at the wintering grounds since the late 1800s, researchers have concluded that the Kirtland's warbler population was highest between 1880 and 1890. About that time, both lumbering and forest fires in the pinelands of Michigan contributed to regenerating young jack pines in the breeding grounds, apparently enhancing the birds' reproduction. In modern times, however, the suppression of forest fires has meant that many of the jack pine stands have aged, outgrowing their usefulness to the warblers.

To satisfy the warbler, the jack pine stands must not only contain sufficient young trees but also cover at least 80 acres and preferably as many as 200. Such stands usually have some deciduous tree (e.g., Hill's oak, red oak, pin cherry, and shadbush) mixed in, but if the proportion of such trees is too high, the Kirtland's warbler will reject the stand.

Furthermore, even though jack pines grow naturally from the maritime provinces of eastern Canada to the upper Yukon Valley in the Northwest Territories and south to Michigan's Lower Peninsula and central Wisconsin, only those in the southernmost part of the range seem to suit Kirtland's warbler.

Forest growth in more northern areas apparently is too lush, while the stands in Michigan are ideal because the dry, nutrient-deficient soil limits the growth of the trees, as well as of the grasses and sedges beneath them, leaving a lot of open sandy areas in between. The soil—almost always a particular type known as Grayling sand—is also very porous, allowing water from heavy June rains to percolate down quickly without flooding the ground nests.

Given the small number of Kirtland's warblers noted in the 1951 census, Michigan took measures in 1957 to protect the bird's breeding habitat on state forestland. A census in 1961 recorded 502 singing males, a slight increase. Two years later, the U.S. Forest Service initiated its own plan, setting aside a 4,010-acre tract of jack pine forest in the Mack Lake area. Management areas were set up so that some jack pine stands would always be at the proper stage for the persnickety warbler. As stands grew too old for the birds, the trees would be harvested and new ones planted. In addition, the Forest Service cut and burned some old jack pine areas in the forest and replanted them with seedlings. In 1966, when federal legislation was enacted to list and protect animal species threatened with extinction, Kirtland's warbler and 33 other species of birds were the first to make the list.

Despite these developments, by the 1971 census, the number of singing male warblers had plunged to 201. Biologists then decided to attack another problem that affected the success of nesting, one first noted in the 1920s: brown-headed cowbirds parasitize the nests by laying their own eggs in them, often destroying some of the Kirtland's warbler eggs in the process. To combat this problem, cowbirds were trapped in the jack pine plains of Michigan and destroyed. More than 50,000 have been killed since 1972, and the threat they pose is now apparently under control.

A recovery team was also organized in 1975 to oversee the measures taken to preserve Kirtland's warbler. This team's work was facilitated by the strengthened Endangered Species Act of 1973, which makes funding available to carry out management programs and to assist federal and state agencies in coordinating their efforts.

Ever since the alarming decline noted in 1971, a census has been taken each summer. The number of singing males appears to have stabilized, with a low of 167 reported in both 1974 and 1987 and a high of 243 in 1980. The total for 1989 was 212. At this point, observers believe that more than enough suitable habitat is available and that the population should eventually increase.

Even so, the management plan for the warbler's breeding grounds may not be enough to ensure the preservation of the species. Since 1972, 800 to 900 warblers have gone south each fall, exiting the United States along the coast of the Carolinas before reaching their wintering grounds in the Caribbean, but only about 400 return to Michigan in June. Something must be happening to the birds during the spring and fall migrations or at wintering grounds, which include Grant Turk Island, South Caicos Island, Hispaniola, and the Bahamas. Little is known about the fate of the birds during those times. Hurricanes, drought, or pesticides may destroy a few birds during migration, but hazards at the wintering grounds, where they spend up to 8 months living amid shrubby, desertlike vegetation, probably take the greatest toll.

Tuttle Marsh

In 1839, land surveyors in eastern Michigan mapped a large peat-filled wetland a few miles west of Lake Huron, noting that "hemlock" trees covered the higher ground. Sometime later, one Jonathan B. Tuttle purchased most of the area, and much of the land was subsequently logged and drained. Cultivation of domesticated crop plants proved uneconomical, however, because of the acidic soil. Eventually the region was managed for blueberries, which grew there naturally. The blueberry growth was enhanced by periodic burning of the vegetation, which robbed the soil of humus.

Things began to turn around in 1929, when the U.S. Forest Service acquired the 5,065-acre property for the Huron National Forest. Surviving wetland was thus saved and ultimately enhanced. In 1961, a modest management plan was adopted that included improving the habitat for waterfowl. Ditches were dug, and at least nine ponds were created; 30 acres were also sown with food plants attractive to wildlife. More intensive management went into effect in 1987. Today the area, known as Tuttle Marsh, contains some 700 acres of thriving wetland, scattered within a slightly more elevated forest of jack pine, red pine, white pine, quaking aspen, red maple, and northern red oak.

The wetland habitats include moist, wooded areas dominated by cone-bearing trees: black spruce, balsam fir, and especially tamarack, or larch. Unlike most other conifers, the tamarack drops all of its needles during the winter. (The only other deciduous conifer in the United States is the bald cypress of the Southeast.) The more obvious wetland habitats, however, are the marshes, most of which lie to the south of gravel Forest Service Road 4546, which bisects Tuttle Marsh diagonally.

The word *marsh* conjures up a mental picture of a soggy, flat area filled with cat-tails and other herbaceous vegetation, with ducks sitting on patches of open water. If the open water is instead continuous and stands throughout most or all of the year, we usually call the area a *pond* or *lake*. If the soggy region has standing trees in it, we call it a *swamp*. If it quakes with layers of sphagnum moss, we probably refer to it as a *bog*. Wetland authority Milton W. Weller offers the following technical definition to distinguish a marsh: a basin that holds water long enough for the germination and survival of water-tolerant, rooted, perennial soft-stemmed plants, such as sedges, cat-tails, and bulrushes. Such plants are often called *emergents*.

Marshes usually form when ponds or river margins fill in. Cat-tails and bulrushes often become rooted in the mud near the bank, where the water is shallow, and creep out farther and farther as the basin accumulates decayed organic matter along with soil washed in from the surrounding area. The plants and added nutrients attract other living organisms. The depth of the water and the extent to which it persists during the year determine the type of vegetation found in different parts of the marsh. The common cat-tail, for example, thrives in shallow water but will not survive in water deeper than 3 feet.

There are 10 different species of cat-tails in the world, 3 of which are native to the United States, including the common cat-tail. All are similar in appearance and structure. Cat-tails have stout, unbranched stems, which grow up to 15 feet tall, and long, narrow, fibrous leaves. Their most rapid growth, however, is below the surface, where a network of underground stems, known as *rhizomes*, may spread over several square feet. The rhizomes send new shoots aboveground, often at close intervals; as a result, cat-tails are usually found in dense colonies, with few other species growing among them.

Cattail rhizomes seem to be stimulated to grow faster and divide more rapidly at higher latitudes, which have a distinct spring season with shorter days and cooler temperatures. Cat-tails are able to grow throughout most of North America; however, unlike many temperate plants, their seeds do not have to pass through a cold dormant period in order to germinate.

Biologists James Grace and Robert Wetzel have noted that cat-tails use much of their energy to proliferate "vegetatively"—that is, to put forth rhizomes and leaves. These plants devote relatively little energy to sexual reproduction. But sometime in late summer, each mature cat-tail usually flowers. Hundreds of male, pollen-producing flowers are crowded into a very slender, cylindrical spike, which sits atop a thicker spike of female flowers. The male flowers are minute, each consisting only of two to seven stamens and a few bristlelike hairs at the base. The female flowers are larger and less numerous; each consists of a single ovary, and it, too, has bristlelike hairs at the base.

The male flowers mature slightly ahead of the female flowers on the same stalk, so that plants usually cross-fertilize. After shedding its pollen grains, the upper spike disintegrates, leaving the larger spike to develop minute seeds. As the seeds mature, the bristles around them enlarge, developing into cottony parachutes that facilitate dispersal.

Because their flowers lack petals and sepals, cat-tails were once believed to be a very primitive type of flower, perhaps not very far removed from the first kind of flower ever produced. Now, however, most botanists think that cat-tails are highly evolved and that the flowers are actually reduced, or simplified, structures that were once more complex.

For a long time, marshes and other wetlands have been regarded as breeding grounds for mosquitoes and places where snakes lurk, and few lamented their loss. U.S. Fish and Wildlife Service biologist Ralph Triner Jr. has noted that between the time the territory of the United States was settled by Europeans and the mid-1970s, 54 percent of all original wetlands were lost to agricultural, urban, and other developments. In 1982, the Michigan Department of Natural Resources estimated that Michigan had already lost 71 percent of its original wetlands.

As Milton Weller has noted, however, wetland habitats serve many vital functions. They control erosion by slowing water movement, protect extensive areas from wind action, collect soil and nutrients, and provide habitat for wildlife. In addition, they improve the water supply by filtering the water that enters streams, enlarging water tables, sealing basins, and storing floodwaters. As indicated by the protection and improvement of the wetlands at Tuttle Marsh, the value of these habitats is being increasingly appreciated.

Manistee National Forest

SIZE AND LOCATION: Approximately 500,000 acres in west-central Michigan, extending from Lake Michigan in the west to Cadillac and Big Rapids in the east. Major access routes are U.S. Highways 10, 31, and 131 and State Routes 20, 37, 55, 82, 115, and 120. District Ranger Stations: Baldwin and Manistee. Forest Supervisor's Office: 1755 S. Mitchell Street, Cadillac, MI 49601, www .fs.fed.us/r9/hrmnf.

SPECIAL FACILITIES: Boat ramps; swimming beaches; winter sports area.

SPECIAL ATTRACTIONS: Lake Michigan Recreation Area; Pere Marquette National Scenic River; Newaygo Prairies; Caberfae Winter Sports Area.

WILDERNESS AREA: Nordhouse Dunes (3,450 acres).

One of the most unusual areas in any of the national forests in the United States is the Nordhouse Dunes (pl. 14) along the eastern edge of Lake Michigan

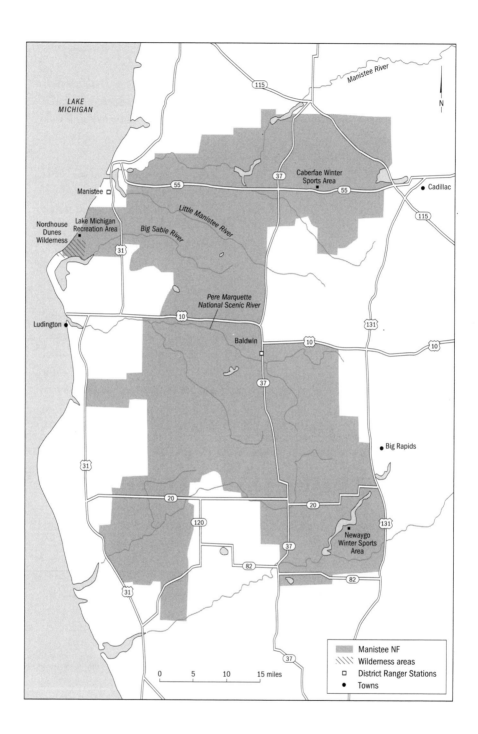

LAKE
MICHIGAN

Manistee River

115

N

Caberfae Winter
Sports Area

37

55

Cadillac

Manistee

55

115

Little Manistee River

Nordhouse
Dunes
Wilderness

Lake Michigan
Recreation Area

Big Sable River

31

Pere Marquette
National Scenic River

10

131

Ludington

10

Baldwin

10

37

Big Rapids

31

20

20

120

131

37

Newaygo
Winter Sports
Area

82

31

82

37

Manistee NF
Wilderness areas
District Ranger Stations
Towns

0   5   10   15 miles

between the towns of Ludington and Manistee. Some of these sand dunes are estimated to be about 4,000 years old and may reach heights approaching 150 feet. The southern end of the Nordhouse Dunes area is a designated wilderness. To the west and north of the wilderness is the Nordhouse Semi-Primitive Motorized Area. Most visitors will opt to drive Forest Roads 5629 and 5972 to the Lake Michigan Recreation Area between two lobes of the wilderness and the semiprimitive area. At the recreation area, where you may get a good look at some of the dunes, are a campground, picnic area, swimming area, hiking trail, and an observation deck for visitors with disabilities.

The Pere Marquette River crosses the Manistee National Forest at about its middle. South of Baldwin along State Route 37 where the Middle Branch and Little South Branch meet at the Forks, the Pere Marquette River has been designated a National Wild and Scenic River east all the way to the State Route 31 bridge at Ludington. This river, tranquil for much of its length, is highly scenic, with high cliffs often lining a part of its route. Canoeists will find the river very relaxing, but those who prefer more adventure will probably consider it tame. Since several highways parallel the river in many places, you have ample opportunities to put your canoe in the river. From east to west, the most popular places are Gleason's Landing, Bowman Bridge, Upper Branch Bridge, Lower Branch Bridge, Walhalla Bridge, and Indian Bridge. From Indian Bridge to Ludington, the Pere Marquette is outside the boundaries of the Manistee National Forest.

Several mammals and birds can be observed along the river. Among the mammals are beaver, mink, river otter, muskrat, and white-tailed deer. Bird enthusiasts should be able to spot wild turkey, grouse, woodcock, blue heron, cedar waxwing, woodpeckers, and warblers. South of the Walhalla Bridge, at Nelans Marsh, mallards, black ducks, wood ducks, redheads, canvasbacks, and scaups are common. Fishermen will enjoy angling for brown trout and other fish.

North of the Pere Marquette River and U.S. Highway 10 are other points of interest in the Manistee National Forest. West of State Route 37 are the Ward Hills, which are interlaced by all-terrain vehicle trails. At the northwest edge of these hills are a number of wetlands in the vicinity of Beaver Pond and McCarthy Lake.

Major rivers in the northern half of the Manistee National Forest include the Big Sable, Little Manistee, Manistee, and Pine. All of these provide opportunities for fishing, canoeing, and exploring. An extensive wetland occurs west of Bass Lake Road, where waterfowl and marsh vegetation may be observed in typical habitats.

The Pine River south of State Route 55 is exceptionally scenic, particularly as it passes along the edge of the Pine River Corridor Semi-Primitive Non-Motorized Area. West of Cadillac and Lake Mitchell is another wild area known as the Brandy Brook Semi-Primitive Motorized Area, with 27 Mile Road and 31 Mile Road forming its western and eastern boundaries, respectively. Other semiprimitive areas are the Manistee where the Manistee River flows through the center, and the Brian Hills at the northern edge of the Manistee National Forest. Winter sports enthusiasts will find the Caberfae Area much to their liking. Caberfae is located 2 miles north of State Route 55 along the west side of 13 Mile Road. The area has snowmobile trails and, for downhill skiers, two chairs, five T-bars, and 18 rope tows.

The area of the Manistee National Forest south of the Pere Marquette River and U.S. Highway 10 includes the Stiles Swamp Semi-Primitive Non-Motorized Area, the Whelan Lake Semi-Primitive Non-Motorized Area, the Highlands Semi-Primitive Motorized Area, the Loda Lake Semi-Primitive Motorized Area, the White River Semi-Primitive Non-Motorized Area, the Loda Lake Wildflower Sanctuary, the Newaygo Prairies, and the Newaygo Winter Sports Area, in addition to myriad small lakes, swamps, marshes, and bogs. All of the semiprimitive areas contain vast wetlands, usually in association with rivers and creeks.

The Loda Lake Wildflower Sanctuary dates back to 1938 when it was started by a group of Michigan garden club members to create and maintain a sanctuary for native plants. The 1,000-acre sanctuary is 1 mile west of State Route 37 on the Loda Lake Road, about 8 miles northwest of White Cloud. Loda Lake is at the center of the sanctuary, with wooded swamps, marshes, a bog, and forested areas at various places near the lake. Several trails have been laid out, including the Marsh Marigold Trail and the Birch Trail.

Northwest of Newaygo, in relatively flat, sandy ground, are remnant prairies that are all that remain of tongues of prairie that at one time extended into central Michigan. Here grow some of the same prairie plants found in the prairies of the Great Plains.

Several vegetation communities exist in the Manistee National Forest. In dry sandy areas are stands of jack pine with an understory that may include rattlesnake plantain orchid, wintergreen, and even an occasional lady's-slipper orchid.

Many of the dry forests, however, are occupied by white oak, northern red oak, and black oak, occasionally mixed with white pine, jack pine, red pine, red maple, bigtooth aspen, sassafras, and witch hazel. Shrubs in these oak forests include huckleberry, narrowleaf blueberry, myrtle blueberry, and a native bush honeysuckle. Common herbs are wintergreen, trailing arbutus, mayflower, cowwheat, starflower, partridge berry, lousewort, and white bluets. Bracken fern is common throughout these woods.

Moister sites often include mesic forests with a rich diversity of species. Trees present are red maple, paper birch, American beech, northern red oak, hemlock, white pine, red pine, yellow birch, and wild black cherry. Mapleleaf viburnum is the common shrub in most of these mesic forests. On the forest floor are several species of ground pines, blue violets, wild sarsaparilla, starflower, dwarf cornel or bunchberry, partridge berry, bedstraws, and wintergreen. Huge cinnamon ferns and spinulose wood ferns are frequent. Flowering plants that lack chlorophyll in these forests include pinedrops, Indian pipe, spotted coralroot orchid, beechdrops, and squaw root.

In even wetter forests along rivers and creeks that may experience seasonal flooding, a forest of red maple, green ash, and bigtooth aspen often occurs. Shrubs beneath these trees usually include winterberry, myrtle blueberry, and mountain ash. Nonwoody plants are marsh fern, interrupted fern, cinnamon fern, sensitive fern, enchanter's nightshade, monkey-flower, smallflower skullcap, clearweed, honewort, water horehound, Jack in the pulpit, and occasional sedges such as hop sedge, Tuckerman's sedge, and three-way sedge. Grasses may include northern mannagrass and broad-leaved wood reed.

Dotted throughout the Manistee National Forest are white cedar swamps, where green ash, red maple, hemlock, paper birch, and yellow birch often join the white cedars. On the forest floor are bluebead lily, Jack in the pulpit, rattlesnake plantain orchid, greenthread, bishop's cap, and enchanter's nightshade. This is a fern lover's paradise, with the usual presence of rattlesnake fern, oak fern, lady fern, cinnamon fern, spinulose wood fern, bulblet bladder fern, and New York fern.

Those visitors interested in wetland plants will have a field day around all the ponds, lake, and marshes in the Manistee National Forest. Many sedges and smartweeds grow here, as well as marsh marigold, bulblet water hemlock, Jack in the pulpit, blue iris, water horehound, swamp candles, clearweed, pink St. John's wort, blue violets, and white violets. Ferns abound, including spinulose wood fern, sensitive fern, cinnamon fern, marsh fern, and Virginia chainfern. Swamp rose is a common shrub.

Acid bogs occur here and there in the Manistee National Forest. Several shrubs occupy this type of habitat, including black chokeberry, leatherleaf, huckleberry, sweet fern, bearberry, raisin tree, narrowleaf blueberry, and myrtleleaf blueberry. Many sedges, including one known as a cottongrass, occur in the bogs. Black spruce, tamarack, and red pine are trees that sometimes grow in the bogs.

Newaygo Prairies

A nearly continuous sea of tall grasses once covered the central region of North America, from the edge of the boreal forests in Manitoba south to Texas, and from the woodland borders in Indiana halfway across the Dakotas,

Nebraska, and Kansas. Pioneers named this landscape a prairie, while the terrain from central Kansas westward to the Rocky Mountains, which was dominated by shorter grasses, became known as the Great Plains. Although much of the original prairie has been plowed under during the past 100 years for such crops as corn, wheat, and soybeans, parts of it survive. Some isolated patches even turn up in (of all places) the Manistee National Forest, amid wooded and cultivated land east of the town of Newaygo.

The Newaygo prairie patches, the largest about 100 acres in size, are remnants of what once was a tongue of grassland that extended north into the forests of central Michigan. Similar extensions (collectively termed the Prairie Peninsula in 1935 by botanist Edward Transeau) penetrated the forests of northwestern Pennsylvania, western Kentucky, and northeastern Arkansas. These, as well as the more extensive sections of the North American prairie, apparently owed their origin to the uplifting of the Rocky Mountains 35 million years ago, which altered precipitation patterns. Winds from the Pacific Ocean were induced to shed most of their water west of the Rockies before reaching the flatter regions to the east. Ecologist Henry Allen Gleason has speculated that the drop in summer rainfall and the even drier winters were responsible for the disappearance of existing forests and their replacement by grassland.

Grasses and other plants that grow on the prairie succeed in part by exposing little surface area in the heat of summer. John Weaver of the University of Nebraska, who devoted a lifetime of study to the prairie, noted that half or more of the typical prairie plant is belowground. In addition, the aboveground parts are usually active only from April to October, dying back for the winter. The rest of the year the prairie is underground, as the root systems of most prairie plants store enough food to start off the spring with a burst of rapid growth. The roots also form a dense network that prevents alien plants from invading the prairie.

Despite a reputation for visual uniformity, prairieland contains grasses and wildflowers varied enough to make the most of the available resources. Some plants have most of their root system in the upper 2 feet of soil, some in soil between 2 and 10 feet deep, and others have roots extending well below 10 feet, so that the soil moisture is tapped at different levels. The energy of the sunlight is captured by plants ranging from small, broad-leaved species that form a ground cover to such grasses as big bluestem, which may grow more than 12 feet high. In addition, some plants have their period of maximum growth in the spring, some in the summer, and a few in the autumn, so that demands for light, water, and nutrients are spread throughout the growing season.

The long, slender grasses that dominate the prairies allow some light to reach the broad-leaved plants below. Still, the growth is so dense that few seedlings are ever able to attain maturity. Prairie plants, therefore, rely heavily on vegetative propagation—reproducing by sprouts from roots or new shoots from the buds on rhizomes (underground stems).

The Newaygo prairies, like most other isolated remnants of the Prairie Peninsula, contain fewer species and support less robust growth than the prairieland to the west, mainly because the moister Michigan climate tends to favor forest species rather than prairie species. However, Kim Chapman and Susan Crispin of the Michigan Natural Features Inventory have recorded as many as 125 plant species in the largest Newaygo prairie patch. Although the dark, rich topsoil typical of Kansas prairieland is replaced at Newaygo by a sandy substratum, the species are much the same. Big and little bluestem dominate among the grasses, with Indian grass and June grass also common. Among the showy wildflowers are the heath aster, prairie cinquefoil, two kinds of blazing stars, and the rare prairie smoke, a member of the rose family.

Nordhouse Dunes

Westerly winds blow across Lake Michigan, shifting the grains of sand that wash up along the eastern shore. Where the wind strikes a plant or another impediment, sand is deposited and begins to accumulate as a dune. An obstacle to the blowing sand, the dune continues to build, changing shape and wandering until vegetation becomes established on it and more or less stabilizes it.

Sand dunes in all stages of development, from embryonic mounds to 150-foot ridges clothed in mature forest, can be seen in the Nordhouse Dunes, located about halfway between the towns of Manistee and Ludington. Part of Manistee National Forest, Nordhouse Dunes occupies more than 4 miles of shoreline, extending inland more than 1 mile. The undisturbed series of sand habitats found there includes, in addition to the dunes themselves, pools edged with rushes and troughs where small flowering plants somehow survive periodic burial in sand.

Although no one knows the original source of the sand deposited in the Lake Michigan basin, apparently it was brought there by glacial action. The bedrock of shale, which at the shoreline is concealed below nearly 500 feet of sand, has been overlain by glacial "drift"—sand and other deposits—ever since a lobe of the Late Wisconsinan glacier covered the area nearly 13,000 years ago. When the glacier receded, the sand deposits were at first inundated by a glacial lake. The water level fluctuated, however, with several successive glacial lakes forming and then receding, periodically exposing large amounts of sand, which was blown eastward by the strong winds. Today, sand on the lake bed continues to be washed ashore by wave action and distributed by the wind.

The dunes are accessible from the Lake Michigan Recreation Area to the north or from a parking lot near the end of sandy Nurnberg Road. From the parking lot, Nordhouse Dunes Trail leads to the shore of Lake Michigan, crossing from the older, stabilized dunes to the younger, ever-changing ones. For three-fourths of its length, the trail winds up and down forested ridges, but the ground underfoot is sand and not the rich organic soil so common beneath other forests.

Where the sandy ridges are moist, there are well-developed stands of red maple, American beech, paper birch, large-toothed aspen, and red oak, punctuated by Canadian hemlock with its short, deep green needles. On the forest floor grow the ghost-white stems of Indian pipe, topped by its single nodding flower. Indian pipe is one of a few flowering plants that completely lack chlorophyll, the green-pigmented compound involved in photosynthesis. As a result, this plant depends on the organic matter in which it grows for its supply of nutrients. Growing with the Indian pipe are other delicate wildflowers of the northern states, including the 4-inch-tall dwarf dogwood, trailing arbutus, wild lily-of-the-valley, and starflower. The tiny but strongly scented wintergreen is abundant along the trail.

On drier, wooded ridges, the trees of the moister soil give way to oaks and pines. White, red, and black oaks predominate, but white pine, jack pine, and red pine are plentiful. Brian Hazlett, a Michigan botanist, has noted that great numbers of young white pines are growing in the drier woods, which suggests that they will dominate this habitat in the future. Because of the reduced moisture, few wildflowers live beneath the trees.

The sand underfoot becomes progressively looser and the vegetation of the ridges sparser as the trail approaches Lake Michigan. Finally, the trail rounds a bend, affording the first glimpse of Lake Michigan and the shifting white sands of the open dunes. The beach proper, which extends inland from the lake for a few feet, is nearly devoid of vegetation because of the constant wave action from the lake. Where a line of driftwood has come to rest, however, a low ridge of sand parallels the lake bank. Known as the *foredune*, this low ridge develops when the wind strikes the driftwood, depositing sand on the windward side. The constantly enlarging foredune provides a place where seeds may become lodged. Although most flowering plants are unable to withstand the desiccating summer heat, two tiny annuals with succulent leaves have adapted to the arid conditions. One is a 3-inch-tall mustard called sea rocket; the other is milk spurge, which lies prostrate on the sand. Only annuals are able to survive in this habitat, because the furious winter storms that churn the sand would destroy the roots of any overwintering perennials.

In some places wave action creates breaks in the foredune, and the sand is blown through these openings to form dunes behind known as *blowouts*. The blowout dunes are constantly changing and shifting as more and more sand passes through the openings in the foredune. Because the plants growing there are farther from shore, the flora of the blowouts is richer than that of the foredunes, but it is still relatively sparse.

During the summer, one of the most conspicuous plants is Pitcher's thistle (fig. 7), with its prickly leaves and beautiful lavender pink flower heads. It is so uncommon in the United States that it is listed as a threatened species by the U.S. Fish and Wildlife Service. The thistle's dense covering of woolly hairs, which impede the evaporation of water from the leaves, enables the plant to avoid desiccation. A similar covering of densely matted hairs also protects the showy orange-flowered puccoon from the wind and sun.

As the blowout dunes build, perennial plants, many of them grasses, gain a foothold on them and serve as additional obstacles to the wind. Most efficient in stabilizing and actually building a dune is marram grass (*Ammophila breviligulata*), which sends out underground stems, called *rhizomes*, in a radial pattern beneath the surface of the sand. The stems and leaves of the grass provide areas where grains of sand become lodged. The radial growth widens the dune. At the same time, the marram grass continues to grow upward to lift itself above the sand, and the dune increases in height. Another grass that aids in the stabilization is sand reed, whose stiff, wiry leaves, stems, and roots resist the mechanical action of the wind. The extensive network of roots and stems below the sand surface binds the sand even after the plant dies.

Eventually the dunes build up into ridges that are receptive to colonization by trees. Once forest is established, some 500 years after the dunes begin to form, the ridges become fairly permanent. Two conspicuous forested ridges, the Algoma and Nipissing, run

Figure 7. Pitcher's thistle, Manistee National Forest (Michigan).

north-south about a mile inland from the bank. Hiking trails follow their crests.

Throughout the complex of sand dunes at Nordhouse are troughs and depressions that have been carved out by the wind. Sometimes the depressions are gouged out to water level and become filled, forming dune pools. Sand-tolerant wetland plants find the edges of the dune pools suitable, and a community of rushes, bulrushes, and horsetails develops. The dry troughs are home to the glasswort, whose fleshy, jointed stems lack leaves, thus preventing excessive water loss.

A number of environmental conditions constrain the growth of vegetation in the various dune communities. The driving winds not only shape the dunes but literally sandblast living and dead trees. (As Henry Cowles, the pioneer plant ecologist at the University of Chicago noted at the turn of the century, mushrooms on the windward side of logs are virtually petrified because sand grains became embedded in the soft, growing tissue.) The sand itself—mostly large-grained, white quartz sand—is a porous substrate, poor in nutrients. Water percolates easily down through the sand, washing away what few nutrients there are and becoming less available to plant life.

The light in the dunes is intense, especially along the banks of Lake Michigan, where the sparse vegetation is surrounded by the reflecting sand. Exposure to strong light can kill chloroplasts, the plant cell structures responsible for photosynthesis. For protection, the leaves of most dune plants are covered by a thick cuticle (wax layer) or by hairs. Because of the paucity of vegetation and high exposure, temperatures on the dunes are more extreme than in surrounding areas—higher in summer and lower in winter. At the same time, the growing season is shortened because the sand does not heat up until late spring and cools rapidly in the autumn.

Despite the limitations of the environment, however, a succession of plant communities succeeds in colonizing and stabilizing the shifting sands of Nordhouse Dunes. The process is constantly being repeated slightly westward, as new supplies of sand are added to the shoreline. As a result, the different stages of ecological change are arranged in orderly fashion across the landscape.

Ottawa National Forest

SIZE AND LOCATION: 989,736 acres in extreme northwestern Michigan in the Upper Peninsula. Major access routes are U.S. Highways 2, 45, and 51 and State Routes 28, 38, 64, and 73. District Ranger Stations: Bessemer, Iron River,

Kenton, Ontonagon, and Watersmeet. Forest Supervisor's Office: East U.S. Highway 2, Ironwood, MI 49938, www.fs.fed.us/r9/ottawa.

SPECIAL FACILITIES: Boat ramps; swimming beaches; winter sports areas.

SPECIAL ATTRACTIONS: Sylvania Recreation Areas; Black River Harbor Recreation Area; waterfalls; Yellow Dog National Wild and Scenic River.

WILDERNESS AREAS: Sylvania (18,327 acres); McCormick (16,532 acres); Sturgeon River Gorge (14,800 acres).

Beautiful forests, wildflowers, woods, gentle lakes, white sandy beaches, clear streams, ample wildlife, and plummeting waterfalls are all parts of the Ottawa National Forest, the northwesterly national forest in Michigan's Upper Peninsula. The forest contains more than 700 lakes and over 2,000

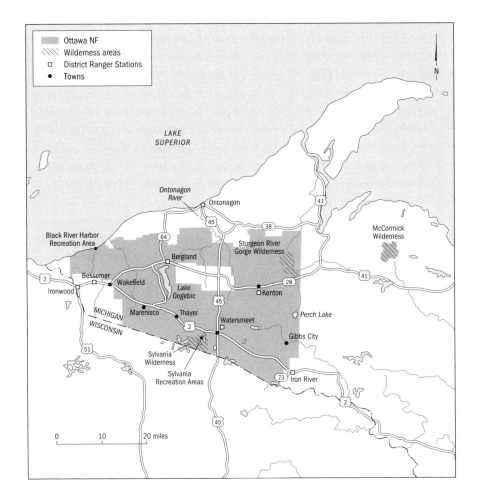

miles of streams. At any season of the year, there is something for everyone. When the snows finally melt in late spring, wildflowers bring a breath of color to the forest floor, and steelhead, brown trout, walleyed pike, northern pike, muskies, bass, and many other species of fish populate the lakes and streams. Snowmelt enhances the flow of waterfalls in the national forest. During summer, water sports bring many people out to water ski, swim, fish, and canoe. During autumn, the leaves of the hardwoods turn color, and Lake Superior steelhead, brown trout, and salmon enter the streams of the national forest on their late-season spawning runs. Winter in the Ottawa National Forest is the time for snowmobiling, cross-country skiing, downhill skiing, and ice fishing. Snowshoeing is becoming increasingly popular, too.

The Ottawa National Forest has frontage on Lake Superior at the Black River Harbor Recreation Area. The area is located at the northern end of County Road 513. The recreation area has a marina, campground, picnic areas, and a swimming beach. A segment of the North Country National Recreation Trail is here. You may hike or drive to several spectacular waterfalls in this area. As you drive north along the Black River toward the Black River Harbor Recreation Area, you will come to the Copper Peak Ski Flying Hill (not a part of the Ottawa National Forest) where downhill skiing is available in season. Proceeding north from Copper Peak, you will come to Chippewa Falls, Algonquin Falls, Great Conglomerate Falls, Potawatomi Falls, Sandstone Falls, and Rainbow Falls, each one worth a visit. The North Country National Recreation Trail begins at the Copper Peak area and winds across the northern part of the Ottawa National Forest all the way to the national forest's eastern edge.

Two major highways cross the Ottawa National Forest from west to east, splitting off from each other at the town of Wakefield, about 12 miles east of Ironwood. State Route 28 is the more northerly of the two routes, with U.S. Highway 2 several miles to the south.

By following State Route 28, you will swing around the north end of Lake Gogebic. A few miles before reaching the lake, where the highway crosses Warbler Creek, is a cedar swamp. The white cedars form a dense canopy so that the forest floor is shaded intensely throughout the day. Mosses carpet the ground, with occasional flowering plants such as goldthread, bishop's-cap (pl. 15), swamp dewberry, starflower, and oak fern growing from them. Although most of the trees in this swamp are white cedars, there are a few black ash, balsam fir, tamarack, and black spruce.

The town of Bergland is located at the northern end of Lake Gogebic, and north of town are the Trap Hills. After about 3 miles on State Route 28, a fine mesic forest is on the west, dominated by large specimens of yellow birch and sugar maple. Other trees present include red maple, northern red oak, hemlock, and white pine. Red elderberry is a common shrub, and violets are abundant on the forest floor in the spring. About 5 miles farther along on Forest Road 400 is Cascade Falls.

State Route 28 continues east from Bergland, leaving the Ottawa National Forest for a few miles. If you take U.S. Highway 45 north at Bruce Crossing, you will reenter the national forest. Soon, the highway crosses the North Country National Recreation Trail. O Kun De Kun Falls is 1.5 miles to the east. Two miles to the west, you will enter a forest dominated by quaking aspen above a shorter layer of balsam fir. In the autumn, bigleaf aster is common on the forest floor. Other species present are red maple, paper birch, bunchberry, goldthread, mayflower, and creeping snowberry.

Forest Highway 16 crosses State Route 28 at Kenton. By driving north on Forest Highway 16, you will have access to the wild and scenic northeast corner of the Ottawa National Forest. As soon as you turn onto Forest Highway 16, you will come to the Sparrow-Kenton Picnic Area. Forest Road 1100 goes west from the picnic area to the Sparrow Rapids Campground on the East Branch of the Ontonagon River. Five miles north along this very crooked river is attractive Onion Falls.

From the Sparrow-Kenton Picnic Area, Forest Highway 16 soon crosses an extensive wetland along Beaver Creek. In a few miles, shortly after Forest Highway 16 crosses the North Country National Recreation Trail, is Bob Lake, with a pleasant campground, picnic area, and boat ramp. Three miles north of Bob Lake, Forest Road 1700 to the east will bring you to Hogger Falls and West Branch Sturgeon Falls, both of them worth a side trip.

Just before leaving the Ottawa National Forest, Forest Highway 16 passes Very Sudden Pond and Sudden Pond and crosses the West Branch of the Firesteel River. Forest Highway 16 dead-ends at State Route 38 just outside the northern edge of the national forest. One and one-half miles west of State Route 38 is a short side road to Courtney Lake, where there are a campground, picnic area, and hiking trail.

Although State Route 38 never actually enters the Ottawa National Forest, Forest Road 2270 south from State Route 38 is a must route for exploring the Sturgeon River Area. In 2.5 miles there is a side road to the dam along the Sturgeon River that forms Prickett Lake. Less than 1 mile beyond this side road to the north is a fine forest dominated by sugar maple with a few large hemlocks, white pines, and yellow birches interspersed. A few mountain maples grow in these woods as well.

Continuing south, Forest Road 2270 is extremely curvy until it crosses the Sturgeon River at the northern edge of the Sturgeon River Gorge Wilderness. This wilderness extends southwest for nearly 9 miles on either side of the very scenic gorge that has been carved by the Sturgeon River. The North Country National Recreation Trail follows the northeast edge of the wilderness. Just within the wilderness boundary on the east side is turbulent Sturgeon Falls.

South of Sturgeon Falls, Forest Road 2270 comes to a T with Forest Road 2200. By taking Forest Road 2200 back to the northeast, you will be on a back road through some of the prettiest woods in the Ottawa National Forest. East of the forest road is a woods dominated by paper birch that are so densely arranged that sunlight is limited on the forest floor. Northern red oak, red maple, and hemlock are also present. Because this woods is on somewhat dry, sandy soil, the vegetation on the forest floor consists of bracken fern, bigleaf aster, mayflower, starflower, and wild sarsaparilla. Hazelnut is the common shrub in these woods. By continuing south on Forest Road 2200, you will stay along the eastern boundary of the wilderness, passing the Sturgeon River Campground.

The southern half of the Ottawa National Forest is accessed off U.S. Highway 2 from Wakefield to Iron River. About 7 miles southeast from Wakefield, U.S. Highway 2 crosses Great Lakes Road. If you take Great Lakes Road south for 1 mile and then turn west on Forest Road 8170, you will come to a beautiful mesic forest dominated by 100-year-old hemlocks. This forest, reminiscent of those found farther north, also contains balsam fir, black spruce, and white spruce. Of the many species of herbs on the forest floor, mayflower, bunchberry, bluebead lily, and several kinds of ground pines are the most common.

At Marenisco, State Route 64 heads south from U.S. Highway 2, crossing the state line into Wisconsin in about 8 miles. The Ottawa National Forest ends at the state line. On either side of this highway are dozens of small lakes interspersed among extensive marshes. Several forest roads provide access to much of the area. Many of the lakes have boat ramps, and campgrounds are available at Henry Lake, Pomeroy Lake, Langford Lake, and Mooseheart Lake.

If you take Forest Road 523 north from Marenisco, you will reach lovely Yondata Falls in less than 4 miles. As you continue eastward on U.S. Highway 2, you will come to Forest Road 527 and Thayer. By taking this forest road to the northeast, you will be able to see Kakabika Falls and Wolverine Falls. In another 2 miles is a magnificent scenic overlook along the road before it comes to U.S. Highway 45. In the vicinity of Forest Road 527 and U.S. Highway 45 are Robbins Pond and Paulding Pond campgrounds.

U.S. Highway 2 east of Thayer comes to Watersmeet in a few miles. South of town is the J. W. Toumey Nursery and the Ottawa National Forest Visitor Center. Be sure to stop at the visitor center to find out about the Sylvania Wilderness, the largest wilderness area in the Ottawa National Forest and one that is filled with lakes of all sizes. Canoeing is popular in the wilderness, and nearly three dozen primitive campgrounds are available here, some of which may be reserved in advance. Fishing for bass, lake trout, walleye, and northern pike is great. Forest Road 535 is a paved route that follows the northern and western boundaries of the wilderness, while Forest Road 6320 is a gravel road that parallels the eastern side of the wilderness. Northwest of Watersmeet a few miles is Ajibikoka Falls.

By continuing eastward from Watersmeet on U.S. Highway 2, you will be near Marion Lake and Taylor Lake campgrounds just north of the highway and Imp Lake Campground about 1 mile to the south. Along the north side of Marion Lake is a mature stand of hemlock, sugar maple, yellow birch, and basswood estimated to be nearly 250 years old. A pleasant hiking trail starts at the Imp Lake Campground.

About 1 mile after U.S. Highway 2 crosses the South Branch of the Paint River is a tamarack–black spruce bog. This swampy area is covered by dense growths of mosses out of which delicate flowering plants such as goldthread and three-leaved Solomon's-seal grow. A dense shrubby thicket is present, consisting of Labrador tea, mountain holly, a blueberry, and leatherleaf. Exploration of this area will reveal a marvelous diversity of flowering plants.

To the east of this bog along U.S. Highway 2 is Golden Lake, with a campground and boat ramp. Forest Highway 16 proceeds north from the campground and will eventually bring you to several points of interest including Tepee Lake and campground, Duppy Falls, and Jumbo Falls.

Just before U.S. Highway 2 leaves the Ottawa National Forest, about 2 miles before the town of Iron River, you may wish to drive County Road 657 north to see the easternmost part of the national forest. Near Gibbs City is the Paint River Campground, and 3 miles north of Gibbs City, between County Road 657 and Golden Creek, is a fine example of a boreal forest. In this forest are balsam fir, white pine, and quaking aspen forming a dense canopy above a moss-covered forest floor where a lot of downed trees make walking very difficult. Starflower, mayflower, and goldthread are some of the typical boreal wildflowers you will encounter.

If you continue north past the boreal forest, you will find a maze of forest roads that will eventually take you to Perch Lake, Lake Sainte Kathryn, Norway Lake, and Lower Dam Lake, all with campgrounds.

West of Iron River via the Lake Ottawa Road are the Lake Ottawa Campground and Hagerman Lake Picnic Area.

Recently incorporated into the Ottawa National Forest is a disjunct parcel of land designated as the McCormick Wilderness, named for three generations of McCormicks, the descendants of Cyrus McCormick, who invented the reaping machine. This area of approximately 27 square miles occurs 10 miles north of Champin and 3 miles east of the Baraga-Marquette county line. The Peshekee River Road passes the southwestern corner of the wilderness. The area contains several creeks and small lakes. It had been logged until the early 1900s, but since that time, a good mixture of conifers and northern hardwoods has developed. Moose have been reintroduced into the wilderness, and loons are often seen. The wilderness contains parts of the headwaters of the Huron, Dead, Peshekee, and Yellow Dog rivers, 18 lakes, and numerous swamps and bogs. The Yellow Dog, designated a National Wild and Scenic River, includes some waterfalls. White Deer Lake Trail is a 3-miler from the lake to County Road 607.

NATIONAL FORESTS
IN MISSISSIPPI

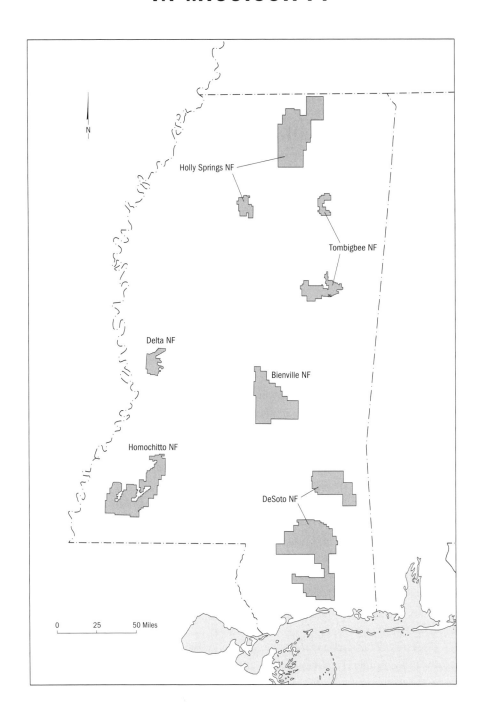

Holly Springs NF

Tombigbee NF

Delta NF

Bienville NF

Homochitto NF

DeSoto NF

N

0   25   50 Miles

Six national forests are found throughout the length and width of Mississippi. All of them occupy lands that were heavily timbered in the past and converted into farmlands. As the farmlands wore out and became severely eroded, the federal government began purchasing this land to be included in the national forest system. All of the national forests of Mississippi are in Region 8 and are administered by one forest supervisor, whose office is at 100 W. Capitol Street, Suite 1141, Jackson, MS 39261.

Bienville National Forest

SIZE AND LOCATION: Approximately 86,000 acres in central Mississippi, between Jackson and Meridian. Major access routes are Interstate Highway 20; U.S. Highway 80; and State Routes 13, 35, 481, 501, and 506. District Ranger Station: Forest. Forest Supervisor's Office: 100 W. Capitol Street, Suite 1141, Jackson, MS 39269, www.fs.fed.us/r8/miss.

SPECIAL FACILITIES: Boat ramp; swimming beaches.

SPECIAL ATTRACTIONS: Bienville Pines Scenic Area; Harrell Prairie Hill Botanical Area.

While the federally threatened red-cockaded woodpecker lives in several southern national forests, the Bienville National Forest is one of the best places to observe this rare bird since the large pines in the Bienville Pines Scenic Area create an ideal habitat for it. Located at the southeastern edge of the town of Forest, on State Route 501, Bienville Pines Scenic Area contains one of the last remaining tracts of virgin pines in Mississippi. An interpretive trail winds through the scenic area, and nest cavities of the red-cockaded woodpecker may be seen.

Less than 2 miles southeast of the Bienville Pines Scenic Area, along Forest Road 518, is the Harrell Prairie Hill Botanical Area. This 150-acre tract preserves one of the few patches of prairie to be found in Mississippi. When one thinks of prairies, Mississippi does not come to mind, but here in the heart of the state is a small tract of prairie. Natural prairies were originally found in the Blackbelt and Jackson Prairie physiographic regions of Alabama and Mississippi, but nearly all of these natural prairies were destroyed by farming.

Harrell Prairie Hill is mostly treeless, although the prairie is surrounded by loblolly pines. Patches of natural prairie occur among clumps of shrubs, and because several nonnative species are present, the prairie has a weedy

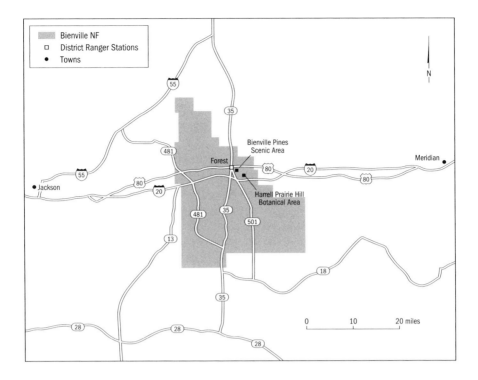

appearance. However, a number of good prairie species occur. Grasses include big bluestem, broomsedge, bushy broomsedge, Indian grass, and switchgrass. Prairie wildflowers present are purple coneflower, blazing star, false boneset, two species of green milkweed, purple prairie clover, white prairie clover, shiny aster, prairie tick trefoil, hairy gentian, oxeye sunflower, mountain mint, and wild bergamot. Woody plants that constantly invade the prairie are roughleaf dogwood, deciduous holly, hawthorn, woolly buckthorn, and vines such as supplejack, snailseed, and sweet potato vine.

The Bienville National Forest has two extensive wildlife management areas: the Bienville Wildlife Management Area in much of the forest's northwestern corner, and the Tallahala Wildlife Management Area in most of the southeastern corner. Just outside the western edge of the Tallahala Wildlife Management Area is Marathon Recreation Area, centered around the reverse J–shaped Marathon Lake. Many recreation features are here, including a campground, boat ramp, swimming beach, and hiking trail. Shockaloe Base Camps I and II are along the 23-mile Shockaloe Horse Trail in the Bienville Wildlife Management Area, and each has a campground, picnic area, and hiking trail.

Shongelo Recreation Area, around 4-acre Shongelo Lake near the southern end of the Bienville National Forest, is a popular destination, too. This region has a campground and a swimming area. A short hiking trail that encircles the lake will enable you to become familiar with the vegetation in this part of the national forest.

Coursey Lookout Tower in the southwest corner of the Bienville National Forest, 3 miles north of State Route 18, is of historical interest.

Six miles south of Marathon Lake is a greentree reservoir along Little Achusa Creek. The Forest Service regulates the amount of water in the area in an effort to enhance waterfowl habitat.

Bienville Pines

In the early 1700s, when pioneer settlers first entered the huge region that stretches from the Carolinas and Florida to eastern Oklahoma and Texas, they found most of it forested with southern pine trees. The oldest trees often had trunks 5 feet thick, but the pioneers cleared much of the land for agriculture. Later, more trees were cut for lumber and to make pulp for the paper industry. In Mississippi's Bienville National Forest, however, 189 acres of old-growth woods are preserved and managed as the Bienville Pines Scenic Area.

That this stand of old trees survived the lumberman's ax is ironic, because the tract was once owned by a lumber company and was adjacent to the sawmill. But the lumber company decided to harvest the pines farthest from the mill first, saving the nearby trees for later. When the U.S. Forest Service acquired the Bienville Pines area in 1935, it decided to leave the trees uncut because the habitat had become a rarity. The area is not pristine, however. During World War II, 150,000 feet of timber were removed for the war effort; in 1952, a tornado further damaged the forest.

A trail winds for 2 miles through the old-growth forest, dominated by loblolly, shortleaf, and longleaf pines. These trees thrive in the well-drained, sandy soils of the southeastern Coastal Plain. They also have thick, insulating bark that protects them from fires, which in the past often raged across the region (with thunderstorms occurring 60 to 90 days a year, lightning strikes are frequent). Some of the loblolly pines are 200 years old and more than 100 feet tall and 3 feet in diameter. Hardwood trees are also interspersed throughout the woods—white oak, winged elm, black gum, and, in moister areas, sweet gum, red maple, and deciduous holly. Unfortunately, poison ivy abounds on the forest floor.

One organism that clings to existence in the Bienville Pines and other surviving old-growth pine areas is the red-cockaded woodpecker. It is on the federal list of endangered species. Ornithologist Jerome A. Jackson has studied this woodpecker in great detail and has fought tenaciously for its protection.

Unlike most other woodpeckers, which carve out cavities for their nests in dead trees, the red-cockaded woodpecker uses living pine trees, which are able to withstand the fires. The trees must be at least 60 years old, the age at which the center of most southern pine trunks begins to rot from the fungus that causes red heart disease. Over a period of months or even years, the woodpecker penetrates the living tissues of the tree until it reaches the rotting center, which is more easily removed to create a nest or roost cavity. The cavities are used year after year.

As a pine grows older and taller, the lower branches drop off, leaving damaged areas where red heart fungus spores can enter the tree. The woodpecker sometimes digs the opening for its nest cavity near these branch scars. The older and taller the tree, the higher up the woodpecker can make its nest, far above any shrubby vegetation growing beneath the tree. Fires also help by killing the lower vegetation. If shrubs do reach up to the nest cavity, the woodpecker usually abandons it. Some suggest that this abandonment may be due to the increased potential for predators, such as the gray rat snake, to reach the cavity.

Gray rat snakes can climb trees rapidly, an ability that may enable them to escape ground fires; however, they have an aversion to the gummy resin that pine stems produce. To capitalize on this, the red-cockaded woodpecker chisels numerous small holes around the opening into the nest cavity. Thick, gummy pine resin exudes from these "resin wells," flowing all around the entrance of the nest cavity. Gray rat snakes climbing a tree with nest cavities stop and try to turn around when they reach the pine gum and typically fall from the tree. They are unable to get past this barrier unless the resin dries out and is not re-covered by fresh resin.

To test the effects of pine resin on gray rat snakes, Jackson captured six snakes and caged them with two pine logs. The snakes climbed over both logs until resin was smeared on one of the logs. The snakes then avoided the resin-smeared log but continued to explore the other one. When Jackson applied a 2-inch band of resin around an upright pine stem, a gray rat snake was able to bridge the ring and then continue its upward journey. Snakes that got fresh resin on their body, however, writhed uncontrollably. While most of the snakes that were coated recovered in 3 or 4 days, one of them died.

Red-cockaded woodpeckers have a distinctive social life. They live in clans of two to nine birds, each member roosting in its own tree cavity. Each clan includes only one breeding pair; the other birds are usually this pair's young male offspring, 1 to 3 years old, which help dig resin wells, incubate eggs, and feed hatchlings. When the breeding male dies, one of the young males in the clan usually assumes the dominant role. According to Jackson, the young females leave the clan by winter and fly around nearby in search of a clan that has lost its breeding female. For a population to survive, then, several clans of woodpeckers must occupy a continuous stand of pines.

Because agriculture and lumbering have broken up the old-growth forests, the number of red-cockaded woodpeckers has plummeted to perhaps 10,000 birds. On top of this, in autumn 1989, Hurricane Hugo knocked down 80 to 90 percent of the nest trees in South Carolina's Francis Marion National Forest, which supported one of the largest populations of red-cockaded woodpeckers.

When a species such as the red-cockaded woodpecker is officially included on the federal list of endangered species, certain positive things are supposed to happen. One is that any listed animal or plant must be protected on federally managed property. Accordingly, most national forests where the red-cockaded woodpecker survives are taking some steps to preserve the trees in which the woodpeckers nest and roost. Another is that funds may be made available for the preparation and implementation of "recovery plans," step-by-step management plans designed by experts to save the species from danger of extinction.

From 1975 until 1982, Jackson headed a recovery team that suggested ways to improve conditions for the red-cockaded woodpecker. One recommendation was to rotate the harvest of pine stands so there would always be ample old-growth stands where woodpeckers could nest, roost, and forage for food. Although the timber industry is generally good about replanting new pine trees after a harvest, they prefer to cut down the new trees long before they become 60 years old (and develop rotted centers). In addition, they prefer to clear-cut all trees in a stand, regardless of age. Jackson's plan called for leaving some old-growth pines in any forest stand from being cut.

As soon as Secretary of the Interior Cecil Andrus approved the recovery plan in 1979, loud protests were heard from the timber industry and foresters. Jackson and his committee were dismissed in 1982, and a U.S. Forest Service biologist was contracted to revise the plan. The U.S. Fish and Wildlife Service approved the essentially new and different recovery plan in 1985. A special committee of the American Ornithologists' Union, however, has criticized the plan as providing at best only the minimum required protection for the red-cockaded woodpecker.

Meanwhile, Jackson has been proposing another way to help the birds: by providing a continuous corridor of old-growth pines along the right of way of interstate highways that cross through red-cockaded woodpecker habitat. Several clans of woodpeckers live near the highways and seem to be little distracted by the traffic and noise. (For example, there is a currently abandoned nesting site on the median of Interstate 20 near the Bienville Pines Scenic Area.) Jackson recommends preserving old pines along the highways, as well as adjacent pine stands, regardless of age, which the birds will use for foraging. He also suggests mowing the understory beneath the old-growth pines to keep down the brushy vegetation (burning would be easier but would create a smoke hazard for motorists).

Besides these strategies, the U.S. Forest Service has instituted several additional ones more recently for saving this species.

Delta National Forest

SIZE AND LOCATION: 59,000 acres in west-central Mississippi north of Vicksburg and east of Rolling Fork. Major access route is State Route 16. District Ranger Station: Rolling Fork. Forest Supervisor's Office: 100 W. Capitol Street, Suite 1141, Jackson, MS 36269, www.fs.fed.us/r8/miss.

SPECIAL FACILITY: Boat launch area.

SPECIAL ATTRACTIONS: Five greentree reservoirs; three research natural areas; Blue Lake Recreation Area.

For hundreds of years, the Mississippi, Yazoo, and Big Sunflower rivers often flowed out of their banks, flooding hundreds of square miles of bottomlands in west-central Mississippi. This area is known as the Mississippi Delta, and it is composed of rich alluvial soils. For centuries this rich bottomland has been home to lush bottomland hardwood forests where trees often reach heights of 100 feet or more. Early settlers, being primarily farmers, cut much of the timber and cleared the land for cultivation of crops, but frequent flooding often wiped out the crops. Much of the land was cultivated long ago but gradually abandoned, allowing the delta to grow back into forests.

Beginning in the mid-1950s, improved flood control, including the building of levees, brought on another rush of land clearing, this time in an effort to grow soybeans. In the meantime, the federal government began buying up the hardwood forests that remained, so that today, the Delta National Forest contains only traces of the magnificent bottomland hardwood forests that once covered the Mississippi Delta.

Today, the Delta National Forest is being managed to attract waterfowl and other wildlife and to preserve the tracts of rich bottomland forests. Four greentree reservoirs have been developed by the national forest, and a fifth on Forest Service property has been constructed by the Mississippi Department of Wildlife, Fisheries, and Parks.

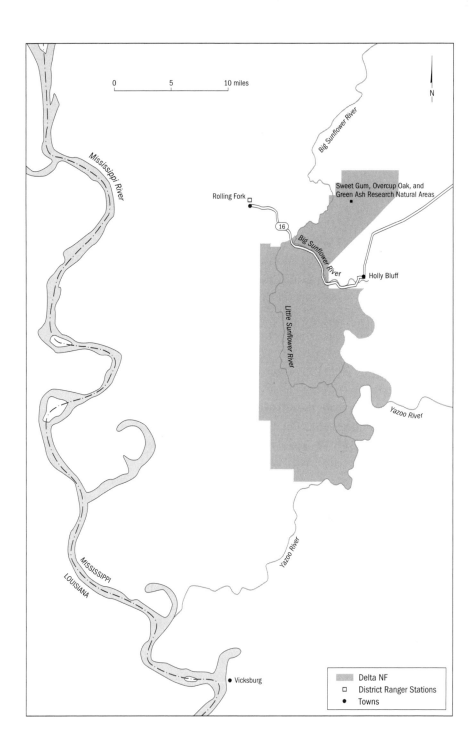

Within the map:

0   5   10 miles

N

Big Sunflower River

Mississippi River

Rolling Fork □
●

16

Sweet Gum, Overcup Oak, and
Green Ash Research Natural Areas
●

Big Sunflower River

Holly Bluff
●

Little Sunflower River

Yazoo River

MISSISSIPPI
LOUISIANA

Yazoo River

● Vicksburg

Delta NF
□ District Ranger Stations
● Towns

Greentree reservoirs are areas where forests of living trees, primarily oaks, are flooded, by either pumping or damming a creek or stream to retain standing water in the forest during the winter months. This provides for enhanced waterfowl habitat. Mallards (pl. 16) and some other waterfowl feed at or just below the surface of the water in search of acorns, as well as insects, worms, snails, and a variety of crustaceans. Nesting productivity of ducks is enhanced by the amount of food consumed by the female ducks on their northward migration. The five greentree reservoirs in the Delta National Forest cover more than 6,400 acres.

Dowling Bayou Greentree Reservoir is at the northern end of the Delta National Forest, on the east side of Dowling Bayou. It is reached from the end of Forest Road 706-D. One mile to the southwest is Green Ash Greentree Reservoir, on the east side of Little Sunflower River. The southern half of this area is bordered by Forest Road 717.

Sunflower Greentree Reservoir on the south end of Big Sunflower River is bordered on the north by Forest Road 715 and on the northeast by State Route 16. Long Bayou Greentree Reservoir is at the western edge of the Delta National Forest, with Forest Road 707 forming the northern and western borders. Near the southern end of the national forest is the South Greentree Reservoir, located between Six Mile Cutoff, Big Sunflower River, and Yazoo River.

Trees that live in the bottomland hardwood forests of the Delta National Forest include Nuttall's oak, overcup oak, cherrybark oak, willow oak, water oak, bur oak, swamp chestnut oak, water hickory, sweet gum, box elder, swamp red maple, American elm, green ash, sugarberry, September elm, honey locust, pecan, black gum, red mulberry, and persimmon. Persimmon, usually a small tree in the forests of the midwestern and eastern United States, may reach heights up to 80 feet in the bottomland forests. A good complement of shrubs consists of swamp privet, green hawthorn, storax, swamp dogwood, palmetto, deciduous holly, and, in slightly drier areas, roughleaf dogwood. The federally threatened pondberry, in the same genus as spicebush, occurs in shallow depressions. Giant cane, a woody grass, is also common. Woody vines are plentiful in the bottomland forests, including poison ivy, Virginia creeper, winter grape, muscadine, cross-vine, supplejack, peppervine, and four species of greenbriers. Also here are several nonwoody vines that die back each winter. Among these are climbing milkweed, bluevine, Mohlenbrock's pea, snailseed, Boykin's gourd, squirting cucumber, climbing hempvine, ladies' eardrops, climbing dogbane, and a blue-flowered clematis.

Herbaceous plants that flower in the spring in the bottomland hardwood forests include white avens, buttercups, butterweed, blue violets, Virginia knotweed, spring corydalis, daisy fleabane, forget-me-nots, bluestar, and several species of sedges in the genus *Carex*. During summer and autumn, conspicuous plants in the herbaceous understory are fernleaf trepocarpus (related to Queen Anne's lace), stingless nettle, stinging nettle, eclipta, false pimpernel, toothcup, ditch stonecrop, camphor weed, boneset, bedstraws, late goldenrod, clearweed, black snakeroot, slender penstemon, and brookweed.

Birds that use the Delta National Forest extensively include red-shouldered hawks, Mississippi kites, yellow-crowned night herons, great blue herons, great egrets, snowy egrets, black-necked stilts, solitary sandpipers, pileated woodpeckers, red-headed woodpeckers, common yellowthroats, indigo buntings, yellow warblers, wood ducks, mallards, gadwalls, wigeons, blue-winged teals, lesser yellowlegs, summer tanagers, wild turkeys, and prothonotary warblers.

Wildlife you may encounter are white-tailed deer, bobcats, river otters, alligators, and several kinds of snakes, including the cottonmouth.

Three significant old-growth tracts of bottomland forest have been designated Research Natural Areas. Sweet Gum Research Natural Area contains huge sweet gum trees, some of them probably at least 300 years old. To reach this area, drive north out of Holly Bluff on State Route 13 for 4 miles. Turn west on Anguilla Road. In a little less than 3 miles, you will come to the Sweet Gum Research Natural Area on the west side of the road.

One-half mile before reaching the Sweet Gum Research Natural Area, turn west off Anguilla Road onto Forest Road 706-I. In about 1 mile, this forest road loops around the Overcup Oak Research Natural Area. This 40-acre site contains large overcup oaks and water hickories estimated to be nearly 200 years old.

Green Ash Research Natural Area is adjacent to the south side of Forest Road 717-B, the road that follows the southern boundary of the Green Ash Greentree Reservoir. This is a 60-acre tract with huge green ash trees that are probably 250 years old. Growing with these mammoth ashes are large specimens of American elm and sugarberry.

The only developed recreation area in the Delta National Forest is Blue Lake. The campground is nestled beneath handsome, wispy bald cypress trees laden with Spanish moss. Fishing in the lake is said to be good for catfish, bass, and bream. A boat ramp is also here.

Sweet Gum Research Natural Area

A vast floodplain, the delta of the Mississippi River covers 37,500 square miles south of Illinois, especially in Louisiana, Arkansas, and Mississippi. The sands and clays transported from the Upper Mississippi basin, an area extending from Pennsylvania to Montana, have made soils in the delta among the richest in the country. When European American pioneers entered the region in the 1820s, they found it mostly covered by mature hardwood forests (deep woods), although Native Americans had cleared openings for agriculture and better hunting grounds. Later in the century, settlers cut down most of the forests for home building, fuel, and various commercial endeavors and subsequently placed the fertile soil under intense cultivation.

When the Mississippi River and its tributaries stayed within their banks, farmers prospered, but more often than not, floodwaters inundated crop after crop. Most farmers soon abandoned the effort to grow crops. Only after the Army Corps of Engineers built an extensive levee system, in the first decade of the 20th century, was much of the cropland reclaimed. In places, the abandoned farmland was never returned to production but was allowed to develop into second-growth forest.

In 1936, the U.S. Forest Service was authorized to purchase 13,200 acres of delta forest in Mississippi, mostly where the trees had never been harvested. This virgin forest area was later rounded out with another 46,650 acres of second-growth forest, and in 1961, the Delta National Forest was officially established. In accordance with its main mission at that time, the Forest Service proceeded to contract with lumber companies to cut down the trees.

All the virgin timber, except for three parcels totaling 160 acres, fell to the ax and the saw. Those remaining acres are now under special management as research natural areas. One of these is the Delta Sweet Gum Research Natural Area, a 50-acre zone located near the confluence of the Yazoo and Big Sunflower rivers. It contains perhaps the only virgin stand of sweet gum (fig. 8) in the world.

Approaching the natural area in mid-April from the tiny village of Holly Bluff, I parked where the road passed by the area's eastern edge and set out on foot. First I had to fight my way through several feet of giant cane. The only bamboo grass native to the eastern United States, giant cane has a hard stem, often fashioned into fishing poles. It often grows 10 feet high or more, and the stems may grow so close together that they form a thicket known as a *canebrake*. Hardly any other vegetation can survive in canebrakes, but a number of animals inhabit them, among them canebrake rattlesnakes. In recent years, the elusive, endangered Bachman's warbler lived here.

After getting through the canebrake, I came to a more open forest and the first sweet gums. Sweet gum, with its star-shaped leaves and spherical, prickly fruits containing numerous tiny seeds, is a common component of bottomland forests from Texas to Florida and ranges as far north as southern Illinois

Figure 8. Sweet gum, Delta National Forest (Mississippi).

and southern New England. Because the wood has been valued for woodenware, general construction, and paper pulp, few of the original sweet gums encountered by the first settlers still stand. The oldest of the denizens in the natural area are estimated to be about 300 years old; one is 130 feet tall with a diameter greater than 5 feet.

Where the sweet gums are densely spaced, few other tree species reach the canopy. Saw palmetto is scattered throughout the understory, while shrubs and small trees include deciduous holly (also know as possum haw), American snowbell, green hawthorn, and swamp dogwood. Here and there are thickets of pondberry, an endangered species of shrub related to the spicebush common in moist sugar maple woods throughout the eastern United States. Wildflowers beneath the sweet gums include hooked buttercup, violets, white forget-me-not, Virginia knotweed, white avens, dwarf nettle, and swamp goldenrod.

Forest Service reports dating back to the 1950s describe this virgin stand as a closed canopy of sweet gums, with no intrusion by other trees. But something must have happened in the late 1950s or early 1960s to change this— perhaps a tornado swept through the area—because today there is a break running through most of the canopy. Some large, dead sweet gums are standing in the forest, and many others are lying on the ground. With the canopy opened up, box elders have filled the void, some of them 60 feet high with girths of 3 feet. There are also hackberries, red maples, and green ash trees, as well as tangled thickets of vines—blackberries, greenbriers, Virginia creeper, poison ivy, and pepper vine.

The terrain in the Sweet Gum Research Natural Area drops almost imperceptibly from east to west. Visiting in April, I found the soil beneath the sweet gums to be moist. As I proceeded west, I encountered standing water, and beyond that point I saw only other types of trees instead of sweet gum. In shallow water, swamp red maple, American elm, and Nuttall's oak were growing; in deeper water, a little farther along, there were occasional bald cypresses.

DeSoto National Forest

SIZE AND LOCATION: 500,500 acres in southeastern Mississippi, between Laurel and Gulfport. Major access routes are U.S. Highway 49 and State Routes 15 and 29. District Ranger Stations: Laurel and Wiggins. Forest Supervisor's Office: 100 W. Capitol Street, Suite 1141, Jackson, MS 39269, www.fs.fed.us/r8/miss.

SPECIAL FACILITIES: Boat ramps; swimming beaches; all-terrain vehicle (ATV) area.

SPECIAL ATTRACTIONS: Black Creek National Recreation Trail; Tuxachanie National Recreation Trail; Gavin Auto Tour.

WILDERNESS AREAS: Black Creek (5,050 acres); Leaf (940 acres).

Mississippi's DeSoto National Forest, the state's southernmost that occupies the southeastern lobe of the state, extending nearly to the Gulf of Mexico, consists of heavily wooded uplands on sandy ridgetops and vast swampy forests in the floodplains along the rivers and streams.

To get into the heart of the DeSoto National Forest, the National Forest has developed several hiking trails, including the 41-mile-long Black Creek National Recreation Trail, the 23-mile-long Tuxachanie National Recreation Trail, and interesting trails at the Big Biloxi and Turkey Fork recreation areas. The forest also has provided two horse trails and a 69-mile Bethel ATV Trail.

The Black Creek National Recreation Trail's western trailhead is at Big Creek Landing on Rock Hill Road, about 4 miles west of U.S. Highway 49 north of Wiggins. The trail accesses a good cross section of the DeSoto National Forest, from forested sandy ridges to dry forest to mesic forest to swampy woods along the creeks. After crossing U.S. Highway 49, the Black Creek Trail meanders through the woods, coming near Moody's Landing on Forest Road 301. Here there is a campground and boat ramp for access to the Black River. The trail stays along the western side of the creek, eventually entering the Black Creek Wilderness, which it crosses diagonally. Leaving the wilderness, the trail continues for 6 miles to Fairly Bridge Landing on Black Creek where the trail ends. Fairly Bridge Landing has a campground and boat ramp. Where the trail is on the dry ridges, the dominant vegetation is loblolly and longleaf pine, with a small tree and shrub layer of flowering dogwood, American holly, yaupon, and gallberry. On midslopes, sweet gum and yellow poplar are plentiful. The bottomland woods consist of sweet gum, sweet bay, willow oak, overcup oak, and red maple above a shrub layer of yaupon, dahoon holly, titi, and giant cane. In the wettest areas where water stands all year are bald cypress and tupelo gum.

Black Creek winds through the heart of the DeSoto National Forest and is a fine creek in which to canoe and kayak. It has beautiful wide, white sandbars. From Moody's Landing, Black Creek has been designated a National Wild and Scenic River for 21 miles. This stretch is often lined by steep bluffs and dense vegetation. When canoeing or kayaking, keep an eye out for stumps and fallen trees in the creek. During the spring, colorful azaleas are common along the creek.

The Tuxachanie National Recreation Trail is a great way to observe most of the plant communities that occur in the DeSoto National Forest. The first 5 miles of the 17-mile trail follow an abandoned logging railroad of the Dantzler Lumber Company. Not only did the railroad haul logs to the sawmills, it also provided transportation for the loggers and their families to the company store. As you hike across some of the creeks along the trail, look around for the remains of old trestles. At the eastern trailhead along U.S. Highway 49 is a group of live oaks planted in 1935 by the chief forester of the Dantzler Lumber Company. Nearby are some of the largest farkleberries you will see anywhere. Farkleberry is a type of highbush blueberry that usually is a shrub, reaching heights of about 10 feet. At the trailhead are small trees of farkleberry, some of them in excess of 30 feet in height. The trail begins in an upland where the common trees you will observe are sassafras, wild black cherry, tulip poplar, southern magnolia, loblolly pine, flowering dogwood, sweet olive, and blackjack oak. Beneath these trees is a shrub layer of large gallberry, American beautyberry, and little-leaved blueberry. Bristly greenbrier and Small's greenbrier are common, prickly vines. The forest floor has a scattering of lyreleaf sage, beefsteak plant, roughleaf goldenrod, Carolina elephantsfoot, purple-leaved woodsorrel, beggarslice, Louisiana shield fern, and Japanese climbing fern. After several hundred feet, the trail drops down to a small creek bordered by a low, swampy woods. In this much wetter habitat is a canopy layer of sweet gum, swamp gum, laurel oak, loblolly bay, sweet bay, and slash pine above shrubs of Virginia sweetspire, deciduous holly, black elderberry, and shrubby St. John's wort. Vines in this wetland include climbing hydrangea, fox grape, and the red-fruited Walter's greenbrier. Two kinds of little white-headed asters, a beggarstick, and a smartweed are the common herbs. In shallow depressions are a beaked rush and Howe's sedge.

In 1 mile, the Tuxachanie Trail crosses West Creek, eventually entering a savanna where longleaf pines are the dominant trees. In the savanna are the yellow pitcher plant and parrot pitcher plant, roundleaf sundew and threadleaf sundew, grass-pink orchid and pale grass-pink orchid, false blaz-

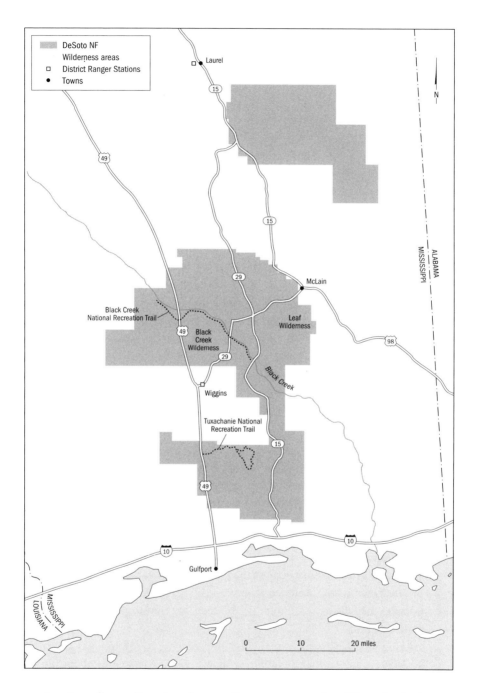

ing star, yellow colicroot, and many other savanna species of flowering herbs. After 4.8 miles of the trail is the Airey Recreation Area where the trail crosses State Route 67. Small Airey Lake is here along with a campground and restroom facilities. The trail curves southward to Copeland Springs and then follows the south side of Boggy Branch Creek for nearly 2.5 miles. At this point there is a 12.5-mile loop. By following the loop, you will cross Boggy Branch Creek and the upper end of Big Fork Creek. After 6.5 miles along the loop, a 0.3-mile side trail leads to the site of a World War II German prisoner of war camp. Nothing remains now except the foundations of a few of the buildings and some old ammunition bunkers. There is a primitive campground here, as well as a small lake. You may also reach the POW camp site by driving Forest Road 402. The return loop lies between Bridge Branch to the west and Spike Back Creek to the east, passing Alligator Pond and Duck Pond before completing the loop.

West of U.S. Highway 49 is the Big Biloxi Recreation Area, which has a large campground, shelters, restroom, and picnic areas on a terrace above Big Biloxi Creek. From the campground is a 1-mile loop hiking trail that winds across an upland woods dissected by a few streams. Walking the trail will reveal an incredible number of species of woody plants. Near the beginning of the trail are sweet gum, southern red oak, longleaf pine, loblolly pine, and water oak. As the trail leaves the campground, red maple, witch hazel, sourwood, black gum, persimmon, American holly, southern magnolia, and wild black cherry are quickly encountered. Perhaps even more striking is the great diversity of shrubs along the first 200 feet of the trail. In this short distance you will find arrowwood viburnum, American beautyberry, Sebastian-bush, wax myrtle, buckwheat tree, farkleberry, little-leaved blueberry, deciduous holly, smooth sumac, winged sumac, yaupon, titi, fetterbush, and Virginia sweetspire.

A depression of about 1,800 square feet adjacent to the campground frequently fills with water and has developed a significant wetland flora. In this small area are six species of rushes, two kinds of beaked rushes, a spikerush, blue sedge, rough panicgrass, two kinds of yellow-eyed grass, a pitcher plant, threadleaf goldenrod, yellow meadow beauty, orange milkwort, and coinwort.

Near the southernmost edge of the DeSoto National Forest is the Turkey Fork Recreation Area, centered around Turkey Creek Reservoir. The recreation area includes a campground, picnic area, boat ramp, swimming beach, and hiking trail.

Along Railroad Creek near the junction of State Route 15 and County Road 450 is a large wetland complex with nearly every wetland shrub known to occur in Mississippi. The dominant shrubs are buckwheat tree or black titi and swamp cyrilla, or southern titi, but other common shrubs are two kinds of gallberries, yaupon, dahoon holly, myrtleleaf holly, and two kinds of wax myrtles. Occasional bald cypress and slash pine trees form a sparse canopy.

At the extreme eastern edge of the DeSoto National Forest, 6 miles southeast of the village of McClain, is the 940-acre Leaf Wilderness. Leaf River flows through this area of extensive bottomland forests with a few upland ridges. Horseshoe Lake, an old oxbow, is at the northeastern corner of the wilderness. A 1.5-mile hiking trail in the Leaf Wilderness is mostly on a boardwalk above the wetlands.

Bigfoot Horse Trail Camp, at the end of Forest Road 401-B near the Tuxachanie Hiking Trail, is a 21-mile loop route that encircles Spike Back Creek. The longer Longleaf Horse Trail begins near the junction of Forest Roads 213 and 218 about 12 miles southeast of the town of Chickasawhay.

The Bethel ATV Trail meanders for 69 miles through mostly pine-covered forests. More experienced ATV enthusiasts can try the steeper hills of the 31-mile Rattlesnake Bay ATV Trail, 25 miles north of Wiggins on Forest Road 312. Little Tiger ATV Trail is in the Chickasawhay Ranger District, accessed from the end of Forest Road 220 just south of Forest Road 202. The Bethel Bicycle Trail has 35 miles of challenging trails for the mountain biker. It is along Forest Road 426, which branches off of State Route 15 about 12 miles north of Interstate Highway 10.

For those wishing to observe a part of the DeSoto National Forest from the comforts of a vehicle, drive the 12-mile Gavin Auto Tour, which has interpretive markers that discuss the past land use of the area and current forestry practices. The tour begins near the Longleaf Horse Trail Camp at the junction of Forest Roads 213 and 218 and follows Forest Road 218 until it meets Forest Road 202. It then follows Forest Road 202 west until it leaves the DeSoto National Forest at the tiny settlement of Ovett.

Five miles southwest of the village of Avent along Forest Road 327 is the second-largest southern magnolia in Mississippi. You will marvel at the size and beauty of this tree.

Of historical interest is the General Jackson Interpretive Trail that describes General Andrew Jackson's march along the Old Federal Road during the War of 1812. The entire march was between Mobile and New Orleans. That part of the trail that passes through the DeSoto National Forest may be accessed from State Route 29 about 10 miles north of Wiggins.

Historic lookout towers are scattered throughout the DeSoto National Forest. Paret Lookout is along Tower Road (County Road 302), about 7 miles south of New Augusta. Maxie Lookout is at the junction of U.S. Highway 49 and Forest Road 320 a short distance north of Wiggins. Hickman Lookout is on the Perkinston Silver Run Road at the extreme edge of the national forest. Airey Lookout is along the Tuxachanie National Recreation Trail. Harrison Lookout is at the end of Forest Road 432-C, 5 miles southeast of the Big Biloxi Recreation Area. Wausau Lookout, in the heart of the Chickasawhay

Ranger District, is located along Forest Road 206. Strong Lookout is along the north side of State Route 42, a few miles northeast of the Turkey Fork Recreation Area.

The Ashe Nursery, constructed by the Civilian Conservation Corps in 1936, produces seedlings and small trees for planting in the national forests of the south. The nursery, located along Forest Road 3008 a few miles southeast of Brooklyn, is open to the public. Nearby is the 8-acre Ashe Lake with a primitive camping area, picnic tables, and a fishing pier. Also within the DeSoto National Forest are the Erambert Seed Orchard between Janice and Oak Grove on State Route 29, and the Black Creek Seed Orchard just southwest of the Ashe Nursery. The Harrison Experimental Forest is 6 miles west of Saucier.

Holly Springs National Forest

SIZE AND LOCATION: Approximately 147,000 acres in northern and north-central Mississippi, extending nearly to the Tennessee state line. Major access routes are Interstate Highway 55; U.S. Highways 72 and 78; and State Routes 4, 5, 6, 30, and 370. District Ranger Station: Oxford. Forest Supervisor's Office: 100 W. Capitol Street, Suite 1141, Jackson, MS 39269, www.fs.fed .us/r8/miss.

SPECIAL FACILITIES: Boat launch areas; swimming beach.

SPECIAL ATTRACTIONS: Chewalla Recreation Area; Puskus Recreation Area.

Beginning only 1 mile from the Tennessee border and stretching southward for 44 miles, the Holly Springs National Forest consists of several parcels of forested lands amid numerous private inholdings. The topography is mostly rolling hills, with some of the higher hills bearing the name of mountains, even though they are only hills. The Little Tallahatchie River separates the northern part of the national forest from the southern, and the areas on either side of the river are mostly swampland.

The drier uplands of the Holly Springs National Forest are covered by forests of white oak, northern red oak, southern red oak, black oak, bitter-nut hickory, mockernut hickory, white ash, slippery elm, and wild black cherry. Flowering dogwood and redbud provide brilliant color to the forests during late April and early May. On mesic slopes, the forest usually consists of tulip poplar, black gum, basswood, and hackberry, while the bottomland forests are dominated by sweet gum, red maple, green ash, and American elm. In areas with standing water for much or all of the year, bald cypress is often found, sometimes associated with pumpkin ash and tupelo gum.

Numerous county and Forest Service roads penetrate the national forest, providing easy access to all parts of the area.

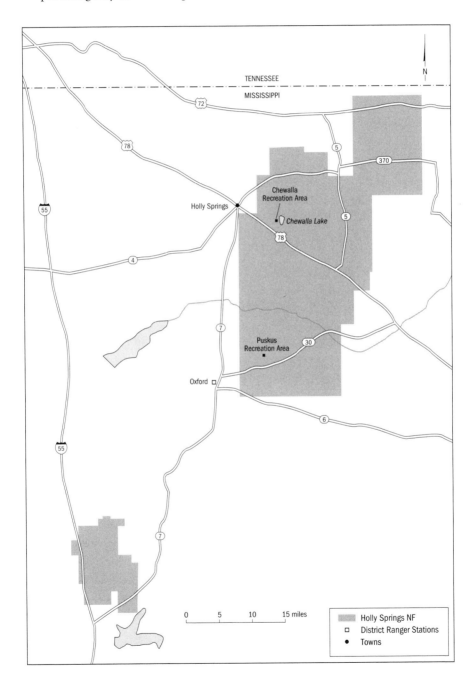

Most of the organized activities in the Holly Springs National Forest center around the Chewalla and Puskus recreation areas. Chewalla Recreation Area lies just 7 miles southeast of Holly Springs and contains Chewalla Lake. A boat launch area and a swimming beach are maintained by the Forest Service, and a nature trail begins at the boat launch area and follows the shoreline of the lake to the swimming beach. Midway along the trail is a wooden overlook deck that provides an unobstructed view of the surrounding area, including outcroppings of red sandstone. The area at one time was home to the Choctaw Indians, and a ceremonial burial mound has been constructed near the lake. The mound is surrounded by a split rail fence. A large campground is along the west side of the lake.

The Puskus Recreation Area is in the southern part of the Holly Springs National Forest, about 10 miles east of Oxford. A boat launch area at the edge of Puskus Lake and a campground near the southeastern corner of the lake are amenities at the recreation area.

Many small lakes are dotted throughout the forest, attracting great numbers of fishermen. Numerous old cemeteries add historic touches to the national forest. The Benton Lookout Tower is located 2 miles south of the Tennessee border a short distance from Walnut Lake.

Another region of the Holly Springs National Forest, the Yalobusha Unit between Oakland and Grenada, is managed by the Tombigbee National Forest. See that unit (p. 187) for a description.

Homochitto National Forest

SIZE AND LOCATION: 189,000 acres in southwestern Mississippi south of Natchez. Major access routes are U.S. Highways 84 and 98 and State Routes 28, 33, and 550. District Ranger Station: Meadville. Forest Supervisor's Office: 100 W. Capitol Street, Suite 1141, Jackson, MS 39269, www.fs.fed.us/r8/miss.

SPECIAL FACILITIES: Boat launch area; swimming beach.

SPECIAL ATTRACTIONS: Clear Springs Recreation Area; Pipes Lake Recreation Area; Mount Nebo Recreation Area.

When the first settlers arrived in the southwestern corner of Mississippi in the early 1700s, they cleared most of the forests so that they could plant corn, cotton, wheat, rye, tobacco, and potatoes. Numerous large plantations developed north of Natchez. From 1795, when the cotton gin was invented,

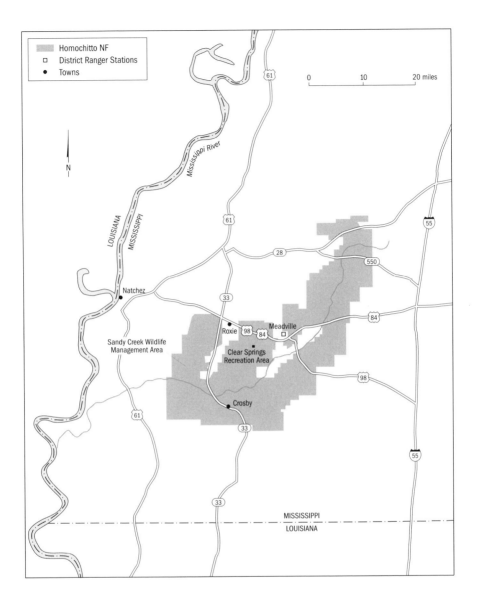

until the Civil War, the area was one of the greatest cotton-producing areas in the country. After the Civil War, the land began to show wear and tear from constant farming, and, when the boll weevil began to decimate the cotton crops in 1908, most of the area was abandoned and left to reforest itself. In 1936, the federal government purchased land that was to become the Homochitto National Forest, and Forest Service management techniques have enabled the forests to recover into attractive woods. Today, exploration for oil and gas is a major endeavor on the Homochitto National Forest, with

86 percent of the producing wells on Forest Service land in Mississippi in the Homochitto National Forest.

The Homochitto National Forest consists today of densely forested hills with slopes that drop abruptly into deep ravines. Loblolly and shortleaf pine are common on the ridges, while the wooded slopes include southern magnolia, American holly, bigleaf magnolia, flowering dogwood, southern red oak, and wild black cherry. In lowlands are bald cypress (many draped with Spanish moss), tupelo gum, and swamp red maple.

One of the most attractive areas in the Homochitto National Forest is Pipes Lake (pl. 17). This lovely lake, whose surface is often covered with duckweeds and small floating leaves of watershield, is at the bottom of a ravine where the surrounding vegetation consists of leathery-leaved southern magnolia and palmettos. In early May, flowering dogwoods on the midslopes are very attractive. Spring wildflowers include several kinds of violets and trilliums, wild geranium, woodland phlox (pl. 18), waterleaf, Jacob's ladder, and wild ginger. A campground, picnic area, and short hiking trail are at Pipes Lake. Cypress-knee sedge and crested fringed orchid are rare species in the Pipes Lake area. Pipes Lake is reached by taking State Route 33 from Roxie, and then a series of narrow, gravelly Forest Service roads for 8 miles. It is at the edge of the large Sandy Creek Wildlife Management Area.

The most popular recreation site in the Homochitto National Forest is Clear Springs Recreation Area, which is 4 miles south of U.S. Highway 84 via Forest Road 102. Centered around Clear Springs Lake, this recreation area has a campground, picnic area, boat launch area, swimming beach, and a hiking trail that encircles the lake. Just behind the picnic area, at the bottom of a deep ravine, are large specimens of royal fern and cinnamon fern.

Mount Nebo Recreation Area is 3 miles north of Crosby at the end of Forest Road 191-A. The small Mount Nebo Lake is a popular fishing spot. A small campground and picnic area are also here.

Most of the southeastern corner of the Homochitto National Forest is within the Homochitto Wildlife Management Area. Many roads and small streams criss-cross this heavily forested region. Between the western edge of the wildlife management area and the town of Crosby is the historic Stephenson Lookout Tower along Forest Road 165.

On Stephenson Ridge is habitat for the federally endangered red-cockaded woodpecker. The driest upland areas of the ridge contain loblolly, shortleaf, and a few longleaf pines. The common shrub in this habitat is farkleberry. At the base of the ridge near the creek is a forest of tulip poplar, American beech, water oak, sweet bay, red maple, and horse sugar. Along the creek are swamp chestnut oak, sweet gum, giant cane, palmetto, arrowwood viburnum, and devil's walkingstick.

At the far western side of the Homochitto National Forest, about 4 miles northwest of Pipes Lake and within the Sandy Creek Wildlife Management Area, is pretty Pellucid Bayou with deep sandy areas at several places along the bayou. An interesting assemblage of plants occurs at this site. Uncommon plants found on the north side of Pellucid Bayou include the heart-leaved climbing hempvine, one-headed pussy-toes, ill-scented trillium, and a beautiful wild camellia. On the south side of Pellucid Bayou is bay starvine, in a family closely related to the magnolia family.

An archaeological site in Franklin County in the vicinity of Meadville was discovered in 1975 by Dale Greenwell of the University of Southern Mississippi. This prehistoric site contains a mound about 400 feet long and 310 feet wide.

Pellucid Bayou

During the Ice Age, which lasted from about 1.6 million to 12,000 years ago, glaciers periodically scoured the northern portions of the North American continent, grinding rock into tiny particles. This dust was picked up by the wind and deposited in the Mississippi floodplain as a fine-grained soil known as loess. The thickest layers accumulated on the eastern side of the Mississippi and Missouri rivers, where the prevailing westerly winds encountered bluffs. Near Natchez, Mississippi, loess deposits are as much as 30 feet deep, although eroded in places because of continuous wind action.

About 25 miles east of Natchez, in the Homochitto National Forest, deposits of loess form gentle ridges adjacent to Sandy Creek and its tributaries, such as Pellucid Bayou. The original deposits are shallower than at Natchez—about 15 to 20 feet thick—but because the wind is less intense, they are less eroded. Plants that grow in the river bottom can also be found on these nearby ridges, alongside upland species, providing a much diversified habitat.

I wandered onto the loess ridges during a pleasant mid-April day. Basswoods and tulip poplars are common upland trees, along with American beech, southern magnolias, occasional upland species of oaks, and the evergreen American holly. In addition, occasional trees usually grow in bottomland forests along the Mississippi River. One of these is Nuttall's oak, a pin oak look-alike with larger acorns that is found only in the forests adjacent to the Mississippi River from New Orleans to southern Illinois. Even more interesting to me was that on the ridges above Sandy Creek and Pellucid Bayou, I saw cherrybark oaks growing next to southern red oaks. This is not only an example of a bottomland species growing next to an upland one; to me it is also evidence that these two trees, which many botanists consider the same species, are really distinct.

I live in an area in southern Illinois where both cherrybark oak and south-ern red oak occur, but they are never found together. The cherrybark oak grows in wet, bottomland forests, whereas southern red oak grows in the dry woods found on ridges. A trained eye can easily tell these two kinds of trees apart. The bark of a mature cherrybark oak, with its smooth patches, resem-bles the bark of a cherry tree (as the name implies), while the more broken-up bark of the southern red oak is characteristically oaklike. Cherrybark oak has leaves with five or seven lobes, the terminal one not decidedly curved; southern red oak has leaves with three or five lobes, with the terminal lobe strongly curved. In addition, the base of a cherrybark oak leaf runs straight across, while the base of the southern red oak leaf resembles an inverted bell.

The problem is that in classifying plants, botanists prefer to rely on re-productive structures, such as the flowers and the fruits. These parts tend to vary less from individual to individual than do nonreproductive structures, such as leaves. For example, anyone who has ever examined the leaves of a northern red oak knows that no two leaves are exactly alike. However, the acorns of one northern red oak are identical to the acorns of any other north-ern red oak.

Many of my colleagues consider the cherrybark oak to be a habitat-induced variant of the southern red oak. The acorns of the two trees are very similar, with only a few, seemingly minor differences. The differences be-tween the leaves and bark, they believe, result somehow from differences in the habitats where the trees grow. I have always been skeptical of this opin-ion, regarding the trees as two distinct species. As I walked along the ridge above Sandy Creek, I came upon a large cherrybark oak growing right next to an equally large southern red oak. The leaf and bark differences between the two couldn't have been more striking, yet the soil they were growing in and their exposure to the sun was essentially identical. This did not prove that they were distinct species, but I felt my view was borne out.

Moving downslope from the ridgetop, I observed umbrella magnolia, an Appalachian species with 2-foot-long leaves, growing with American elm, sweet gum, and hop hornbeam. As I walked farther down into a ravine (locally called a *hollow*) between adjacent ridges, I found myself in lush veg-etation. Clumps of Christmas ferns covered the slopes, interspersed with maidenhair fern (rare this far south) and a plant I had never seen before, fetid trillium, an evil-smelling plant with three rusty-maroon petals.

Other plants in the hollow included a white-flowered violet and a blue-flowered violet; climbing hydrangea and bittersweet (both are vines); and doll's eyes, a member of the buttercup family whose black-dotted, round, glossy white fruits resemble the eyes of a porcelain doll. Continuing downs-lope, I observed two very uncommon plants, a beautiful flowering shrub known as the silky camellia and a vine related to the magnolia family called bay starvine because of its 0.5-inch-wide, star-shaped, strawberry pink flowers. Finally, I came to Pellucid Bayou, beneath whose clear water I could see the sandy-colored loess of the stream bottom.

Pipes Lake

Lake beds form naturally in various ways—through the sinking of underlying rock, abandonment of channels by meandering rivers, movements of the Earth's crust, gouging by glaciers, and the creation of volcanic craters. Pipes Lake, in the heart of Mississippi's Homochitto National Forest, arose relatively recently when a colony of beavers dammed up a stream flowing through a valley.

The clear waters of Pipes Lake reflect the outline of the surrounding trees—an assortment of loblolly and longleaf pines, southern magnolia, and American holly. Wisps of Spanish moss draped on branches and occasional saw palmettos in the understory are a reminder that the area supports subtropical vegetation. The wooded slopes drop abruptly to the edge of the lake in some places; here, upland species such as the narrowleaf ironweed and wild verbena may grow at the shore. Elsewhere the lake merges almost imperceptibly into the valley, in a gradual transition from open water to marsh to valley forest. The varied marsh vegetation includes sedges, rushes, and bur-reeds.

Around the periphery of the lake are shallow zones where such rooted aquatic plants as arrowheads, beaked rushes, and flat sedges have become established. Scattered among these plants and growing in deeper water are floating plants, or *floaters*. As in most lakes and ponds, the floaters at Pipes Lake consist of a variety of flowering plants, including duckweeds.

Duckweeds are tiny, generally flat and roundish plants that grow in quiet or slightly moving water in all parts of the world. A few species approach 0.5 inch in size, but most are much smaller, some only 0.03 inch in diameter. The body of a duckweed, which is not differentiated into leaves and stems, is a green structure referred to as a *frond*. At one or two places on each frond is an area of small, actively dividing cells that produce new fronds. Plant physiologist William S. Hillman, who has spent much of his life studying duckweeds, has found that each frond may produce 10 or more fronds before it dies. As new plants are formed, they adhere in colonies before eventually breaking apart.

Duckweeds grow at least twice as fast as most plants, owing in part to their small size and relatively simple body form. With very few conducting structures to move water and nutrients through their tissues, and no mechanical cells to provide rigidity, duckweeds devote almost all their energy to photosynthesis. Cell for cell, duckweeds produce more food than corn and other crop plants. The cells of some duckweeds contain as much as 20 percent protein and are a major source of food for ducks—hence the name—and other birds and for fish and muskrats. In some countries, duckweeds are cultivated in ponds without fertilization, and they are harvested every 3 or 4 days for most of the year. Although duckweeds are sometimes consumed by people, they are mostly fed to livestock and poultry.

Duckweeds are particularly abundant in organically rich water, even thriving near sewer outlets. Hillman has shown that unlike most other flowering plants, all duckweeds can absorb large organic molecules, such as carbohydrates and some amino acids, without having to wait for other organisms to break them down. This permits duckweeds to scavenge complex nutrients from the water.

Duckweeds often tolerate copper, zinc, and boron, and other nutrients in concentrations that would curtail growth in most other species, prompting interest in their potential use in wastewater treatment facilities. In addition, duckweeds provide favorable conditions for aquatic animals that also help purify water by consuming blue-green algae and noxious bacteria. On the other hand, fisheries biologist William Lewis has found that duckweeds sometimes grow in such thick layers that oxygen in the water beneath them is depleted, killing fish. Although duckweeds in the uppermost layer give off oxygen to the atmosphere, those in the lower layers are so shaded that they cannot carry out photosynthesis and thus do not release oxygen into the water.

Duckweed flowers are so minute and appear so sporadically that most people, including biologists, have never seen them. A flower consists of a pistil (where the seeds ultimately develop) and from one to three pollen-producing stamens. Petals are absent, although sometimes a membranous sheath partly surrounds the pistil and stamens. Hillman suggests that unknown changes in the composition of lake water may induce flowering. Several species may flower at once in one lake, while the same species in nearby ponds are devoid of blooms.

The 30 or so duckweed species in the world (family Lemnaceae) are usually grouped into four genera. *Lemna*, the genus with the most species, has a single rootlet projecting from the lower surface of the frond. *Spirodela* has the largest fronds and usually bears two or three rootlets. *Wolffia* contains the smallest species. Often no larger than a pencil point, they are more or less spherical and have no roots. *Wolffiella*, also rootless, has a strap-shaped frond. At Pipes Lake, I observed species of all the genera except *Wolffiella*.

Perhaps if I had skimmed my hand across the surface of the water enough times, I might have found it as well.

Tombigbee National Forest

SIZE AND LOCATION: 66,000 acres in east-central Mississippi, including the Yalobusha Unit of the Holly Springs National Forest. Major access routes are Interstate Highway 55; the Natchez Trace Parkway; U.S. Highway 82; and State Routes 7, 15, 25, and 41. District Ranger Station: Ackerman. Forest Supervisor's Office: 110 W. Capitol Street, Suite 1141, Jackson, MS 39261, www.fs.fed.us/r8/miss.

SPECIAL FACILITIES: Boat launch areas; swimming beaches.

SPECIAL ATTRACTIONS: Owl Creek Archaeological Site; Lake Tillatoba Recreation Area; Choctaw Lake Recreation Area; Davis Lake Recreation Area.

When the Tombigbee National Forest was established more than seven decades ago, the forest acquired severely eroded farmland and cut-over woods from the previous private landowners. Through sound forest management techniques, the forested land has made a modest comeback, with nice woods now occurring on dry ridges, mesic slopes, and wet bottomlands. Prior to the coming of European settlers, the land served as the hunting and agricultural grounds for Chickasaw Indians between 300 and 1500 A.D.

The most important archaeological site is the Owl Creek area 3 miles west of the Natchez Trace Parkway and about 23 miles southwest of Tupelo. The site consists of five flat-topped mounds that were probably part of a ceremonial center. The two largest mounds probably supported a temple and a house for one of the more important chiefs. The major occupation of the site by the Indians was around 1000. The Owl Creek Archaeological Site lies on a small terrace in the alluvial floodplain of Chuquatonchee Creek, near where Goodford, Owl, and Davis creeks empty into the Chuquatonchee. Corn was apparently cultivated in the alluvial floodplain. Wooden steps allow one to climb to the top of the largest mound.

One and one-half miles southwest of the Owl Creek Archaeological Site is the Davis Lake Recreation Area, which features a campground, boat launch area, swimming beach, and nature trail. The 300-acre Davis Lake is serenely set among a forest of pines and hardwoods. This section of the Tombigbee

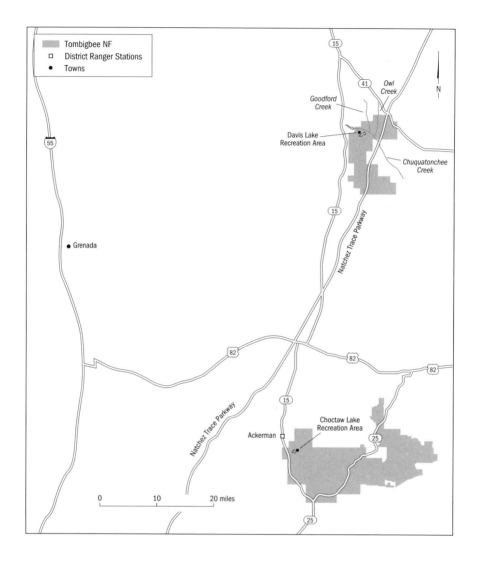

Tombigbee NF
□ District Ranger Stations
• Towns

National Forest is the most northeastern of the three units and is crossed diagonally by the Natchez Trace Parkway. Less than 1 mile east of the Parkway is the Witch Dance Lookout Tower. The Witch Dance Horse Trail is here, too. This unit of the Tombigbee lies between Tupelo and Houston.

South of U.S. Highway 82 is the southern district of the Tombigbee National Forest. Ackerman is only 1 mile from the western edge of this unit. Choctaw Lake Recreation Area, 1 mile east of State Route 15, is popular with campers, fishermen, and boaters, and amenities are present to satisfy all of these groups of recreationists. Three miles east of Choctaw Lake is Tombigbee Lookout Tower, and at the extreme southern end of this unit is the historic Winston Lookout Tower, only 1 mile north of Louisville. The eastern edge of this unit of the national forest abuts the Noxubee National Wildlife Refuge.

A short distance to the east of Interstate Highway 55, between Oakland and Grenada, is the Yalobusha Unit of the Holly Springs National Forest, but the unit is administered by the Tombigbee National Forest. Lake Tillatoba Recreation Area, the popular attraction in this unit, features a campground and boat launch area. Two miles to the northeast is the Coffeeville Lookout Tower. Walker Lake, at the northern end of the unit, and Texas Lake, at the southern end, are popular with fishermen.

NATIONAL FOREST IN
NEW HAMPSHIRE AND MAINE

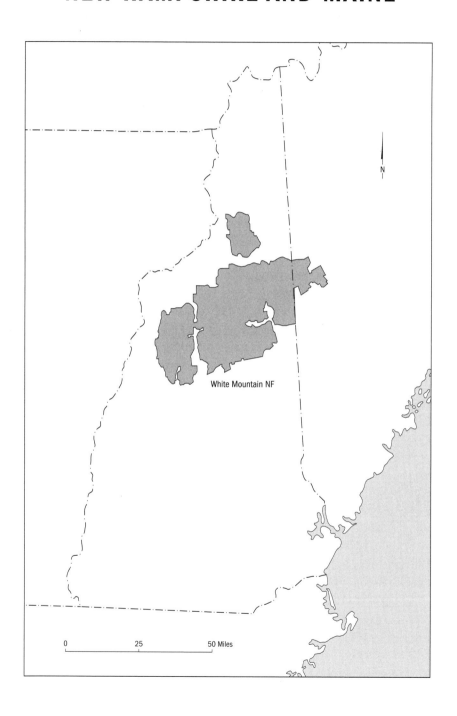

White Mountain NF

N

0   25   50 Miles

Most of the White Mountain National Forest is in New Hampshire, although a small part extends into western Maine. This national forest is in Region 9 of the United States Forest Service.

White Mountain National Forest

SIZE AND LOCATION: Approximately 770,000 acres in east-central New Hampshire and adjacent western Maine. Major access roads are Interstate Highway 93; U.S. Highways 2, 3, and 302; New Hampshire State Routes 16, 25, 25C, 26, 49, 110, 112, 116; and Maine State Route 113. District Ranger Stations: Bethlehem, Conway, Gorham, and Plymouth. New Hampshire. Forest Supervisor's Office: 719 Main Street, Laconia, NH 03246; www.fs.fed.us/r9/wmnf.

SPECIAL FACILITIES: Boat ramps; swimming beaches; winter sports areas.

SPECIAL ATTRACTIONS: Numerous designated scenic areas; Kancamagus National Scenic Highway.

WILDERNESS AREAS: Caribou–Speckled Mountain (12,000 acres); Great Gulf (5,552 acres); Pemigewasset (45,000 acres); Presidential Range–Dry River (27,380 acres); Sandwich Range (25,000 acres).

Usually during the first 2 weeks in October, the White Mountain National Forest attracts millions of visitors to see the brilliant autumnal coloration of sugar maples, red maples, and other hardwood trees. In addition to this marvelous display of colors, the White Mountain National Forest has much more to offer during all seasons. From high-elevation mountaintops with alpine flora, swift-flowing streams, dramatic waterfalls, clear mountain lakes, and tracts of virgin forest, the national forest is one of the natural treasures of the eastern United States.

Near the center of the national forest is the Presidential Range of mountains, topped by the incomparable Mount Washington. Darby Fields was the first white person to climb to the summit of this 6,288-foot mountain in 1642, and since that time, thousands of persons have made it to the top by hiking, driving, or taking the remarkable cog railroad. The summit of Mount Washington is bleak and harsh nearly every day of the year, and the mountain holds the world record for the highest wind velocity when on April 12, 1934, the wind was clocked at 231 miles per hour. These climatic conditions, along with the rocky summit, prevent much vegetation from growing, yet a

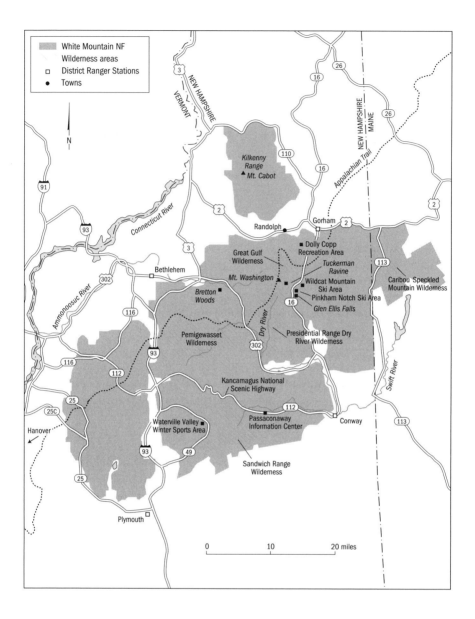

small but hearty group of plants makes their home on the top of Mount Washington. The cog railroad climbs the western side of the mountain from a station off Forest Road 4. The Mount Washington Auto Road, 8.5 miles in length, begins from the Glen House on State Route 16, a few miles north of Pinkham Notch.

To the north of Mount Washington is the Great Gulf Wilderness, while to the south is the Presidential Range–Dry River Wilderness. Mount Washington is the only presidential mountain not in these wildernesses, although the boundaries of these wilderness areas include portions of Mount Adams, Mount Quincy Adams, Mount Madison, Mount Jefferson, Mount Eisenhower, Mount Pierce, Mount Monroe, and Mount Jackson. Except for Mount Washington, however, all of these presidential mountains must be hiked to if one were to reach the summits. The area between the two wildernesses is the Pinkham Notch Scenic Area. A stop at the visitor center along State Route 16 will provide information about the area; it is open daily for meals, lodging, and day use.

The scenic area, crossed by State Route 16, contains such popular destinations as Mount Washington, Tuckerman Ravine, Huntington Ravine, Glen Ellis Falls, Crystal Cascade, and the Pinkham Notch Ski Area. Tuckerman Ravine is a large glacial cirque on the eastern flank of Mount Washington. Skiers climb the Tuckerman Ravine Trail to get to a popular ski run area, and hikers use the Tuckerman Ravine Trail as the best walking route to the summit of Mount Washington. The trail up Huntington Ravine is one of the most treacherous in the White Mountain National Forest and should be attempted only by the experienced hikers. After crossing the Cutler River, the Huntington Ravine Trail crosses several brooks, meanders past broken rocks known as the Fan, and passes several rock formations, including the Pinnacle. Between Tuckerman Ravine and Huntington Ravine is the narrow Raymond Cataracts, a waterfall dropping 400 feet down the mountain slope. The 1.5-mile trail to this spot begins from the parking lot of Wildcat Mountain Ski Area. Also reached by trail from this parking lot is the Thompson Falls, descending nearly 200 feet, including three major drops.

More spectacular and easy to get to because it is alongside State Route 16 is Glen Ellis Falls, which drops 64 feet in a horsetail fashion. The trail from the parking lot to the falls passes under the highway. Northwest of the visitor center are the 80-foot Crystal Cascades, one of the prettiest in the White Mountain National Forest. Also in the Pinkham Notch Scenic Area is the Chudecoff Falls off George's Gorge Trail.

The Great Gulf Wilderness north of the Pinkham Notch Scenic Area was the first designated wilderness in the northeastern United States when it received this status in 1964. Mounts Washington, Jefferson, Adams, and Madison surround the largest glacial cirque in the White Mountains, the Great Gulf. The headwall of the mountains is up to 1,500 feet higher than the gulf. The cirque drains into the West Fork of the Peabody River.

South of Pinkham Notch Scenic Area is the Presidential Range–Dry River Wilderness, which includes Mounts Monroe and Eisenhower, as well as others such as Mount Jackson, Mount Isolation, and other lesser mountains. The Dry River flows through the northwestern side of the wilderness, closely followed by a hiking trail.

Wildcat Mountain Ski Area is between the Pinkham Notch Scenic Area and the Mount Washington Auto Road where all winter sports may be enjoyed. State Route 16 continues to parallel the Peabody River to the Dolly Copp Recreation Area. Located on the site of two early farms adjacent to the Peabody River—the Copp and the Culhane—the recreation area has a campground, picnic area, nature trail, a memorial to Dolly Copp, and several old farm foundations. Culhane Brook, which empties into the Peabody River, flows through the recreation area. You may also take several longer trails from the campground. Northeast of Dolly Copp along Forest Road 2 are three cascades that comprise Triple Falls.

Beyond Triple Falls, Forest Road 2 ends at U.S. Highway 2 near Randolph at the northern edge of this district of the White Mountain National Forest. Just inside the forest boundary is the Snyder Brook Scenic Area, on either side of lovely Snyder Brook. The area is surrounded by the rugged gray Northern Peaks. Along Snyder Brook are three waterfalls, all reached by hiking the Brookbank Trail. Gordon Falls is the northernmost falls, dropping 20 feet over rocks in a beautiful setting. The middle falls in the scenic area is Salroc Falls. Although the drop is only about 10 feet, it forms a pretty fan. Near the south end of Snyder Brook Scenic Area is Tama Falls, another fan but with a longer drop of 25 feet.

West of Snyder Brook, Forest Road 4 to the south follows the South Branch of the Israel River and is a few miles west of the Great Gulf Wilderness. The Cap Bridge Trail to the wilderness begins from the forest road and eventually connects with the Appalachian Trail. The forest road then swings westward and parallels Jefferson Brook. Just after the forest road crosses the Ammonooosuc River is the station for the Mount Washington Cog Railroad.

Just before Forest Road 4 reaches U.S. Highway 302, between the highway and the western edge of the Presidential Range–Dry River Wilderness, is Gibbs Brook Scenic Area. This 1,650-acre scenic area, in the drainages of Gibbs Brook and Elephant Head Brook, preserves an old-growth red spruce–balsam fir forest in the higher elevations. Lower down, along the brooks, are majestic northern hardwood trees and hemlocks. Some of the latter are thought to be more than 500 years old. Near the summit of Mount Pierce on the eastern edge of the scenic area are small patches of subalpine vegetation. The Crawford Path, constructed in 1819 by Abel Crawford and his son Ethan Allen Crawford, follows Gibbs Brook in the scenic area. Forest Road 4 then ends at U.S. Highway 302.

By taking U.S. Highway 302 north, you will come to the resort area known as Bretton Woods. Just beyond this are Zealand and Sugarloaf campgrounds on either side of the Ammonoosuc River. After U.S. Highway 302 leaves the national forest, U.S. Highway 3 enters the forest and passes the Beaver Brook

Winter Sports Area. West of the sports area is the Churchill Mountain Overlook. U.S. Highway 3 then joins Interstate Highway 93.

Although the interstate highway is not in the White Mountain National Forest, it enters Franconia Notch with national forest land, all mountainous, to both the west and the east. The western segment includes Cannon Mountain, Mount Pemigewasset, Mount Moosilauke, and Kinsman Mountain with its North and South peaks. All but Mount Pemigewasset are above 4,000 feet. Mount Pemigewasset is well-known because of the rock formation known as Indian Head that is readily visible from the interstate highway. Cannon Mountain is a short distance south of the town of Franconia. To the west is Bridal Veil Falls, a 50-foot falls above a 30-foot slide.

Between Cannon Mountain and the North Peak of Kinsman Mountain are the Cannon Balls, three humps on the ridge between the mountains. The Appalachian Trail connects Kinsman Mountain with Kinsman Notch on State Route 112. To the east of the trail is another Gordon Falls, this one with a nice drop of about 35 feet. From Kinsman Notch, you may take State Route 112 west through more of the national forest until the highway exits the forest before coming to Woodsville, or you may take the highway south and then east to Interstate Highway 93 at the town of North Woodstock. Just northwest of town is an interesting formation known as Balanced Rock.

There is land to explore south of Kinsman Notch, however, and State Route 118 provides access to some of the areas. Along this highway is the Jackman Brook Overlook, just north of Hubbard Brook Experimental Forest. Not only are current forest management practices developed here, but the large area is scenic as it is bisected by Hubbard Brook. Forest Road 22 and several hiking trails cross the area.

At the far southwestern end of the White Mountain National Forest, on the west side of Carr Mountain, is Waternomee Falls on Clifford Brook.

A vast region of the White Mountain National Forest occurs between Interstate Highway 93 on the west and U.S. Highway 302 on the east, and much of this area is composed of the Pemigewasset Wilderness. The 45,000-acre area includes what many believe to be the most remote part of the national forest. The wilderness is bounded by the steep and narrow Franconia Range on the west and by Mount Garfield, Gale Head Mountain, and Mount Guyon, all above 4,000 feet, on the north. The Franconia Ridge consists of Mounts Lafayette and Lincoln, both over 5,000 feet above sea level, as well as two 4,000-footers: Mounts Liberty and Flume. The 4,403-foot Mount Hancock is along the southern border. Also within the wilderness are Thoreau and Franconia falls; Black, Ethan, and Shoal ponds; and several mountain peaks in the central part of the wilderness. Of the latter, Mount Bond, at 4,698 feet, and The Cliffs, at 4,265 feet, are the highest. Just outside the northern edge of the wilderness are Hawthorne and Zealand falls. The eastern end of the wilderness has flatter terrain. The Appalachian Trail stays near the northwestern and northern border of the Pemigewasset Wilderness, and numerous other hiking trails are within the wilderness.

The Kancamagus National Scenic Byway from Lincoln to Conway cuts across the White Mountain National Forest from west to east and lies only a short distance south of the Pemigewasset Wilderness for part of its way. Rough and narrow Forest Road 87 north of the byway follows the Pemigewasset River and enters a narrow cleft between two lobes of the wilderness and provides access to Franconia Brook Campground near Franconia Falls and Black Pond. The Lincoln Woods Visitor Center is at the junction of the Kancamagus Highway and Forest Road 87.

Outside the northwestern, northeastern, and southeastern corners, respectively, of the Pemigewasset Wilderness are the Lafayette Brook, Lincoln Woods, and Nancy Brook scenic areas. Lafayette Brook Scenic Area, located near Franconia Notch, consists of the steep, V-shaped Lafayette Valley on the rugged slopes of Mount Lafayette. The gray, treeless summit of this mountain towers to an elevation of 5,249 feet above sea level, and supports an alpine flora. Halfway down the mountain's western side is a rocky landmark known as Eagle Cliff. The Eagle Lakes are nearby. Areas of old-growth red spruces and balsam firs occur in the scenic area.

Lincoln Woods Scenic Area protects fine stands of forests, and the trail that runs through it is the major route to reach the Pemigewasset Wilderness.

Between the southeastern edge of the Pemigewasset Wilderness and U.S. Highway 302 is Nancy Brook Scenic Area, situated in a depression between Mount Bemis to the north and Duck Pond Mountain to the south. Pick up a brochure at a ranger station to learn the sad story about a young girl named Nancy for whom the area is named. The closest way to this area is via a 3-mile trail from U.S. Highway 302, but the hike is well worth it. The scenic area contains waterfalls, landslides, clear-water glacial ponds, virgin red spruce forests, and other forest types. Nancy and Norcross ponds are both within the scenic area.

A short distance northwest of North Conway and due west of Intervale and U.S. Highway 302 is an interesting series of water-filled depressions at the base of a 20-foot-high falls. Known as Diana's Baths, the pools are a popular destination, particularly during the summer. You can reach them by taking an easy 0.5-mile trail from a parking area along West Side Road in North Conway.

The Kancamagus National Scenic Byway has numbered mileposts, from 64 to 100, with the lower number at the eastern end near Conway. The highway climbs to nearly 3,000 feet at Kancamagus Pass. Beginning at Conway,

Figure 9. Rocky Gorge, White Mountain National Forest (New Hampshire/Maine).

you will soon come to the Sago Ranger Station where you may obtain information about the scenic byway. The start of the highway parallels and eventually crosses the Swift River. A side road north of Blackberry Crossing Campground comes to the Covered Bridge Campground near one of the interesting covered bridges still in the region. Vehicles more than 8 feet tall cannot pass through the covered bridge. Next is a picnic area near the Lower Falls of the Swift River (pl. 19). The scenic byway comes to the Rocky Gorge Scenic Area (fig. 9) where the Swift River drops about 20 feet over rocky boulders into a solid granite rock gorge. The area also contains 6-acre Falls Pond with an average depth of 11 feet. The pond was formed naturally when a narrow ridge of glacial debris was laid down by a retreating glacier. The pond is surrounded by majestic red spruces and white pines.

Between the Rocky Gorge Scenic Area and the Jigger Johnson Campground is a hiking trail to the south that quickly comes to the picturesque Champney Falls. Beyond the falls, the trail continues into the Mount Chocorua Scenic Area dominated, of course, by Mount Chocorua. Five hundred yards west of the Jigger Johnson Campground is the historic George

House dating back to 1810. The house is used now as the Passaconaway Information Center. The Rail 'n River Nature Trail wanders for 0.5 mile through a forest at the information center.

At the Jigger Johnson Campground is the junction of Forest Road 26 with the Kancamagus Highway. By following this road to the north, you will come to the southern end of the Bartlett Experimental Forest where research is being conducted on how to manage timber and wildlife habitat.

West of the Passaconaway Information Center is Sabbaday Falls, one of the nicest and most heavily visited waterfalls in the national forest. Actually a series of cascades in a narrow flume, the falls can be reached via a 0.5-mile trail. South of the falls, Sabbaday Brook enters the Sandwich Range Wilderness, with dense forests, clear streams (some with waterfalls), and six mountain peaks that are 4,000 feet above sea level or nearly so. Outside the southeastern corner of the wilderness, in a large cirque between Mount Wonalancet and Mount Whiteface, is the Bowl Natural Area.

West of the Sandwich Range Wilderness, near the community of Waterville Valley, is the Waterville Valley Winter Sports Area. In the vicinity are such attractions as the Norway Rapids, The Ledges, and the Fletcher Cascades.

The Kancamagus National Scenic Byway climbs to the Sugar Hill and Kancamagus overlooks before reaching Kancamagus Pass. About 1 mile south of the pass is Greeley Ponds Scenic Area. Here are 810 acres of red spruce and balsam fir forest and two pretty ponds nestled beneath rugged cliffs and mountain peaks. Upper and Lower Greeley Ponds are south of the Mad River Notch that divides the East Peak of Osceola and Mount Kancamagus. The Greeley Ponds Trail goes through the heart of the scenic area and past the ponds. On the east side of the East Peak, about 1 mile from the Greeley Ponds, are huge erratic boulders—Davis Boulders and Goodrich Rock.

Before the Kanacamagus Highway reaches Lincoln, it passes Otter Rocks Roadside Rest Area, Big Rock and Hancock campgrounds, and the Loon Mountain Winter Sports Area.

The eastern portion of the White Mountain National Forest extends from State Route 16 east and into the western edge of Maine. Although this is a vast area of mountains and brooks, there are fewer recreation areas and fewer trails, although the Appalachian Trail does cut across the northern end of this region. Most of the points of interest in this district are on either side of Maine State Route 113 as it stays alongside Cold River approaching Evans Notch. At the notch is a panoramic view across the Cold River to East Royce and West Royce mountains, which rise to 3,114 and 3,116 feet, respectively. Nearby on the west side of State Route 113 are the Cold River and Basin Brook, each with a campground adjacent. The Royce Trail north from the Cold River Campground brings you near Mad River Falls in about 2 miles.

Near the north end of Basin Brook is lovely Hemit Falls. North of Evans Notch, the Wild River Road (Forest Road 12) branches off State Route 113 and heads southwestward into the hinterland. At the end of the road is the Wild River Campground.

East of Evans Notch is the Caribou–Speckled Mountain Wilderness in Maine. About one-fourth of the White Mountain National Forest that is in Maine is in the wilderness area. Several mountains rise here, the highest being Speckled Mountain with an elevation of 2,906 feet. Many streams eventually flow into either the Saco or Androscoggin rivers.

Several trails lead from State Route 113 and into the wilderness. Red Rock Trail climbs over Speckled Mountain in the southern part of the area; Haystack Notch Trail crosses the middle part, skirting the southern end of Haystack Mountain; the Caribou Trail, at the north end of the wilderness, climbs to the flat granite top of Caribou Mountain, after passing 15-foot Kees Falls on its way to the summit.

The highest elevations in the Caribou–Speckled Mountain Wilderness support a grand forest of balsam fir, white spruce, white pine, red pine, white cedar, and hemlock, interrupted by deciduous forest trees. The slopes of the mountains are covered by a heavy concentration of sugar maple, red maple, American beech, three kinds of birches, and several other species. In autumn, the orange leaves of sugar maple, red and yellow leaves of red maple, and golden brown leaves of American beech leave an indelible picture in the minds of the viewers. On the bare, granite top of Caribou Mountain and on adjacent granite ledges is a unique flora of low-growing plants that are able to survive in this xeric habitat. Most significant of the plants in this habitat is the White Mountain silverling, a plant known from only about 200 populations in the Caribou Mountain area. Often growing with the White Mountain silverling is the rare mountain sandwort.

West of State Route 113 and in New Hampshire is Mountain Pond. Near the pond is an old-growth northern hardwood forest with an abundance of basswood, white ash, American beech, yellow birch, and sugar maple, some of them nearly 300 years old.

At the extreme eastern side of the White Mountain National Forest in Maine is the Patte Brook Multiple Use Management Area. Patte Brook, a tributary of Crooked River, contains Broken Bridge and Sunken ponds and the 45-acre Patte Mill Waterfowl Impoundment. A good diversity of trees grows in the area, including red spruce, balsam fir, white cedar, hemlock, white pine, red pine, sugar maple, red maple, yellow birch, paper birch, gray birch, white ash, American elm, northern red oak, and American beech. Common shrubs or small trees include striped maple, hobblebush viburnum, alder, beaked hazelnut, and witch hazel. The management area was set aside to produce maximum benefits that can be derived from the forest. Just outside the western edge of the Patte Brook Management Area is Crocker Pond, which has a nice campground.

Two other areas of the White Mountain National Forest are not contiguous with the main forest area. One is a few miles north of U.S. Highway 2 between that highway and New Hampshire Route 110 that includes the Kilkenny Range. The other is a narrow corridor on either side of the Appalachian Trail.

In the Kilkenny Range, Mount Cabot is the highest elevation at 4,080 feet above sea level. Numerous trails cross the region, but the only developed recreation area is South Pond near the northern end of the district. A short distance southeast of South Pond, however, is an interesting area known as the Devil's Hopyard. Located in a scenic gorge, the hopyard has a small stream with moss-covered boulders and overhanging cliffs.

Southwest of the large area of the White Mountain National Forest, the Appalachian Trail makes its way from Hanover to Moose Mountain, then over Smart Mountain where there is a lookout tower, and eventually to State Route 25C where it enters the main part of the national forest. The corridor on either side of the Appalachian Trail is included in the White Mountain National Forest. Then, after crossing the national forest, the corridor along the trail picks up again east of Gorham and passes over the Mahoosuc Range to the Maine state line. This part of the corridor is also included in the national forest.

Alpine Garden

When he stood atop Mount Washington, P. T. Barnum remarked, "This is the second greatest show on Earth." Highest of the 11 peaks that make up the Presidential Range, Mount Washington lies in the heart of White Mountain National Forest. The 6,288-foot summit extends well above the region's tree line, and bleak tundra covers nearly eight square miles of the surrounding territory. Here, 1,000 feet down and east of the mountaintop, is a natural area known as the Alpine Garden. Two-thirds of the garden's 110 kinds of flowering plants are alpine species whose major range is far to the north. Many plant species that live here are more common in Labrador, Newfoundland, and areas in the Arctic. Alpine azalea, moss cassiope, mountain heath, alpine bearberry, and two-eyebrights range from Newfoundland and the mountains of Quebec south only to the Presidential Range. Alpine bluet grows just on the islands of Saint Pierre and Miquelon and on Mount Washington.

There are three ways to reach Mount Washington's summit and the adjacent Alpine Garden. For the nostalgic, a cog railroad makes the climb to the top in 70 minutes, as it has since 1869. The 8.5-mile automobile toll road takes a more circuitous route. The third way to go is to hike one of the several trails. As a botanist, I prefer the last method because it affords a close look at a variety of habitats en route.

It was a crisp and sunny late June morning when I hiked away from Pinkham Notch toward Tuckerman Ravine, on the eastern flank of Mount Washington. But at about 4,800 feet, I left the ravine forest and came out into the open, where sharp gusts of bone-chilling wind nearly took my breath away and a heavy, wet fog hovered above me.

I felt as if I had been translocated into the Arctic, and the types of plants around me contributed to the illusion. I had broken into the *krummholz*— "crooked wood," in German—that zone of gnarled dwarf trees a few hundred feet below the treeless tundra. Because of the fierce climate, which reduces the rate of photosynthesis, black spruces and balsam firs are only a few feet tall, despite their considerable age. Frigid and furious winds prune away all branches of these hardy plants that protrude above the layer of winter snow.

At about 5,100 feet, I passed the last of the krummholz plants, and the solemn-looking, rock-strewn slope of the Alpine Garden lay ahead. The garden surface has been shaped by frost through the centuries. Permanent moisture and fluctuating low temperatures have created the subsurface soil known as *permafrost*, which has remained frozen for thousands of years. Above the permafrost is a layer of water-saturated soil that repeatedly freezes and thaws, heaving rocks and pebbles out of the ground. Sometimes the rocks are deposited in rings and polygons, sometimes in small piles. In places, thawing, saturated soil creeps downslope over bedrock (a process known as *solifluction*), eventually forming very low, undulating ridges called solifluction terraces.

Plants in the Alpine Garden contend with bitter conditions. The mean annual temperature on the tundra is only 27 degrees Fahrenheit, with July the warmest month at 48 degrees. The winds surpass 100 miles per hour every month of the year, with a world-record high of 231 miles per hour noted in 1934. Clouds block out the sun about 75 percent of the time, mostly in the form of wet fog, especially on the western slopes. Winter snow piles up an average 195 inches; during the winter of 1957–1958, 344 inches fell on Mount Washington.

The plants that have adapted to life in the tundra make the most of the natural features. Something as seemingly unimportant as the lee side of a rock may give just enough protection for a seedling to grow into a mature plant. The nearly imperceptible ridge of a solifluction terrace may provide the right exposure to the sun for seeds to germinate and develop. Snowbanks built up by the freezing, hurricane-force winds insulate the parts of plants that otherwise would perish during the winter. Natural springs from high up on Mount Washington flow out across the Alpine Garden in clear, cold streams, offering the bordering plants the water they need to complete their life cycle during the short growing season.

Most plants on Mount Washington are perennials, with taproots that penetrate more than a foot into the soil. Because they have a stable root system, they needn't perpetuate themselves every year by flowering and forming fruit and seed. If the conditions are especially harsh, a plant need only form a few new leaves to replenish its supply of stored food.

Since the growing season in the Alpine Garden seldom lasts more than 100 days, plants begin their life cycle as early as possible. The three-toothed cinquefoil, for example, begins to flower as soon as the temperature climbs above freezing. The flower buds in such species are formed during the previous season and kept safely protected by surrounding tissues over the winter.

To avoid the intensity of the wind, most alpine plants are very short, and often they are plastered against the ground. While most willows are upright trees or shrubs, the Alpine Garden's bearberry willow grows in dense mats only 2 inches high. Other plants, like the diminutive moss campion, have branches tightly woven into a rounded cushion that surrounds a deep taproot. At the center of the cushion, where the branches and leaves are densely packed, the temperature is warmer than in the surrounding environment. The threat of desiccation is also minimized, because the wind tends to blow over the rounded form.

Small, compact leaves, like those of the moss cassiope, are common in alpine species because they require less energy for their maintenance. Plants with grasslike leaves, such as sedges and rushes, are also prominent members of the alpine flora. Their slender leaves are less likely to be shredded by the strong winds. Occasional patches of sphagnum (peat moss), as well as a few other plants, are colored by anthocyanin, a reddish-blue pigment. Anthocyanin absorbs rays of light, helping to keep plant tissues warm.

Although at first the Alpine Garden appears uniform, discrete plant communities can be recognized. On most of the slopes that grade down to the level terraces, clumps of alpine rush are surrounded by lower, compact growths of mountain cranberry and bog bilberry, two heaths with pink, bell-shaped flowers and small, leathery leaves. By late July, the mountain cranberry (pl. 20) forms bright red, acidic berries, while the bilberry bears sweet, black fruit. Other plants in this community include the mountain sandwort, whose bright white flowers enliven the landscape. The sandwort's penchant for growing along the edges of trails makes hiking in the dense fog somewhat easier.

In shallow depressions and on the lee side of rock piles, a dwarf shrub heath community often develops. Low sweet blueberry, Labrador tea, and the alpine bearberry, as well as the mountain cranberry and bog bilberry, form dense mats as much as 12 inches deep. Within the boundaries of rock polygons, solid masses of diapensia often grow so densely that they prevent other flowering plants from becoming established. Diapensia is a cushion plant that resembles members of the heath family except for its flowers.

The streams that radiate across the Alpine Garden are lined with dark green bands of vegetation. The streamsides are the most interesting of the plant communities because they contain a diversity of species. One is the tealeaf willow, which looks like a normal willow in miniature, having a stature of no more than 2 feet. Another curious plant, the alpine bistort, forms several pale pink flowers, but none of the flowers set seed. Instead, tiny balls of plant tissue, called *bulblets*, develop below the base of the flowers. After dropping to the ground, the bulblets can grow into new plants without benefit of pollination and fertilization.

The range of some of its alpine species sets the Alpine Garden apart from similar habitats elsewhere in New England. At one time scientists thought that arctic plants reached the White Mountain region when they ranged southward in advance of continental glaciation. In this view, Mount Washington and adjacent peaks became areas of refuge as the glaciers marched past at lower elevations. However, glacial till has been discovered on the mountain summits, showing that they, too, were covered over. Biologists speculate that the arctic plants found a haven farther south and migrated into the area as the glaciers began to recede.

I was only able to stay about 2 hours on my first visit to the Alpine Garden because the weather was nearly unbearable. On a subsequent trip, however, the fog was gone and the winds were moderate. The view was spectacular. It must have been on such a day that P. T. Barnum stood atop Mount Washington.

Devil's Hopyard

Having driven to the northern tip of New Hampshire's White Mountain National Forest, I stopped at South Ponds, only 30 miles from the Canadian border. A sign that pointed down an inviting trail to the Devil's Hopyard immediately intrigued me. Following a ravine forested with American beeches, sugar maples, and other hardwood trees, and a sprinkling of hemlocks, red spruces, and balsam firs, I walked for 0.5 mile, occasionally crossing a meandering rivulet on partly submerged, slippery stones. Then the unexpected happened: around a sharp curve, the slope on either side abruptly encroached upon the trail, narrowing the canyon to only 75 feet. Scattered over the canyon floor were countless boulders, debris left by Ice Age glaciers. Some were 10 feet high, others smaller, and each was covered by the most prolific growth of mosses I had ever encountered outside a tropical rainforest. The lower parts of most of the tree trunks in the canyon were also covered by mosses.

The prevailing moss was a rather large species, *Hylocomium splendens*, whose segments were arranged in flattened sprays one above the other. Howard Crum and Lewis Anderson, in their encyclopedic work on mosses of the United States, propose "stairstep moss" as the common name for this species. Each spray represents 1 year's growth. Unlike most mosses, stairstep moss can grow rampantly if conditions are favorable; in the forests of northern Sweden, it often forms dense carpets.

Like other green plants, mosses manufacture their own food through photosynthesis. But mosses lack roots and a vascular system. At the base of each moss plant are branched filaments, called *rhizoids*, which are nothing more than hairlike tubes, or chains of cells, that anchor the moss to whatever it is growing on. There is little evidence that these rhizoids are responsible for the mosses' intake of water and dissolved minerals. Instead, this function is left to the mosses' "leaves" (technically termed *microphylls*), which are mostly only one cell thick.

Because moss leaves lack the waxy cuticle that covers the leaves of most other plants, they are able to absorb even tiny amounts of moisture, such as a drop of dew. But without the protection of a cuticle, they are also in constant danger of drying out. Most mosses thus do best in moist, shaded areas. The arrangement of their leaves maximizes their ability to capture moisture as well as the light needed to carry out photosynthesis.

A moist habitat is also vital for reproduction in mosses, which form swimming sperm cells. Usually a drop of dew is all that is required for the sperm to reach the egg, since both are formed at the base of the uppermost leaves of the plant. Like ferns, mosses alternate between a sexual generation, which reproduces through male and female gametes, and an asexual generation, which reproduces via spores. The familiar, conspicuous ferns belong to the asexual generation, and the sexual plants are tiny and inconspicuous. Among mosses, however, the familiar plants are sexual, while the spore-producing stage is usually reduced to a stalk and a spore-bearing capsule that protrude from the top of the plant.

Plants with vascular tissues are able to conduct water and minerals for considerable distances—more than 300 feet in the case of the tallest redwood trees. Since mosses lack these special conducting tissues, they can transport materials only very short distances. Consequently, they can attain only a

"Lilliputian form of growth," as described by John Bland in his popular book *Forests of Lilliput: The Realm of Mosses and Lichens*. Entire moss plants may range from only 0.06 inch to just a few inches in height. Little wonder that they rarely dominate natural habitats.

The unusual abundance of mosses at Devil's Hopyard may be due to a number of favorable conditions. Nancy Slack of Russell Sage College, studying mosses in similar habitats in the Adirondack Mountains of New York, discovered that mosses tend to do well on acidic substrates where there is much shade, late snow cover, and high humidity. At Devil's Hopyard, the needles that drop from the red spruce, balsam fir, and hemlock trees provide an acidic environment, and the density of the trees in the narrow ravine provides the deep shade required for lush growth. Snow, which often blankets the area until late spring, increases the humidity. Finally, the far end of Devil's Hopyard is closed by ever-larger and more numerous boulders, forming a box canyon. With the ravine closed in on three sides, high humidity is maintained throughout the year.

White Mountain Meanderings

Spreading across northern New Hampshire, with some overlap into Maine, White Mountain National Forest encompasses much of the White Mountain range, so named because its peaks are often enshrouded in dense fog or capped with snow. Among the roads that provide a scenic itinerary through this region is the Kancamagus Highway, named after a 17th-century Native American leader. Kancamagus kept a confederacy of 17 tribes together from 1684 to 1691, until aggressive European settlers forced him northward out of the region. A section of New Hampshire State Route 112, the Kancamagus Highway covers the 34 miles between Conway and Lincoln. From Conway it follows the Swift River for several miles to the Passaconaway Valley, ascends to nearly 3,000 feet along the north flank of Mount Kancamagus, and then descends toward Lincoln, twisting and turning as it follows the Hancock Branch and then the East Branch of the Pemigewasset River.

Eight miles west of Conway, one stretch of the Swift River has been designated the Rocky Gorge Scenic Area. Here the clear mountain stream plunges 20 feet into a gorge of solid granite. A short trail from the parking area passes through a forest dominated by red spruce and white pine to pastoral Falls Pond. The 6-acre pond formed during the Ice Age, when a melting glacier left behind a ridge of sand and gravel that dammed up a tributary of the Swift River.

A little west of Rocky Gorge, just beyond the turnoff for Bear Notch Road, is the Passaconaway Historic Site. Pioneer Austin George established the settlement of Burton Intervale here in 1800, and his cabin, built in 1810, is now used by the Forest Service as a visitors' information station. Several trails can be followed from here. The Rail 'n River Trail is an easy 0.5-mile loop through a forest of red spruce, balsam fir, and northern hardwoods. The rugged 4.5 miles of Sawyer Pond Trail eventually ascends to Owl's Cliff, providing an outstanding view of Sawyer Pond and the forest below. The 46-acre pond contains crystal-clear water averaging 44 feet deep. It is abloom with white waterlilies during the summer. The Church Ponds Loop Trail leads to a wetland that includes two ponds and the largest open bog in the White Mountains, all surrounded by a forest of spruce, fir, and white pine. Bogs often form in lakes where poor drainage permits the accumulation of organic matter, especially sphagnum moss, which builds up in thick layers of peat. Much of the Church Ponds bog is a floating mat of vegetation, so walking on it is hazardous to the hiker, as well as deleterious to the plants.

The Kancamagus Highway reaches its high point at a mountain pass, where most of the forest consists of red spruce and balsam fir, with scattered heart-leaved paper birches. Near one of the sharp curves on the descent to Lincoln, a trail a little more than 2 miles long follows a branch of the Pemigewasset River to Lower and Upper Greeley ponds. Although the surrounding spruce and fir woodland is not virgin forest, some of the spruces are estimated to be nearly 200 years old.

From Lincoln one can follow U.S. Route 3 north to Franconia Notch, one of the most rugged and scenic passes in the White Mountains. Some of the high rocky outcrops have been given such fanciful names as Indian Head and Cannon Rock. Just north of Franconia Notch, Route 3 crosses Lafayette Brook, which has meandered down from the slopes of Mount Lafayette. A trail along the lower reaches of the brook passes through a forest of red spruce, balsam fir, and paper birch. Before reaching the summit of Mount Lafayette, the trail passes the lovely, high-mountain Eagle Lakes.

Farther north, Route 3 intersects U.S. Route 302, which one can follow southeast back to Conway. This highway goes near two mountain brooks, Elephant Head and Gibbs, which are surrounded by a forest of virgin red spruce, with a healthy mix of balsam fir, yellow birch, and paper birch. A trail follows Gibbs Brook and then rises steeply as it climbs up the northwestern flank of Mount Pierce. Near the summit, the spruces and firs are dwarfed and scrubby because of the harsh, gale-force winds that blow across the mountaintop.

Paralleling the Saco River, Route 302 continues through Crawford Notch, a pass located in a state park of the same name. South of the park, a few miles short of Bartlett, is a trail to serene, shallow Nancy Pond, a 4-acre glacial pool. The 3-mile trail, along Nancy Brook, is often muddy in places. Common trees are red spruce and balsam fir, interspersed with such broad-leaved species as American beech and sugar maple. High above the pond are the Nancy Cascades, a series of four waterfalls with a combined drop of 275 feet.

Descending into Bartlett, Route 302 passes through a broad river valley to the town of Glen and a junction with State Route 16. From here, you can either return south to Conway or turn north on Route 16, which winds through some of the most beautiful scenery in the White Mountains. Along the way are several covered bridges, including a much-photographed red bridge on a side road to Jackson. The main highway, following Ellis Creek through a narrow passage in the mountains, goes near Glen Ellis Falls, a silvery, 60-foot cascade. The highway then climbs over Pinkham Notch. Beyond are the mighty summits of the Presidential Range: Mounts Washington, Jefferson, Adams, and Madison. (A toll road leads to the top of 6,288-foot Mount Washington.) Several mountaintops in this range support a tundra-like flora during a growing season that lasts no longer than 100 days.

Venturing farther north on Route 16 to Gorham, you can follow U.S. Route 2 eastward into Maine and then take Maine Route 113 south, crossing through that state's relatively small portion of White Mountain National Forest. Short of the Evans Notch pass, a trail leads east from the highway, skirting scenic Kees Falls before ascending Caribou Mountain. Because of the cold and wind, this granite summit is nearly devoid of vegetation; plants grow only in the thin gravel lodged in crevices. Two species found here, the White Mountain silverling and the mountain sandwort, are among the rarest in the country.

Hardwood–spruce forests, consisting of deciduous hardwoods mixed with red spruce and some balsam fir, occupy most of the slopes below 2,500 feet. The major hardwoods are sugar maple—whose brilliant fall foliage attracts numerous visitors in October—American beech, and yellow birch. Paper birch is usually found in wetter areas. Because the canopy is somewhat open, an understory of smaller trees is commonly present, with striped maple, mountain maple, mountain ash, and pin cherry. Two prominent ferns are intermediate fern and hay-scented fern; the latter is especially common in disturbed areas. Wildflowers include mountain sorrel (an evergreen), false lily-of-the-valley, bluebead lily, and White Mountain aster.

Spruce–fir forests appear at higher elevations, where the deciduous trees begin to drop out. Red spruce and balsam fir are the major species here. Several wildflowers of the hardwood–spruce forest are also present, along with cucumber root, foamflower, downy Solomon's-seal, twisted stalk, yellow violet, and blue cohosh. Maidenhair fern is scattered throughout.

White pine forests are most common in well-drained sandy soil or on moist, gentle slopes, although white pine trees may grow in other forests as well. Balsam fir, basswood, hemlock, sugar maple, yellow birch, and American beech are also found here. Jack in the pulpit, bunchberry, partridgeberry, mountain sorrel, purple trillium (pl. 21), and ground pine grow on the forest floor.

Balsam fir forests grow at the highest elevations. In addition to balsam fir, they often host red spruce, white spruce, mountain ash, and heart-leaved paper birch. Wildflowers such as goldthread, creeping snowberry, and large-leaf goldenrod, among others, grow in these forests, along with many species from lower elevations.

Open bogs, several of which are scattered through White Mountain National Forest, contain a great number of plant species. In the shrubby category are bog rosemary, leatherleaf, Labrador tea, sweet gale, sheep laurel, bog kalmia, wild raisin, large cranberry, black chokeberry, and shining willow. Among the wildflowers are calla lily, swamp candles, bog green orchid, swamp pink orchid, twin-flower, and many kinds of sedges. Even sundew and pitcher plant can be found.

Marshes are found adjacent to bogs and along some of the streams. Common plants include sweet flag, water plantain, purple avens, blue iris, marsh marigold, sundrops, water smartweed, and several kinds of white-flowered aster.

Lake aquatic plants include the common bladderwort and whitewater crowfoot (a type of buttercup), both of which grow entirely underwater except for their blooms. In other species, the leaves emerge above the surface as well. Among these are white waterlily, water shield, and bullhead lily (spatterdock).

Plate 1 (left). Purple pitcher plant, Conecuh National Forest (Alabama).

Plate 2 (right). Tulip poplar, William B. Bankhead National Forest (Alabama).

Plate 3. Bridge at Leon Sinks, Apalachicola National Forest (Florida).

Plate 4. Hammock Sink, Apalachicola National Forest (Florida).

Plate 5. Flowering dogwood, Ocala National Forest (Florida).

Plate 6. Yellow lady's-slipper orchid, Chattahoochee National Forest (Georgia).

Plate 7. Sosebee Cove, Chattahoochee National Forest (Georgia).

Plate 8. Timber rattlesnake, often seen in Shawnee National Forest (Illinois).

Plate 9. LaRue Swamp, Shawnee National Forest (Illinois).

Plate 10. Redbud, Hoosier National Forest (Indiana).

Plate 11 (top). Angel Windows, Red River Gorge, Daniel Boone National Forest (Kentucky).

Plate 12 (right). Beggarticks, Daniel Boone National Forest (Kentucky).

Plate 13. White cedar, Hiawatha National Forest (Michigan).

Plate 14. Nordhouse Dunes, Manistee National Forest (Michigan).

Plate 15. Bishop's-cap, Ottawa National Forest (Michigan).

Plate 16. Mallard, often seen in Delta National Forest (Mississippi).

Plate 17. Pipes Lake, Homochitto National Forest (Mississippi).

Plate 18. Woodland phlox, Homochitto National Forest (Mississippi).

Plate 19. Lower Falls, Swift River, White Mountain National Forest (New Hampshire/Maine).

Plate 20. Mountain cranberry, Mount Washington, White Mountain National Forest (New Hampshire/Maine).

Plate 21 (top left). Purple trillium, White Mountain National Forest (New Hampshire/Maine).

Plate 22 (top right). Venus flytrap, Croatan National Forest (North Carolina).

Plate 23 (left). Stream in Joyce Kilmer Memorial Forest, Nantahala National Forest (North Carolina).

Plate 24. Catawba rhododendron, Pisgah National Forest (North Carolina).

Plate 25. Hills covered bridge, Wayne National Forest (Ohio).

Plate 26 (left). Allegheny Reservoir, Allegheny National Forest (Pennsylvania).

Plate 27 (bottom left). Coqui, Caribbean National Forest (Puerto Rico).

Plate 28 (bottom right). Hooded warbler, often seen in Francis Marion National Forest (South Carolina).

Plate 29. Bald cypress swamp, Francis Marion National Forest (South Carolina).

Plate 30. Shell midden, Sewee Shell Mound, Francis Marion National Forest (South Carolina).

Plate 31. Cherohala Skyway in the autumn, Cherokee National Forest (Tennessee).

Plate 32 (top left). Upper Tellico River, Cherokee National Forest (Tennessee).

Plate 33 (top right). Forest floor near Stratton Pond, off Long Trail, Green Mountain National Forest (Vermont).

Plate 34 (right). View from Blackies Hollow, George Washington National Forest (Virginia).

Plate 35. Marbled salamander, George Washington National Forest (Virginia).

Plate 36. James River Face Wilderness, Jefferson National Forest (Virginia).

Plate 37. Dolly Sods, Monongahela National Forest (West Virginia).

Plate 38 (top). Wood sorrel, Monongahela National Forest (West Virginia).

Plate 39 (right). West Virginia northern flying squirrel, Monongahela National Forest (West Virginia).

Plate 40. Loon, often seen in Nicolet National Forest (Wisconsin).

NATIONAL FOREST
IN NEW YORK

Finger Lakes NF

0   25   50   75 Miles

The Finger Lakes National Forest, the only national forest in New York, is
the newest region to gain national forest status. It is administered by the forest supervisor of Vermont's Green Mountain National Forest. Finger Lakes
National Forest is in Region 9 of the United States Forest Service.

Finger Lakes National Forest

SIZE AND LOCATION: 16,032 acres in west-central New York, between Seneca Falls and Elmira. Major access roads are State Routes 79, 227, and 414 and County Road 1. District Ranger Station: 5218 State Highway 414, Hector, NY 14841. Forest Supervisor's Office: 151 West Street, Rutland, VT 05701, www.fs.fed.us/r9/gmfl.

SPECIAL FACILITIES: Horseback trails; snowmobile trails.

SPECIAL ATTRACTIONS: Blueberry Patch Recreation Site; North Country National Recreation Trail.

The Finger Lakes of central New York are named for several long and narrow, finger-shaped lakes that trend north-south and are nearly parallel with each other. The lakes are located between Interstate Highway 90 to the north and Interstate Highway 86 to the south. The largest of the Finger Lakes are Seneca Lake and Cayuga Lake, and on the ridge known as the Hector Backbone between the southern end of these two lakes is the Finger Lakes National Forest. Originally known as the Hector Land Use Area, the region was given national forest district status in 1985. Although a separate national forest, the Finger Lakes National Forest is administered by the forest supervisor of Vermont's Green Mountain National Forest.

The area occupied by the Finger Lakes National Forest at one time was home to the Iroquois Indian Confederacy until the Indians were pushed out of the region in 1779. European settlers soon moved into the region and began farming operations, but the poor soil conditions were not conducive to farming, and the land quickly became depleted. In 1934, the federal government began to acquire the land and start a reforestation program.

Today, the area is very scenic, with pastoral woodlands that are particularly pretty in the spring when azaleas and numerous wildflowers are in bloom. Some of the more attractive wildflowers are the pink lady's-slipper orchid, Jack in the pulpit, false Solomon's-seal, Solomon's-seal, wild geranium, starflower, and mayapple. In moist areas are blue iris. Poison ivy is also plentiful.

Several town roads cross the national forest, usually following township lines. The southern end of the Finger Lakes National Forest lies about 4.5 miles north of the town of Watkins Glen. County roads form the western and northern borders of the national forest, and State Routes 79 and 227 lie along its southeast corner. A number of parcels of private land are interspersed within the forest boundaries.

The North Country National Recreation Trail runs through the southern part of the national forest. The only significant recreation area is the Blueberry Patch Recreation Site, also located in the forest's southern end. A small campground and picnic area are nestled under trees, adjacent to an extensive patch of high- and lowbush blueberries. North of this area is the Potomac Group Campground with several wildlife ponds nearby, and the Backbone Campground is available for horse users. An area known as the Caywood Property accesses Seneca Lake.

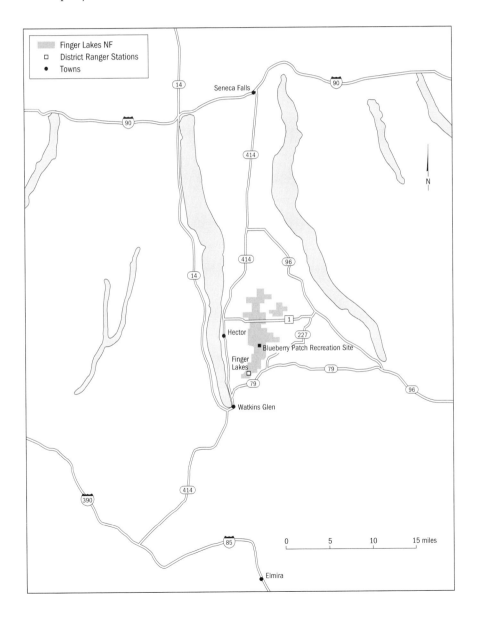

Several miles of hiking and horseback trails criss-cross the national forest. The Interlocken Trail runs the entire length of the forest, climbing over Hector Backbone. The Backbone Trail also crosses the Hector Backbone from east to west. The Gorge Trail enters a small gorge, and the Ravine Trail drops into a low area. Other trails are the Southslope Trail near the southern end of the national forest and the Burnt Hill Trail south of the Blueberry Patch Campground. The Gorge Trail begins at a parking area adjacent to the pair of Gorge Ponds. It is a pleasant trail with the beginning flanked by a forest on the right and a pond on the left. The forest consists of northern red oak, shagbark hickory, black walnut, witch hazel, sugar maple, white ash, yellow birch, with gray dogwood and elderberry in the shrub layer. On the forest floor are Christmas fern and lady fern. During the autumn are goldenrods, asters, and white snakeroot. Black willows occur along the shore of the pond, with soft rush, spikerushes, dark green bulrush, bur-reed, and narrowleaf cat-tail plentiful. In low, moist areas near the pond are New England aster, spotted joepyeweed, and flat-topped aster.

NATIONAL FORESTS IN
NORTH CAROLINA

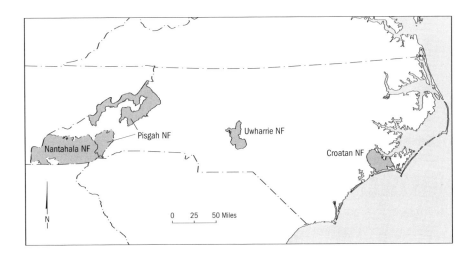

The national forests in North Carolina are the Croatan, Nantahala, Pisgah, and Uwharrie. The Croatan is in the Coastal Plain Province of the state, the Uwharrie is in the Piedmont Province, and the Nantahala and Pisgah are in the Mountain Province. All of the national forests are administered by one forest supervisor whose office is at 160A Zillicoa Street, Asheville, NC 28802. The national forests of North Carolina are in Region 8 of the United States Forest Service.

Croatan National Forest

LOCATION AND SIZE: Nearly 160,000 acres along the southeastern North Carolina coast, between New Bern and Morehead City. Major access routes are U.S. Highways 17 and 70 and State Routes 24, 58, 101, and 1004. District Ranger Station: New Bern. Forest Supervisor's Office: 160A Zillicoa Street, Asheville, NC 28802, www.cs.unca.edu/nfsnc.

SPECIAL FACILITIES: Boat ramp; swimming beach.

SPECIAL ATTRACTIONS: Cedar Point Tideland National Recreation Trail.

WILDERNESS AREAS: Catfish Lake South (8,530 acres); Pocosin (11,709 acres); Pond Pine (1,685 acres); Sheep Ridge (9,297 acres).

Although consisting of only 160,000 acres, the Croatan National Forest must be considered one of the unique national forests in the country. Most of the national forest is made up of an unusual boggy type of habitat known as a *pocosin*. It is one of two national forests with frontage on the Atlantic Ocean. It has more different species of carnivorous plants than any other national forest; it has one of the greatest number of poisonous reptiles and alligators. There are 31,000 acres of wilderness in the Croatan National Forest, but not a single trail penetrates them! Moreover, the forest is surrounded on all sides by water—Bogue Sound and the Atlantic Ocean, White Oak River, Trent River, Neuse River, Newport River, Brice Creek, Clubfoot Creek, and Harlowe Creek.

Cedar Point is a recreation area at the mouth of the White Oak River where tidal marshes exist along with hardwood forests and forests of pine. The water is an estuary where fresh water and salt water mix, with the tides providing a major influence. The Cedar Point Tideland Trail, partly on a boardwalk, will allow you to examine an estuary.

At Flanner Beach and Fishers Landing along the Neuse River, the Forest Service has a campground and swimming area, and Pine Cliff has picnic areas. Also at Pine Cliff is the beginning of the 21-mile Neusiok Trail that winds through stands of hardwood trees and pines. The trail begins in a cypress-lined sandy beach on the Neuse River, passes through hardwood forests and a pocosin, and ends at an estuary of the Newport River. Along Brice Creek just south of New Bern is a picnic area, boat ramp, and self-guided canoe trail.

An interesting woods occurs along Island Creek where there is a beautiful mixture of southern trees with trees that are also found in forests of the Midwest. Bald cypress draped with Spanish moss, umbrella magnolia, and loblolly pine occur with trees such as northern red oak, southern red oak, white oak, American beech, black walnut, tulip poplar, and flowering dog-wood. The Croatan National Forest and the Trent Woods Garden Club of New Bern have established a trail through this woodland.

Most of the Croatan National Forest is covered by pocosins, boglike areas where great amounts of organic matter have accumulated, usually on small rises in the topography. Meaning "swamps on the hill," the pocosins are strongly acidic and poor in nutrients. About 80 percent of the soil is muck ranging in depth from a few inches along the border to several feet in the center. Dense vegetation usually occurs in the shallower muck, including pond pine, red bay, loblolly bay, sweet bay magnolia, fetterbush, honeycup, and titi. There are great entanglements of greenbrier vines. The honeycup seems to be restricted to the pocosins. Of particular interest are the carnivorous plants. The four wildernesses in the Croatan National Forest preserve the best of the pocosins.

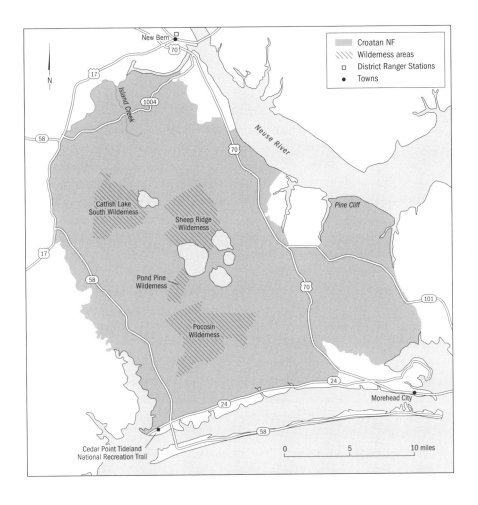

There is a strong interest in the animal life in the Croatan National Forest. Alligators are present, along with cottonmouths, copperheads, eastern diamondback rattlesnakes, canebrake rattlesnakes, and pigmy rattlesnakes. The rare Atlantic sturgeon has been found in the waters near the national forest. Among the birds are bald eagle, osprey, peregrine falcon, woodcock, and the endangered red-cockaded woodpecker.

Along State Route 24 midway between Cedar Point and Morehead City is the Patsy Pond Nature Trail, a project of the North Carolina Coastal Federation and the Croatan National Forest. This self-guiding trail loops through longleaf pine flatwoods and passes four small ponds and a large diversity of plants.

Cedar Point

At less than 250 square miles, Croatan is one of the smallest U.S. national forests, but it contains many biological attractions. It is the northernmost national forest where the American alligator and the Atlantic white cedar have been recorded. It is home to copperheads, cottonmouths, and three kinds of rattlesnakes. And its pocosins, or hillside swamps, harbor 11 different kinds of carnivorous plants. Located on the coast of North Carolina between New Bern and Swansboro, Croatan also embraces Cedar Point, where plants and animals thrive in the salt water of a tidal estuary.

At Cedar Point, seawater from the Atlantic Ocean meets and mixes with fresh water from the White Oak River, whose lower reaches are a mile wide. The gentle, sandy ridges and dry woods of Cedar Point are broken up by lower areas of moist woods and brackish marsh that adjoin inlets of the estuary. The U.S. Forest Service has constructed an interpretive trail that meanders through this varied habitat. It is 1 of about 800 National Recreation Trails designated by the federal government.

From a parking lot, the trail leads through a woods where loblolly pines, water oaks, and sweet bay magnolias grow among shrubs of bayberries, highbush blueberries, yaupon hollies, and Hercules' clubs. Other plants in this heavily disturbed area include giant cane (a type of bamboo), roundleaf greenbrier, muscadine grape, and poison ivy.

As the trail drops imperceptibly toward the White Oak River, a marsh comes into view. A boardwalk through it permits close inspection of the vegetation. Among the species adapted to the brackish water are saltmarsh cordgrass; a dark brown to blackish plant known as needlerush; seaside goldenrod; and the shrubby saltmarsh elder, a greenish flowering plant related to giant ragweed. Growing with these are two kinds of plants that do equally well in freshwater areas: switchgrass, a large, clump-forming grass that may grow 6 feet high, and southern red cedar, a small to medium-sized tree that probably gives Cedar Point its name. Southern red cedar is related to the more common eastern red cedar, which is also present, but not in the brackish areas.

The trail advances into a sandy upland woods supporting red bays, wax myrtles, loblolly pines, water oaks, and scraggly leather-leaved oaks above a grassy understory of little sea oats, hairy panicum, and melic grass. The influence of the salt water is evident in the encroachment of needlerush and seaside goldenrod.

Another boardwalk takes the hiker into a large, brackish marsh where open water and little streams surround clumps of saltmarsh cordgrass and needlerush. Several dead snags stand in the marsh, the remains of pines that could not tolerate the salt water. Some of the birds that frequent the marsh are great blue heron, snowy egret, kingfisher, semipalmated plover, osprey, willet, and black duck.

Here the ebbing tide exposes patches of soil where fiddler crabs come out to feed. The male fiddler crab is the one with the large claw, which it waves to attract the females' attention. Plants tolerant of salt water also inhabit these patches of ground and survive through the fluctuations of the tide. They include sea oxeye, which has small sunflower-like heads; a couple of short, narrow-leaved, wiry grasses known as saltgrass and saltmarsh dropseed; a perfectly branched, nearly leafless plant with papery flowers called sea lavender; and sea purslane and glasswort, which are both succulents. Sea purslane has flat, fleshy leaves and five-petaled, purplish flowers. Glassworts, also called pickleweeds, have upright, succulent, jointed stems without leaves. Their flowers usually form in groups of three in the hollow of the uppermost joints. In late summer, the stems turn red.

After leaving this part of the marsh, the trail meanders in and out of a diverse upland forest that contains black gum, tulip poplar, American holly, water oak, loblolly pine, white oak, chestnut oak, wild black cherry, and sassafras. The trail swings back near the marsh one more time and then enters moist woods dissected by a small stream. These woods are home to sweet gum, southern magnolia, red maple, white alder, and witch hazel. Along the stream are such wetland species as buttonbush, poison sumac, and cinnamon fern.

Biologists estimate that approximately 95 percent of all ocean life originates in or is dependent on estuaries, which are extremely high in nutrients. They consider tide-washed brackish marshes, such as that at Cedar Point, to be more productive than the richest farmland. One source of nutrients are the microscopic organisms known as plankton, which stream into the estuary with high tide, providing food for fish and shellfish. In addition, as larger plants and animals that live in the marsh die, their detritus (decaying organic matter) is broken down into nutrients by bacteria, fungi, and algae, which in turn are the major source of food for snails and crabs. Ultimately, the fish, snails, and crabs are eaten by shorebirds and mammals such as raccoons, opossums, and, of course, humans.

Island Walk

If you drive along the sandy back roads of North Carolina's Croatan National Forest—being careful not to get mired in the soft shoulders—you will see a mostly dry pineland of shortleaf and loblolly pines. But this 250-square-mile forest, located on the Atlantic coast and extending inland for 30 miles, encompasses several other habitats. Tidal marshes include Cedar Point, at the southern end of the forest, which harbors plants that thrive in brackish water and others that do not tolerate it. Pocosins, which are usually flat, occur at elevations slightly higher than the surrounding land. As mentioned earlier, the black, nutrient-poor, acidic soil of the pocosins supports 11 different kinds of carnivorous plants. One of the prettiest places is a dense forest made up of hardwood trees, shrubs, and wildflowers that are found regularly in the Appalachian Mountains but rarely on the Atlantic Coastal Plain.

The Coastal Plain consists mostly of flat, sandy terrain that is subjected to the salt spray that blows in from the Atlantic Ocean. It supports plants that grow in relatively open areas; even where forest grows, the trees are spaced far enough apart to create an open canopy. In contrast, the mountainous Appalachian landscape, with its clear streams and deep coves, is optimal for the dense growth of shade-loving species.

Situated 7 miles southwest of New Bern and bordered by Island Creek, Long Branch, and Highway 1004, the Appalachian enclave has been set aside for special management. With the assistance of the Trent Woods Garden Club of New Bern, the United States Forest Service has developed an interpretive trail providing access to a cross section of vegetation. The circular trail begins and ends at a small roadside parking area.

The slightly more elevated, drier areas near the trailhead are dominated by species of oak, walnut, hickory, and ash that are major components of hardwood forests from the Appalachians to the Mississippi River. As the trail descends and eventually follows a terrace above Island Creek, these species give way to beech and other trees that require a richer, moister soil.

Below the terrace and adjacent to the creek is a bottomland that is subject to occasional minor flooding. Fed by minor drainage from several pocosins, the water in the creek is stained dark brown by minerals picked up from the soil. Several large bald cypress trees grow in the bottomland. (According to the Forest Service, the largest and oldest cypresses were removed in the past, but otherwise the forest has not been logged.)

The long wisps of Spanish moss that hang from many of the trees are a reminder that this is still the Coastal Plain, since this plant does not usually appear in the Appalachians. Another clue is the scattering of shortleaf and loblolly pines among the hardwoods. These were probably the dominant species in this area until 500 years ago. The growing conditions may have been altered gradually by the influence of the nearby Trent and Neuse rivers, which created a milder and more temperate climate. Other major rivers that traverse the Coastal Plain probably had similar effects, but the forests that grew near them were not spared from logging.

The drier forest in this region consists of white oak, northern red oak, southern red oak, black walnut, shagbark hickory, white ash, and basswood. These trees tower above a midlayer of hop hornbeam, flowering dogwood, red mulberry, witch hazel, and beautyberry (with its bright magenta berries). An abundance of wildflowers and ferns, such as bloodroot, meadow parsley, alum root, May apple, wild sage, heartleaf aster, and Christmas fern, are in the understory.

The moist forest is dominated by the stately American beech, but tall, straight tulip poplars, sugar maples, umbrella magnolias with their giant leaves, black gum, and the evergreen American holly are also quite common. Spicebush and American hornbeam, or musclewood, are abundant woody plants in the midlayer. The ground-level plants include rattlesnake fern, broad beech fern, creeping partridge berry, and a dazzling array of wild-flowers: false Solomon's seal, wood betony or lousewort, wild ginger, hepatica, and rattlesnake plantain orchid and spotted wintergreen, two species whose mottled green-and-white leaves are more conspicuous than their flowers.

The bottomland forest is made up of bald cypress, red maple, American elm, sweet gum, and overcup oak, all species that can stand periodic inundation. The understory in this community is sparse, but an occasional sensitive fern and a few kinds of grasses and sedges intermingle with the cypress knees.

Plants in the Pocosins

On casual inspection, Croatan National Forest is unalluring. This small pineland on North Carolina's Coastal Plain contains large expanses of bog-like habitat with scrubby vegetation. But in these areas, where the soil is generally too poor to support trees, grows the widest selection of carnivorous plants found in any U.S. national forest.

The alternate lifestyle of carnivorous plants arouses almost everyone's curiosity. It is not true, as botanist G. E. Rumphius reported from Malaysia in 1783 (and as *American Weekly* magazine told its readers in the 1920s), that vicious plants grow in Madagascar, the Malay Archipelago, and the Philippines that are capable of killing and even swallowing humans. But some 350 different kinds of flowering plants in the world do trap and digest insects and other small animals. Since their victims include small amphibians and reptiles and an assortment of invertebrates, they are referred to as carnivorous, rather than insectivorous, plants.

Botanists recognize 45 species of carnivorous plants in the United States, and at least 11 of these live in Croatan. The Venus flytrap (pl. 22) is the most famous, probably because the movements of its inch-long traps are easily observed. The plant's traps are twin-lobed leaves that are hinged between the lobes. Each lobe is rimmed with narrow, stiff teeth and bears three slender, supersensitive hairs on its inner surface. If two of these hairs are touched within a period of about 30 seconds, the two lobes of the trap close, and the teeth along their edges become intermeshed.

The requirement that two hairs need to be touched to trigger the trap is a way the plant saves energy. If an inanimate, and therefore probably nonnutritious, object brushes a hair by chance, the trap remains open. But an insect crawling on the trap is apt to brush against two of the hairs, triggering the trap. At first, the teeth along the edges do not intermesh completely, allowing very tiny insects, which have little food value, to escape. If there is no substantial prey, the trap then reopens, but if the insect is large and continues to wriggle, the trap closes completely. Once entombed, the insect is digested in 24 to 36 hours by enzymes. Afterward, the trap reopens and lies in readiness for its next meal.

Venus flytrap and other carnivorous plants possess chlorophyll and thus are capable of manufacturing their own nutrients; their carnivory is an adaptation to nutrient-poor environments. In Croatan National Forest, the habitat in which they grow are pocosins, flat, slightly elevated areas with poor drainage, ranging up to several hundred acres in extent. They seldom flood, but small pools may appear during the winter.

The pocosin's ooze of black soil looks rich but is highly acidic. Soil acids combine with bases to create water-soluble salts; in this form, many essential soil minerals are washed away by the frequent drenching rains. The soil gets its deceptively rich appearance from carbon—charcoal left by spring fires. Spring is a dry season in Croatan, and plant debris from the previous growing season dries out rapidly. Fires commonly break out and remove the litter, inhibiting the growth of shrubs and aiding the carnivorous plants, which thrive in sunlight. When natural fires fail to suppress shrubs, U.S. Forest Service personnel sometimes supervise controlled burning.

In addition to the Venus flytrap, 10 other species of carnivorous plants are found in the Croatan pocosins: two pitcher plants, two sundews, a butterwort, and five bladderworts. A pitcher plant has one or more tubular or funnel-shaped leaves with an opening at the upper end known as the mouth. Projecting upward beyond the mouth is a lid, or hood. Early naturalists, including Linnaeus, thought that the lid was hinged and could be opened or closed by the plant to regulate the amount of water that accumulated inside the pitcher. Actually, the position of the lid is fixed. Eighteenth-century naturalist Mark Catesby speculated that the pitchers were used as hiding places for frogs and other small animals. It wasn't until 1817 that James Macbride, a physician, realized that the pitchers were able to attract flies with a sweet secretion. He noted that once it has fallen into the bottom of a pitcher, a fly cannot escape because downward-pointing hairs line the leaf's inner walls.

Like flowers, most pitchers are brightly colored to attract nectar-seeking insects. The thick rim, which secretes nectar, is colorful, shiny—and slippery. The visiting insect eventually slips on the rim surface and falls into the bottom of the pitcher, which contains a mixture of rainwater and digestive enzymes. A pile of undigested insect skeletons eventually accumulates within the plant.

Two of the 120 kinds of sundews in the world live in very damp areas in Croatan. Sundews are perky-looking little plants whose crowded rosettes of leaves form at ground level. All the leaves secrete droplets of a colorless, sticky liquid at the tips of countless slender hairs. Insects that venture on them are unable to pull free and eventually perish. Sundew leaves also have enzyme-producing glands, mostly along the center of each leaf, and an insect caught in this area is engulfed quickly. Should an insect land near the edge of the leaf, the hairs will bend inward, transporting the captive to where the enzymes are concentrated.

The blue butterwort, or *Pinguicula*, is another kind of "flypaper" plant. Although it is not related to the sundews, the butterwort also has all its leaves in a basal cluster. A sticky substance covers the upper surface of these leaves, but not the tips of hairs, as in the sundews. If an insect gets mired down on one of the gummy leaves, the leaf curls slightly so that the surface becomes concave, forming a basin in which the digestive enzymes accumulate.

Numerous depressions with shallow standing water dot Croatan, and most of them support still another group of carnivorous plants, bladderworts. Five kinds of bladderworts grow in or along the margins of these depressions, their intricately branched stems floating in the water. All along the stems are transparent air-filled bladders about 0.16 inch long. Each bladder has a small mouth at one end, but the mouth is normally closed by a trap door. When a tiny water animal swims into contact with one of two trigger hairs at the entrance to the mouth, the trap door suddenly swings inward, and the inrushing water sweeps the animal inside. The water filling the bladder forces the trap door closed, and the prey is condemned to digestion by enzymes secreted within the chamber. After several hours, the water is forced out of the bladder, and the trap is set again.

Charles Darwin, one of the first scientists to consider the biology of carnivorous plants, was aware that pitcher plants, sundews, and others of their kind were adapted to poor soils. But what nutrients are missing from lands like the pocosins? Because plant growth is often inhibited by a lack of assimilable nitrogen, scientists had long assumed that plants developed carnivory specifically to supplement their nitrogen sources. However, botanists Lionel Eleuterius and S. B. Jones Jr., working with carnivorous plants in Mississippi, found that the soils that supported these plants were not deficient in nitrogen, phosphorus, or potassium. When fertilizers that contained these elements were applied to the soil, the carnivorous plants actually did poorly. Recently, zoologist George Folkerts, a former student of mine, suggested that a shortage of trace elements—particularly molybdenum, which is scarce in acidic soils—may have been the key to the development of carnivorous plants. The Fish and Wildlife Service has been studying pocosins for their value as a wetland resource. The results of their soil analyses may shed further light on the relationship between carnivorous plants and their characteristic habitat.

Nantahala National Forest

SIZE AND LOCATION: Approximately 555,000 acres in western North Carolina. Major access routes are U.S. Highways 19, 21, 64, 129, and 441; and State Routes 28, 106, 107, 143, 281, 1140, 1319, 1326, 1342, 1365, and 1448. District Ranger Stations: Franklin, Highlands, Murphy, and Robbinsville. Forest Supervisor's Office: 160A Zillicoa Street, Asheville, NC 28802, www.cs.unca.edu/nfsnc.

SPECIAL FACILITIES: Boat ramps; swimming beaches; off-road vehicle areas; whitewater rafting areas; lodge.

SPECIAL ATTRACTIONS: Mountain Waters National Scenic Byway; Joyce Kilmer Memorial Forest; Beech Creek Seed Orchard; Nantahala Gorge; Cullasaja Gorge; Whiteside Mountain; Whitewater Falls; Cherohala Skyway.

WILDERNESS AREAS: Ellicott Rock (8,274 acres, partly in the Chattachoochee and Sumter national forests); Joyce Kilmer–Slickrock (17,394 acres, partly in the Cherokee National Forest); Southern Nantahala (23,473 acres, partly in the Chattahoochee National Forest).

The Nantahala National Forest contains many deep, narrow gorges where the sun reaches the ground only when it is directly overhead at noon. Thus the word *Nantahala*, Cherokee for "land of the noonday sun," is appropriate for the river, gorge, and national forest of the same name. This huge forest, stretching east to west for 75 miles to the Tennessee border, and north to south from the Great Smoky Mountains National Park to Georgia, is one of five national forests in the beautiful southern Appalachian Mountains.

The Nantahala River bisects the national forest nearly in half, beginning just north of the Georgia border northwest of Dillard, Georgia, passing the Standing Indian Recreation Area, then going between the Nantahala Mountains to the east and the Tusquitee Mountains to the west, and, finally, being dammed to form Nantahala Lake. Northwest of Nantahala Lake, the river enters the spectacular 9-mile-long Nantahala Gorge on either side of U.S. Highway 19 and finally converges with the Little Tennessee River where the two rivers form Fontana Lake. Standing Indian Mountain, at 5,499 feet, is the highest of several peaks that ring the Standing Indian Basin. The basin contains several hiking trails, including the Appalachian Trail that follows the high ridge around the basin. The Standing Indian Campground and Kimsey Creek Group Camp are here, as well as the Hurricane Creek Primitive Camping Area. The recreation area may be reached by Forest Road 67, a short paved road south of U.S. Highway 64 and 12 miles west of Franklin.

About 3 miles south of the Standing Indian Campground is the northern edge of the Southern Nantahala Wilderness, which extends into the Chattahoochee National Forest in Georgia. By following Forest Road 67 south from the campground, you will come to Big Falls and then Mooney Falls, both just outside the western boundary. Forest Road 71, which leads south of U.S. Highway 64, winds to the edge of the wilderness at Deep Gap where the Appalachian Trail may be accessed.

The Southern Nantahala Wilderness is in steep, rugged mountain terrain, and that part of the wilderness in the national forest is remote and little used. Only the 15 miles of the Appalachian Trail along the boundary of the wilderness receives much visitation. The Tallulah River crosses the entire length of the area from north to south. On either side are deep hardwood coves, rocky cliffs, and treeless balds.

West of the Southern Nantahala Wilderness is large Chatuge Lake, which laps over into Georgia. The Jackrabbit Mountain Recreation Area is the only

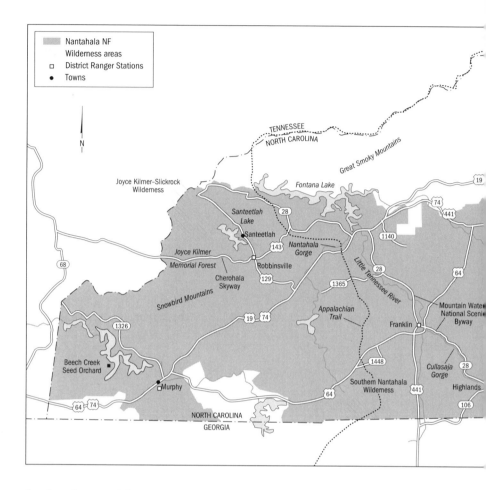

developed national forest recreation area around the lake. The lake is popular for canoeing, kayaking, water skiing, sailing, and fishing; it also features a campground, boat ramp, and swimming beach. Two loop hiking trails totaling 3 miles go around the mountain, following the lakeshore.

North of the Southern Nantahala Wilderness is the small but interesting Chunky Gal Mountain. Although the mountain is only about 10 miles long, it is rugged, particularly along Big Tuni Creek. U.S. Highway 64 crosses the mountain at Chunky Gal Gap. Chunky Gal Trail crosses the Appalachian Trail at Chunky Gal and then continues south to White Oak Swamp at the Southern Nantahala Wilderness western edge. At the northern end of Chunky Gal Mountain is Big Tuni Creek south of Nantahala Lake.

Between U.S. Highway 19 at Andrews and U.S. Highway 64 at Hayesville are the Tusquitee Mountains and the exciting Fires Creek area. This area is sanctuary for black bears and is densely forested, although a number of balds

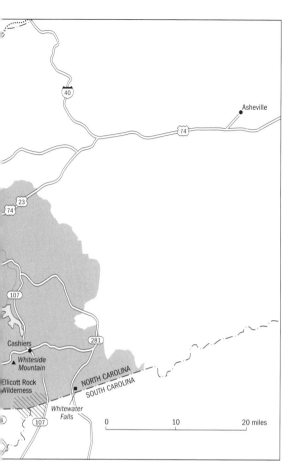

occur on the eastern side of the mountains. The Rim Trail is a rugged 25-mile-long trail that follows the ridge circling the Fires Creek basin. Most people begin the Rim Trail from Fires Creek Picnic Area. A shorter partially paved trail is the Leatherwood Loop Trail that begins and ends at the Fires Creek Picnic Area. The Leatherwood Falls is a series of cascades, with a shrub known as leatherwood fairly common nearby.

Nantahala Lake, the highest lake in North Carolina at 3,013 feet, may be fished for largemouth bass, smallmouth bass, and walleye. The Bartram Trail may be accessed at the lake; a 9-mile trail to London Bald is another option.

Nantahala Gorge extends as a narrow defile for 9 miles, beginning at Wesser at its northern end. During the summer, thousands of people canoe, kayak, and whitewater raft in the Nantahala River. There are three river accesses in the gorge. Many of the rapids are rated as Class II and III, with Nantahala Falls having a Class IV status. Wesser Falls is Class V. In some places in the gorge, sheer cliffs rise more than 1,000 feet above the river. The Appalachian Trail crosses the Nantahala River at Wesser.

At the northern edge of the Nantahala National Forest, the Little Tennessee River has been dammed to form Fontana Lake, whose waters are on either side of the North Carolina–Tennessee border and adjacent to the Great Smoky Mountains National Park. Part of the Tennessee Valley Authority's system of lakes, Fontana Lake consists of 11,685 acres and was constructed in 1945. Nestled between the Great Smoky Mountains to the north and the Cheoah Mountains to the south, Fontana Lake is in a beautiful setting. Cable Cove Recreation Area is situated along the southern edge of this lake, and the Tsali Recreation Area is near the eastern end. Both recreation areas may be reached by short roads north of State Route 28. Lemmons Branch Boat Ramp near the Tsali Recreation Area permits boaters to reach the lake, even during normal drawdowns during late summer. The Appalachian Trail, having crossed the Cheoah Mountains and passing such attractive areas as Sassafras Gap, Cheoah Bald, and Stecoah Gap, circles around the western end of Fontana Lake and into the Great Smoky Mountains National Park.

One of the best paved scenic routes in the Nantahala National Forest that receives relatively little traffic is the Wayeh Road, State Route 1310, which connects Beechertown at the south end of Nantahala Gorge with the town of Franklin. It is part of the Mountain Waters National Scenic Byway. This curvy, mostly mountain highway stays next to bubbling Nantahala River as far as Nantahala Lake. It then takes an eastward course along Jarrett Creek and Wayeh Creek, where there is a campground. A side road to the north, Forest Road 69, ends at Wayeh Bald, at one time a treeless knob at an elevation of 5,342 feet. Trees have encroached on the bald, with only a small, mowed grassy area left at the turnaround on the summit. Vegetation near the summit is still interesting, with white rhododendron, a white-flowering azalea, flame azalea, and mountain laurel prominently blooming in June. Other trees at the summit include black oak, northern red oak, white oak, yellow birch, and basswood. A very short hike to the stone lookout is worth taking. The square tower dates back to 1937, and the views from the top are incomparable on clear days. Both the Appalachian Trail and the Bartram Trail cross Wayeh Bald, and the shorter Rufus Morgan and Shot Pouch trails are accessed from the road to the summit. On the way to the bald, the now-abandoned Wilson Lick Ranger Station may be observed. Built in 1916, this was the first ranger station in the Nantahala National Forest. The station is surrounded by a forest of oaks, hemlock, and white pine.

Between Wayeh Crest and Franklin, along the Wayeh Road, is Arrowwood Glade Picnic Area in a scenic setting along a mountain stream. South of Arrowwood Glade and west of Milksick Knob is 60-foot Rough Fork Falls, reached from the Rufus Morgan Trailhead off Forest Road 388.

Mountain Waters National Scenic Byway continues on U.S. Highway 64 from Franklin to Highlands. It goes through Cullasaja Gorge where several waterfalls are easy to observe. Cullasaja Falls is where the Cullasaja River drops dramatically for 250 feet into a deep, rocky gorge. A little farther south is Dry Falls with an adjacent parking area. Several stone steps lead to this mighty falls that plummets for 40 feet, surrounded by lush vegetation. The trail permits visitors to walk behind the falls to another observation point on the other side. In fact, this great falls is called Dry Falls because you can walk behind it and stay dry, but that is not always the case. Still father south on State Route 28 is Bridal Veil Falls. A spur road lets you drive behind this misty falls.

South of Cashiers, State Route 28 continues into South Carolina, crossing into the Sumter National Forest. East of the highway are several waterfalls, including Silver Run Falls and the incomparable Whitewater Falls. Whitewater Falls, which drops for 411 feet in two bursts, is the highest waterfall east of the Rocky Mountains. A 0.2-mile hike from a large parking area goes to an observation deck across from the falls. If you want a closer look, take the 154 wooden steps down to another observation deck. The more hearty can continue to hike downward on the Foothills Trail to the Whitewater River. You can hike the Foothills Trail for 80 miles into the Sumter National Forest of South Carolina.

Forest Road 1100 is a gravel road west off State Route 28 that in a few curvy miles comes to the Chattooga Wild and Scenic River. The Chattooga is a first-class whitewater rafting stream, and only the experienced rafter should try it (fig. 10). It goes through the Ellicott Rock Wilderness and into the Sumter National Forest in South Carolina.

Figure 10. Rafting on the Chattooga River, Nantahala National Forest (North Carolina).

A few miles south of Highlands, via State Route 1603, is the northern terminus of the trail to Ellicott Rock. One and one-half miles farther south on the forest road is an iron bridge over the Chattooga River where there are two hiking trails. (See Sumter National Forest.)

State Route 106 west of Highlands and U.S. Highway 64 east have plenty to offer in the Nantahala National Forest. The 10 miles of State Route 106 from Highlands to the Georgia border at Sky Valley goes past Glen Falls, Blue Valley Vista, and Ozark Mountain Vista. Glen Falls is the result of the East Fork of Overflow Creek cascading down a rocky escarpment. The short trail to view the falls is steep, rocky, and often muddy. Blue Valley, seen from an overlook along State Route 106, is a wild area of mountain ridges, deep hardwood forest coves, and rushing mountain streams. The Blue Valley Experimental Forest conducts research on white pine and various hardwood species.

Between Highlands and Cashiers, U.S. Highway 64 provides access to Whiteside Mountain. The east side of this marvelous mountain rises as much as 750 feet above the valley floor to an elevation of 4,930 feet. Light-colored feldspar and quartz alternates with bands of dark granite, forming a picturesque pattern on the mountain cliff faces. The peregrine falcon (fig. 11) was introduced onto the mountain in 1985. The Whiteside Mountain National Recreation Trail is a loop, about 2 miles long, that climbs to rocky cliff faces and through a forest dominated by red oak, black oak, white oak, and chestnut oak.

The 39,000-acre Roy Taylor Forest occupies the northeastern corner of the national forest and includes Blackrock Mountain and Rich Mountain. Panthertown Valley along Panthertown Creek is south of Blackrock Mountain, and several scenic waterfalls may be hiked to in this region, including Schoolhouse Falls, Granny Burrell Falls, and Greenland Creek Falls. Southwest of the Tuckasegee River and reached via State Route 281 is the spectacular Tuckasegee Gorge, with huge boulders and several cascades. Deep hardwood and hemlock forest coves are along the river. Rich Mountain, northeast of State Route 281, is highlighted by Rich Mountain Bald, Charley Bald, and Gray Bald. Balsam Lake Lodge is secluded at the end of State Route 1756. The lodge, situated on tiny but picturesque Balsam Lake, offers fully accessible accommodations and good fishing.

West of the Nantahala River and U.S. Highway 129/19 are other major attractions in the Nantahala National Forest. Santeetlah Lake is beautifully

Figure 11. Peregrine falcon, often seen in Nantahala National Forest (North Carolina).

situated in a basin surrounded by the Cheoah Mountains to the east, the Snowbird Mountains to the south, the Great Smoky Mountains to the north, and the Unicoi Mountains to the west. The Forest Service has three boat ramps into the lake, one at Cheoah Point Campground along the northern side near the dam, one at Massey Branch at the southern end of the lake near the Cheoah Ranger Station, and one at Avey Branch at the northwest fringes of the lake near the Joyce Kilmer Memorial Forest.

The Joyce Kilmer Memorial Forest (pl. 23) at the edge of the Joyce Kilmer–Slickrock Wilderness is a prized natural area of virgin forest. Consisting of 3,800 acres, the memorial forest commemorates Joyce Kilmer, author of the poem "Trees" and a former journalist for the *New York Times*; he was killed in action in France during World War I. Two miles of trails, often muddy in places, wind through this forest of gargantuan trees. Huge tulip poplars, maples, Ohio buckeyes, hemlocks, and pines are plentiful, as are large specimens of silverbell trees. The canopy is dense, forming great amounts of shade that are optimal for the growth of mosses, ferns, and wildflowers. Black bears are plentiful. Northwest of the memorial forest is the remainder of the wilderness, a part of the rugged Unicoi Mountains. Several hiking trails are in the wilderness, including one that climbs to 5,341-foot Stratton Bald, the highest spot in the area.

The Cherohala Skyway travels the crest of the Unicoi Mountains south of the wilderness and climbs through Stratton Gap on the state line with Tennessee and into the Cherokee National Forest. The views from the road are spectacular with the highest overlook on the skyway, Santeetlah, at 5,390 feet. Short trails along the skyway provide a glimpse of the northern hardwood forest of American beech, wild black cherry, and yellow birch. You may walk to the top of Huckleberry Knob, covered with grasses and wildflowers, for a panoramic view from 5,560 feet. The Snowbird Mountains to the south are steep and rugged, with several waterfalls. Several attractive waterfalls occur along Sassafras Creek, reached by a hiking trail that follows the creek for several miles.

Northwest of Murphy is the major part of the Tusquitee Ranger District with a series of three interconnected lakes. The Hiawasse is the middle lake and by far the largest, with tiny Cherokee Lake to the south and the somewhat larger Appalachia Lake to the west, extending to the Tennessee state line. Hanging Dog Recreation Area, 4.5 miles west of Murphy on State Route 1326, has a campground, a boat launch area, and two hiking trails. The Hanging Dog Interpretive Trail is about 2 miles long and follows a ridge on Ramsey Bluff. Cherokee Lake has a picnic area and a 0.25-mile trail from the picnic area to the dam.

North of Appalachia Lake is the large Upper Tellico Off-Highway Vehicle

Area. It is known as an extreme challenge for large four-wheel-drive vehicles such as Jeeps and sports utility vehicles.

West of Murphy is the Beech Creek Seed Forest where trees of white pine, shortleaf pine, Virginia pine, black cherry, tulip poplars, and oaks are grown from seed and then transplanted into forests of the southern Appalachian Mountains.

Pisgah National Forest

SIZE AND LOCATION: Approximately 500,000 acres in the Appalachian Mountains of western North Carolina. Major access routes are Interstate Highway 40; U.S. Highways 19W, 23, 25, 70, 221, 276, and 321; and State Routes 143, 181, 209, and 215. District Ranger Stations: Burnsville, Hot Springs, Nebo, and Pisgah. Forest Supervisor's Office: 160A Zillicoa Street, Asheville, NC 28802, www.cs.unca.edu/nfsnc.

SPECIAL FEATURES: Boat ramps; swimming beaches; off-road vehicle (ORV) areas.

SPECIAL ATTRACTIONS: Cradle of Forestry in America; Forest Heritage National Scenic Byway; Linville Gorge; Roan Mountain Gardens.

WILDERNESS AREAS: Linville Gorge (12,002 acres); Middle Prong (7,460 acres); Shining Rock (18,483 acres).

The science of forestry in the United States had its beginning in what is now the Pisgah Ranger District of the Pisgah National Forest when George Vanderbilt, who owned 80,000 acres of land near the Biltmore Estate at Asheville, North Carolina, hired Dr. Carl Alwin Schenck from Germany's Black Forest to establish the first School of Forestry in the United States in 1898. During the school's operation between 1898 and 1913, more than 350 students learned the principles of forestry. This Biltmore Forest School, as it was called, was purchased by the U.S. Department of Agriculture Forest Service in 1914 from Vanderbilt's widow, Edith, and it became the focal point for an area known as the Cradle of Forestry in America, established by Congress in 1968. Located along U.S. Highway 276 about 4 miles south of Blue Ridge Parkway and 12 miles north of Brevard, North Carolina, the Cradle of Forestry in America includes the very informative Forest Discovery Visitor Center, the remaining buildings of the school, and the Forest Festival Trail. Nearby is a picnic area and trail complex in an area called the Pink Beds.

Start your visit to the Cradle of Forestry in America at the modern Forest

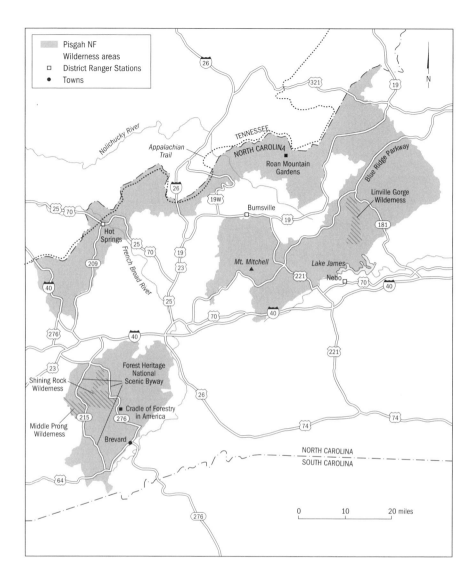

Discovery Center, which features numerous hands-on exhibits, the Carl Schenck Education Wing, a well-stocked gift shop, and the beginning of a pair of mile-long, paved trails. One trail winds through the old Biltmore Campus where you will see a schoolhouse dating back to 1906; the commissary, which served as post office and general store; a ranger's dwelling built in 1882; Dr. Schenck's office furnished with a desk and an old typewriter; the Black Forest Lodge made from native chestnut; and a student dormitory referred to as Hell Hole (fig. 12). The hiking trail that connects these buildings passes through a good example of the vegetation in this part of the Appalachian Mountains.

The Forest Festival Trail passes a seedling nursery; a Norway spruce plantation more than 100 years old; a portable, steam-powered sawmill; an old logging locomotive; and an American log loader that could lift more than 1,000 logs per day up a forested slope by a steam-driven cable. Just north of the Forest Discovery Center is the Pink Beds picnic area, where a trail leads into an area vivid in the spring with blossoms of rhododendron, azaleas, and mountain laurel.

Figure 12. Dormitory, Cradle of Forestry in America, Pisgah National Forest (North Carolina).

The Cradle of Forestry in America is just one point of interest along the 79-mile loop Forest Heritage National Scenic Byway, which utilizes U.S. Highways 276 and 64 and State Route 215. By continuing south along U.S. Highway 276 from the Cradle of Forestry in America, you will come to Sliding Rock, where a fast-flowing mountain stream washes over a gentle slope of glasslike rock for about 50 feet into a pool of water 7 feet deep. Between Memorial Day and Labor Day, scores of people in swimsuits slide down this attractive falls. A lifeguard is on duty between 10 A.M. and 5:30 P.M. when the area is open to the public for a small fee. Another 2 miles south along the byway, Looking Glass Creek plummets 60 feet over a rocky cliff into a splash pool. The falls, visible from the highway, is 30 feet wide in a forested setting of hemlocks, maples, rhododendrons, and mountain laurels. Rock steps from an observation point along the highway lead to the base of the falls, but the steps are uneven and often wet from the spray from the falls.

West of U.S. Highway 276 across from Looking Glass Falls is Looking Glass Rock Scenic Area, where several hundred acres are covered by beautiful forests that are full of stately trees and colorful wildflowers. The center-piece of the area is Looking Glass Rock, a granite monolith always wet from seeps, that glistens mirrorlike in the sunlight. Forest Road 475 follows the southern edge of the Looking Glass Rock Scenic Area, and Forest Road 475B goes north along the western edge. A hiking trail off Forest Road 475 will bring you to the famous rock.

Forest Road 475 also leads to the Pisgah Center for Wildlife Education and the Pisgah Forest Fish Hatchery, operated by the North Carolina Wildlife Resources Commission. The center presents a movie about fisheries, and visitors may observe different trout species in the hatchery. The Davidson River is nearby where you may try your luck at trout fishing. A trail from the center to the south goes to John Rock Scenic Area where another rocky landmark lies within gorgeous scenery.

Continuing south along the scenic byway toward Brevard, you will come to Coontree and Sycamore Flats picnic areas and Davidson River Campground, all situated along scintillating Davidson River. Tubing is popular on the river; you may rent tubes where U.S. Highway 276 enters Brevard. The Pisgah Ranger Station visitor center is across the highway from the Davidson River Campground, where there is also a 0.7-mile self-guided Andy Cove Nature Trail. The visitor center is filled with interactive exhibits to help you plan your recreation experience in the Pisgah Ranger District.

From Brevard to Rosman, U.S. Highway 64 is part of the Forest Heritage Scenic Byway but outside the boundaries of the national forest. At Rosman, State Route 215 continues the Scenic Byway north and immediately enters the Pisgah National Forest, following the North Fork of the French Broad River for many miles. After State Route 215 crosses the Blue Ridge Parkway, it goes through a narrow corridor separating Shining Rock Wilderness to the east from Middle Prong Wilderness to the west. This part of the Scenic Byway follows the West Fork of the Pigeon River, emerging from the national forest at the Sunburst Campground.

Shining Rock Wilderness takes its name from the huge, spectacular, glistening white crystalline feldspar rocks that outcrop on the top of Shining Rock Mountain. The Art Loeb Trail from Iverson Gap is the best way to get to the Shining Rocks. Be aware that in this wilderness there are no trail signs or blazes, and campfires are not allowed. Come prepared with a good map and a compass, and know how to use them. Iverson Gap is 2.5 miles north of Exit 420 on the Blue Ridge Parkway. The trail follows Shining Rock Ledge for 1.5 miles to Shining Rock Gap. At this point, you have three choices. You may follow the trail east along the North Prong of Shining Creek for 3 miles to U.S. Highway 276. You may also hike west, eventually coming alongside the Little Fork of the Pigeon River, exiting the wilderness in 3 miles at the Daniel

Boone Boy Scout Camp. Or you may continue north on the Art Loeb Spur Trail for 3.5 miles, climbing over Stair Mountain, through The Narrows and Deep Gap, and up Cold Mountain with its cone-shaped peak, the highest point in the wilderness at 6,030 feet. In any case, you will be treated to magnificent scenery and marvelous Appalachian forests. At the higher elevations are forests of Carolina hemlock and Fraser fir, while lower elevations support rich hardwood forests of red oak, scarlet oak, Ohio buckeye, tulip poplar, and various species of birches and maples. Rhododendrons, azaleas, mountain laurels, and serviceberries add color in the spring and early summer. Bountiful wildflowers include trilliums, Solomon's-seals, phloxes, wild gingers, and many others in the spring, and fire pinks and turk's-cap lilies in the early summer.

While hiking in the mountains, you will notice occasional low areas where moisture accumulates, sometimes giving rise to boggy habitats where sphagnum moss forms a dense cover. In some of these bogs are two attractive but rare plants that bloom during July. Swamp pink, in the lily family, has thick spikes of handsome pink flowers on a stalk that arises from the center of a cluster of large, basal leaves. Robin runaway (*Dalibarda*), a member of the rose family, is a delicate, evergreen herb with violet-shaped leaves and small, white, five-petaled flowers borne singly at the end of a long stalk. In small depressions on some exposed granite in the Pink Beds area, a grassy-looking tuft with drooping leaves may be the rare Biltmore sedge.

Middle Prong Wilderness is a smaller, similar area west of Shining Rock Wilderness and separated from it only by State Route 215. The southwestern edge of the wilderness is formed by the Blue Ridge Parkway, which dips through several gaps and passes Richland Balsam Mountain, the highest elevation in the Pisgah District at 6,410 feet.

Outside the southern boundary of Shining Rock Wilderness is the popular hiking area known as Graveyard Fields, a meadowlike region dotted with occasional shrubs and small trees. The appearance of Graveyard Fields is due to extensive fires that burned the area several times in the past few decades.

The Blue Ridge Parkway, from its junction with U.S. Highway 276 to its junction with State Route 191 just before reaching Asheville, allows visitors to explore the remainder of the Pisgah District. This portion of the national forest contains Mount Pisgah; the scenic Stoney Fort Picnic Area along State Route 151; Lake Powhatan with swimming, fishing, hiking, and camping available; and the Bent Creek Experimental Forest, the oldest experimental forest in the South, dating back to 1927. Research at the station centers around fire, insects, diseases, timber, wildlife, and water in the southern Appalachians.

Two districts of the Pisgah National Forest lie north of Interstate Highway 40 and extend to the Tennessee border. The Appalachian District is on the North Carolina side of the state line, directly across from Tennessee's Cherokee National Forest. This district is crossed by the Pigeon and Nolichucky rivers. Interstate Highway 40 parallels the Pigeon River through the western edge of the district and provides access to the Sutton Top Lookout via a circuitous mountain road off the interstate.

The French Broad River passes through the community of Hot Springs. If you take State Route 1304 (River Road) from town, staying along the north side of the river, you will come to a rocky escarpment called Paint Rock. By exploring on these historic cliffs, you may see an interesting array of vegetation including an uncommon shrub known as piratebush, or buckleya. South of Hot Springs via State Route 209 is Rocky Bluff Recreation Area where there is a campground, hiking trails, and the old Brooks Cemetery. Fishing for trout and bass in Spring Creek is popular.

Across the French Broad River east of Hot Springs is the Silvermine Group Camp and a trailhead. From here you can hike to Lovers Leap overlooking the river or access the Appalachian Trail.

Floating is popular on the French Broad River and the Nolichucky River. On the French Broad River, most floaters enter the river at Barnard and can float for 5 miles to Stackhouse or 8 miles to Hot Springs. The put-in for the Nolichucky is at Poplar Boat Launch east of Erwin, Tennessee.

Several hiking trails are accessible near Hot Springs. One trail popular with mountain bikers follows an old railroad grade for 36 miles along the Laurel River from the junction of U.S. Highway 25/70 and State Route 208 to the ghost town of Runion. On Rich Mountain is the 7.5-mile Golden Ridge Bike Trail. The 4-mile Mill Ridge Bike Trail is in the vicinity as well. About 17 miles south of Hot Springs, Max Patch is a high grassy bald offering 360-degree views from its 4,629-foot summit. The Appalachian Trail crosses the bald.

The east sides of the Appalachian and Grandfather ranger districts lie between Asheville and Blowing Rock and are separated by the Blue Ridge Parkway. The Appalachian District, which lies north of the Parkway, contains such spectacular areas as Roan Mountain, Elk Falls, the Black and Craggy mountains, and several beautiful regions containing Carolina hemlock forests.

Elk Falls is at the end of State Route 1305 just north of the village of Elk Park. A hike of about 500 feet will take you to the top of the impressive falls where Elk River plunges 30 feet. To get to the catchpool at the base of the falls, continue 300 feet down rocky steps.

Roan Mountain may be the highlight of the Pisgah National Forest, at least during mid-June, when the incomparable Catawba rhododendrons (pl. 24) bloom. The greatest concentration of these is in the Rhododendron Garden, a nearly treeless, parklike setting on top of Roan Mountain. Treeless mountaintops such as these are known as *balds*, and the absence of trees is due to past climatological and cultural events, not because the area is above treeline. Several rare plants grow on or near the bald, including the Roan lily, which is a 6-foot-tall type of turk's-cap lily but with gorgeous red flowers; bent avens, a white-flowered herb in the rose family; and a sedge known as Fernald's hay sedge. If you follow the trail to the rocky cliffs of Roan High Bluff, you may also encounter the rare foot-tall Blue Ridge goldenrod and the Roan Mountain roseroot, a succulent member of the sedum family. The town of Bakersville, North Carolina, holds a Rhododendron Festival the third weekend of June each year. You may check with the Burnsville office for a blooming update of the magnificent display of Catawba rhododendrons.

The same Catawba rhododendron also grows in the Craggy Mountains, but because of the lower elevation, it blooms a week or two earlier than its counterpart on Roan Mountain. While the rhododendrons on Craggy Mountain are best viewed at Craggy Gardens on the Blue Ridge Parkway (a National Park Service area), they also grow in the adjacent Craggy Mountain Scenic Area in the Pisgah National Forest. On the west side of the Craggies, you can hike into the Pisgah National Forest to two significant waterfalls. Walker Falls drop twice for a total distance of 50 feet; Douglas Falls plummet 70 feet into a 250-foot-wide catchpool. Behind the catchpool and under the falls is a picturesque alcove. Carolina hemlocks provide added beauty to the area around the falls.

Mount Mitchell, named for pioneer botanist Elisha Mitchell, is the highest mountain in the eastern United States at 6,684 feet. Mount Mitchell is the centerpiece of Mount Mitchell State Park accessed by State Route 128. The eastern flank of the mountain is in the Pisgah National Forest. There are hiking trails to the top and several campgrounds along Forest Roads 472 and 2074, the latter reached off of the Blue Ridge Parkway or from U.S. Highway 19E and State Route 80 out of Burnsville.

Carolina hemlock forests are throughout this part of the Pisgah National Forest, including at the Carolina Hemlock Campground northeast of Mount Mitchell. The campground, set among a virgin stand of the hemlocks, is adjacent to the South Toe River, where there is a swimming beach. You may fish for trout in the river. It is 5.5 miles by trail from Black Mountain Campground to the summit of Mount Mitchell.

On the upper slopes of Black Mountain, near Mount Mitchell, is the steep, 17.5-mile Buncombe Range Horse Trail. The Black Mountain Crest Trail is a rugged hiking trail that connects some of the highest peaks in the eastern United States.

Grandfather Ranger District lies south of the Blue Ridge Parkway and includes the Linville Gorge Wilderness, Brown Mountain, and several waterfalls in the vicinity of Wilson Creek. The Linville River has carved a 12-mile-long, deep gorge from the community of Linville Falls to within a couple of miles of Lake James. The area on either side of the 2,000-foot-deep gorge is a wilderness, but 39 miles of trails permit visitors to explore much of the area. Linville Mountain forms the western side of the wilderness, and the Kistler Memorial Highway stays on the crest just outside the boundary of the wilderness. At Wiseman's View, you may look straight down into the gorge or across to Jones Ridge, which forms the wilderness's eastern border. The Jones Ridge Road goes along part of the ridge to Table Rock Picnic Area. From this road you will see rock formations such as Setting Bear Mountain and Hawksbill. From Table Rock you may hike to The Chimneys. Rhododendrons, azaleas, and mountain laurels are plentiful, while on the rocky ledges above the gorge, Heller's gayfeather and the federally endangered mountain golden heather are rare.

North of Linville Falls, State Route 181 exits off the Blue Ridge Parkway. If you take the highway south, you will come to the Brown Mountain Overlook. Brown Mountain, 5 miles to the east, has mysterious flickering lights observed by the early Indians and the settlers since. The overlook is a good place from which to look for these lights. Farther south along State Route 181 is a trail to 30-foot Upper Creek Falls. There is a designated off-road vehicle area on the south slope of Brown Mountain with 33.5 miles of trails mostly for trail bikes and all-terrain vehicles.

State Route 1511, which leads from the Blue Ridge Parkway, will take you to the village of Edgemont in the midst of the Wilson Creek drainage, a National Wild and Scenic River. From Edgemont you will have access to several waterfalls. South Harper Creek Falls slides for 100 feet over bedrock, while North Hope Creek Falls slides for 25 feet over bedrock. Mortimer Campground offers camping, picnicking, and fishing near the site of a logging town (1900–1917) and a Civilian Conservation Corps camp (1934–1942).

Paint Rock

From its origin in southern North Carolina, the French Broad River flows northwest, joining the Holston River at the eastern edge of Knoxville, Tennessee, to form the Tennessee River. On its tortuous journey, it winds through territory of both North Carolina's Pisgah National Forest and

Tennessee's Cherokee National Forest. Just where the river crosses state lines, in the Unaka Mountains, a bright sandstone bluff rises 70 feet above the north bank. This conspicuous landmark is Paint Rock; the boundary between the two states runs from the shore and up along its sharp east-west ridge.

While the bluff south of the ridge is bare, the north-facing slope, on the Tennessee side of the border, supports a dense, shady forest of hemlock, red oak, tulip tree, striped maple, and many other species. In the understory are myriad wildflowers, many of which display vivid colors during the spring.

The crest itself, which is flat enough for walking, is strewn with hundreds of loose boulders. This exposed area supports a community of plants, including Virginia pine and rock chestnut oak, that grow well under very dry conditions. Because of the limited moisture, these trees grow no more than 20 feet tall and are spaced fairly far apart. As a result, sunlight strikes the ground virtually unimpeded, limiting the understory to a sparse community of grasses, scattered wildflowers, and crinkly reindeer lichens. At the western end, however, where the ridge slopes down to the river, there is more shade and soil. Sugar maples and Carolina buckthorns grow here, with bladdernuts filling in the shrub layer. Blue-flowered phacelias and yellow trilliums bloom here in the spring. Notable for its seemingly wilted leaves is the clammy-leaved cup plant.

In 1816, pioneer botanist Thomas Nuttall traversed the north side of the French Broad River, collecting plant specimens as he proceeded past Paint Rock. One of the plants he collected was a spindly shrub with pale green leaves. He recognized his find as a species of parasitic plant new to science, but he felt it belonged to an existing genus. Botanist Samuel B. Buckley, exploring the crest of Paint Rock in 1843, rediscovered Nuttall's shrub and, intrigued by it, send a specimen to John Torrey, a prominent New York botanist. Torrey named the plant *Buckleya distichophylla*, creating a new genus within the family of parasitic flowering plants known as the Santalaceae, or bastard toadflax family.

Botanists have since discovered that Paint Rock and its immediate area are home to a number of plant species rare for this part of Tennessee and North Carolina, and among them, *B. distichophylla*—or piratebush, as it has come to be known—is one of the rarest plants in the world. Populations are known only in North Carolina, Tennessee, and Virginia, and specifics of the plant's method of growth are still incompletely understood.

In a recent study, botanist Thomas Mowbray found that piratebush tolerates great variations in moisture, being found anywhere from the banks of streams to dry ridgetops, but is intolerant of shade. All the plants identified in the wild are established either under a very open canopy or in openings created by rocks and cliffs. The largest and healthiest specimens are located in areas receiving the most direct sunlight.

The obvious feature of the piratebush is the pale green of its leaves, a sign that they contain less chlorophyll than those of most flowering plants. Photosynthesis supplies only part of the plant's nutritional needs; the remainder comes from parasitizing other species. Beneath the ground, the piratebush puts forth structures, known as *haustoria*, that attach themselves to the roots of other kinds of plants and draw off nutrients from them.

For 150 years after Nuttall's discovery, botanists thought that the survival of the piratebush depended on its making connections with hemlock trees, since it usually is found in their vicinity. But in 1965, botanist Martin Piehl observed that in nature, the piratebush may attach itself to a number of different woody and herbaceous plants. In addition, botanist Lytton Musselman and forester William Mann Jr. experimented by planting piratebush fruits in pots by themselves and with each of 18 different species of trees. After 5 months, they found that haustoria of the piratebushes had made successful connections with every tree species tested except the tulip tree. They also discovered that piratebush seedlings had developed and survived without making any connections with other plants, so that at least in its early stages of growth, the piratebush can be self-sufficient.

Why the piratebush is so rare, even though there are many open, rocky habitats throughout the Appalachians, continues to puzzle scientists. William Carvel and Hardy Eshbaugh, of Miami University, have noted that piratebush grows most successfully in association with hemlocks and that it also requires direct sunlight, a combination of conditions not commonly available. They also speculate that the plant's pattern of seed dispersal and germination (concerning which little is known) may not enable piratebush to spread to isolated spots where it might otherwise survive.

Roan Mountain

An early name for this area was Cloudland. From a broad base 2,500 feet above sea level, Roan Mountain rises to 6,285 feet, forming a rounded divide that extends nearly 5 miles along the North Carolina–Tennessee line. On either side of the divide, rounded spurs alternate with V-shaped valleys. Oak and chestnut forests once covered the lower slopes, but these were cleared many years ago for cultivation and timber (a blight that struck the chestnuts in the late 1800s helped to speed the process along). Above 3,500 feet, however, large tracts of natural habitat persist, including Catawba rhododendron gardens that offer a spectacular show in June. The rhododendron communities and other treeless areas (balds) seem inexplicable breaks in an otherwise forested area.

In a band about 3,500 to 5,000 feet above sea level, where the deep, rich soil is relatively free of surface rocks and outcrops, American beeches and sugar maples may stand 70 feet tall. Yellow birch, apparently because it is better adapted to cooler temperatures, becomes common near the upper limits of this forest community, which also has a scattering of shorter trees—sweet birch, Ohio buckeye, mountain and striped maples, and smooth serviceberry. There is a shrub layer composed of hobblebush viburnum, alternate-leaf dogwood, and elderberries; beneath it, a number of wildflowers span the growing season. Jack in the pulpit blooms in early spring, followed by the great star chickweed, with its large, white flowers. Orange jewelweed blooms in summer, and the branched white aster in the autumn.

Upslope, the beeches and sugar maples become smaller and are more widely spaced as they reach their environmental tolerance limits. Above 5,400 feet and all the way to the summit—where the climate is cooler, snow more plentiful in winter, and the winds stronger—they are replaced by a coniferous forest of red spruce and Fraser fir. Scattered among the cone-bearing evergreens, which raise their needle-clothed branches to nearly 90 feet, are such deciduous trees as yellow birch, striped maple, mountain ash, and sweet buckeye. Beneath the canopy trees there is a modest shrub layer, which includes an occasional rhododendron. Aided by thick summer fogs, which hover over the mountain, mosses and lichens cover more than half the ground surface, often edging their way up the tree trunks for 4 or 5 feet, and wispy strands of *Usnea* lichens hang from the tree limbs. A few flowering herbs, such as mountain sorrel, foamflower, and purple turtlehead, rupture the moss cover.

The spruce–fir forest occupies Roan Mountain's most rugged terrain, where numerous exposed rock outcrops occur. This stony topography is particularly evident at Roan High Bluff, a scenic overlook reached by trail from the rhododendron gardens at Roan High Knob. Trees toppled by wind or scarred severely by lightning testify to the harsh climate.

But the plant communities known as balds evoke the most interest on Roan Mountain. Some area residents still refer to them as *slicks* because of their shiny appearance from a distance. A. F. Marks, who has studied the southern Appalachian balds thoroughly, defines a *bald* as "an area of naturally occurring treeless vegetation located on a well-drained site *below* the climatic treeline in a predominantly forested region."

Three kinds of balds occupy the upper slopes of Roan Mountain, all in the zone normally inhabited by the spruce–fir forest. Grass balds are the most common, generally occurring between 5,500 and 5,700 feet. Because they are devoid of most shrubs, they are sometimes called *mountain meadows*. Mountain oat grass forms a cover through which sheep sorrel, ticklegrass, and wild strawberry occasionally protrude. Some small mounds, called *moss hummocks*, support the growth only of mosses. At roughly the same altitude are also found the rarest of the Roan Mountain balds, the alder balds, which seem to have developed where there is more available soil moisture. These thick, shrubby stands of 3- and 4-foot-tall green alders are known from no other area in the Carolinas.

The showplaces of Roan Mountain, however, are the rhododendron balds, located mostly on the southeastern side of the divide at between 6,000 and 6,150 feet. At their peak in mid-June, these balds contain hundreds of brilliant rose pink Catawba rhododendrons, interrupted occasionally by patches of grassland. Here and there, a stray red spruce or mountain ash has established a foothold and overtops the rhododendrons, but few plants grow beneath the shrubs because their densely crowded, thick leaves let little light reach the ground. The rhododendrons also produce a network of feeder roots in the upper 2 inches of the soil, excluding most herbaceous plants. The Pisgah National Forest has provided parking areas near Roan High Knob, where trails permit visitors to wander through 600 acres of rhododendrons.

Biologists have proposed many theories to explain the origin of the balds. The first suggestion, made near the end of the 1890s, was that balds were areas where the trees had been damaged and ultimately killed by winter ice storms. Although ice does damage some of the trees on Roan Mountain, large acreages of trees would probably not be wiped out while the surrounding forest remained intact. Years ago a prominent Nebraska botanist suggested instead that fire was the primary cause of the development of the balds. In the southern Appalachians, however, fires are normally followed by the invasion of fire cherries, which do not seem to have been an important part of the Roan Mountain flora. The treeless glades and hill prairies of the Midwest are accounted for, at least in part, by their western exposures, which leave them with soil so dry that trees cannot grow. Although many of the balds on Roan Mountain are on southwestern slopes that receive drying winds, their soil is not significantly drier than in the adjacent forests, and trees can grow there if given a chance. Still another suggestion is that Indians cleared the region of trees and maintained the treeless areas for campgrounds and grazing by animals, but we have little evidence to support this theory.

The most plausible explanation for the origin of balds involves past climatic conditions and the reaction of vegetation to them. Typically, the upper elevations of the southern Appalachians support a coniferous spruce–fir forest. Below 5,000 feet, deciduous beech–maple forest prevails. About 4,000 years ago, a lengthy hot and dry period affected the region. The spruces and firs growing near their lower limits of tolerance gradually succumbed to the hot, arid climate and were replaced by beeches and maples. Subsequently, a cooling trend, which has continued to the present, made life difficult for the beeches and maples that had occupied this upper range. Many trees perished during the cold winters, and others fell because of severe winds, ice damage, and possibly even lightning fires.

The spruce–fir forest generally failed to reestablish itself in this zone because any seedlings that germinated soon perished in the absence of a protective tree canopy. The treeless zone was invaded instead by grasses, and the grass balds came into existence. The uppermost balds nearest the spruce–fir forest were eventually populated by rhododendrons, whose seeds came from the rhododendrons in the forest understory. Fir and particularly spruce seeds also fell into these adjacent balds, but the conditions necessary for their germination and survival apparently did not prevail. Those balds that have remained grass balds probably harbor very few rhododendrons because there is not a large rhododendron seed source nearby.

Today, the balds of Roan Mountain are unstable communities. Some are expanding; others are shrinking. Where there is expansion, it is because mature trees that grow at the edge of the balds are exposed to more severe environmental conditions than those within the forest. In historical times, also, local settlers who used the balds as grazing areas for their domestic animals cut the bordering trees to provide more space. On the other hand, the edges of the balds are constantly threatened by the invasion of woody plants. The spruces and firs are potent seed sources. A seed may be blown beneath a clump of rhododendrons, whose shade permits the conifer seedling to develop. On occasion, this seedling will grow and break through the shrub canopy, ultimately shading out the rhododendrons and killing them. A continuing ecological war thus ensues between the balds and the adjacent forests.

Uwharrie National Forest

SIZE AND LOCATION: 50,189 acres in the Uwharrie Mountains of central North Carolina. Major access routes are State Routes 24, 27, 109, and 1150. District Ranger Station: Troy. Forest Supervisor's Office: 160A Zillicoa Street, Asheville, NC 28802, www.cs.unca.edu/nfsnc/recreation/index.

SPECIAL FACILITIES: Boat ramps; swimming beaches.

SPECIAL ATTRACTIONS: Uwharrie National Recreation Trail; Badin Lake.

WILDERNESS AREA: Birkhead Mountains (5,160 acres).

The Uwharrie Mountains are a low range of mountains in the Piedmont region of central North Carolina, trending southeast to northeast between Albemarle and Asheboro. The Uwharries are ancient volcanoes, perhaps the oldest on the North American continent and maybe as high as 20,000 feet at one time. They have eroded until they are less than 1,000 feet above sea level today.

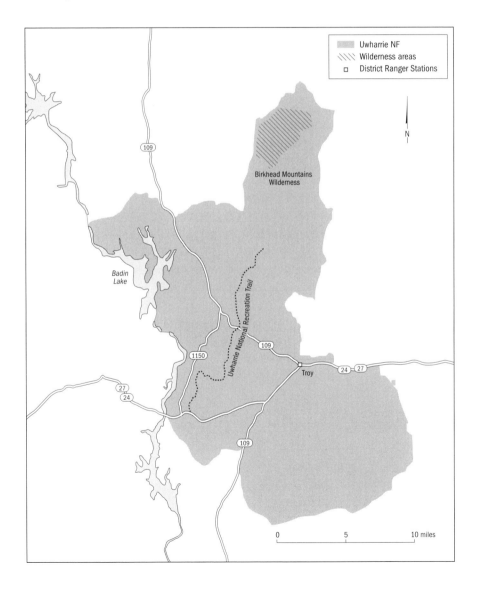

Most of the activities in the Uwharrie National Forest are concentrated on the east side of Badin Lake. This large reservoir was formed when the Yadkin River was dammed just above its confluence with the Uwharrie River. There is a boat ramp at the Cove near the Arrowhead Campground and at the Deepwater Trail Camp. From the Badin Lake Campground is a 6.5-mile loop trail around a peninsula that protrudes into the lake. There is also an off-road vehicle (ORV) area nearby. Horseback riding is popular in the Uwharrie National Forest, and Canebrake Horse Camp caters to riders and their horses.

Forest Road 553 goes through a part of the Uwharrie National Forest between Falls Mountain and Gladtop Mountain and is a good place to join the Rocky Mountain Loop Trail and the Dickey Belt Trail. Both trails are open to ORVs.

Northwest of Troy along State Route 109 is the Back Mountain Lookout Tower. About 1 mile south of the lookout, the Uwharrie National Recreation Trail crosses State Route 109. The northern terminus of this 20-mile trail is midway between the Ophir and Immer churches, while the southern end is on State Routes 24 and 27 about 1 mile south of the Wood Run Horse Camp.

At the northern end of the Uwharrie Mountain Range are the Birkhead Mountains, around which a 5,000-acre wilderness has been designated. The Birkhead Trail crosses the area north to south between the North Prong and Robbins Branch streams. Bare rock on dry, west-facing cliffs support a scrub vegetation of post oak, blackjack oak, and red cedar, with prickly pear cactus in the understory. Rich, moist forests contain American beech, tulip poplar, and sugar maple. Most of the dry woods consist of black oak, red oak, shagbark hickory, and mockernut hickory. Tall sweet gums and sycamores grow along many of the streams.

Near Badin Lake and the Arrowhead Campground is an ORV area with 17 miles of trails, ranging in length from 0.8 mile to 3.5 miles. Horseback riders enjoy 40 miles of trails. Arrowhead Campground offers electric hook-ups at 33 of the 50 sites—an unusual feature for national forest campgrounds.

Twenty-two miles of mountain bike trails are part of the Wood Mountain Trail System near Troy.

NATIONAL FOREST IN OHIO

Ohio's only national forest is the Wayne, located in the south-central and eastern parts of the state. For a short time it had been combined administratively with the Hoosier National Forest in Indiana. The Wayne National Forest is in Region 9 of the United States Forest Service.

Wayne National Forest

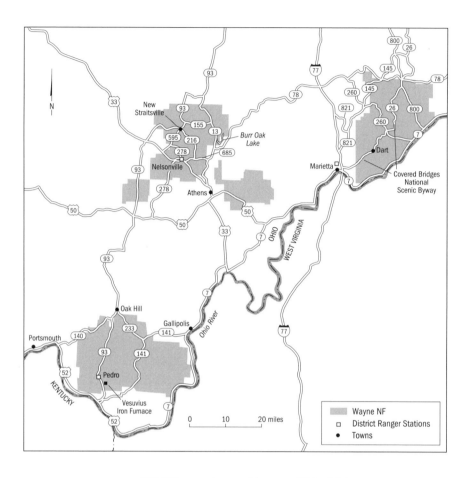

SIZE AND LOCATION: 176,000 acres in southeastern Ohio. Major access routes are Interstate Highway 77; U.S. Highways 50 and 52; and State Routes 7, 13, 26, 33, 78, 93, 140, 141, 145, 155, 216, 233, 260, 273, 278, 595, 685, and 800. District Ranger Stations: Marietta, Nelsonville, and Pedro. Forest Supervisor's Office: 13700 U.S. Highway 33, Nelsonville, OH 45764, www.fs.fed .us/r9/wayne.

SPECIAL FACILITIES: Boat ramps; swimming beaches.

SPECIAL ATTRACTIONS: Vesuvius Iron Furnace; Covered Bridges National Scenic Byway; Buffalo Beats.

In the Hill Country of southeastern Ohio are three small and separate units of the Wayne National Forest. By driving the back roads in this rural part of

Ohio, travelers will be treated to rolling hills, pastoral scenes, and fine colorful foliage during the autumn.

Southeastern Ohio forests are prime examples of mixed mesophytic forests. Plant ecologists generally recognize seven types of deciduous forest vegetation in the eastern United States. The mixed mesophytic association forms a narrow band that extends from western Pennsylvania, through most of West Virginia, southeastern Ohio, the eastern half of Kentucky, eastern Tennessee, and the extreme north-central part of Alabama, with extensions into cove forests of the southern Appalachian mountains. The mixed mesophytic association is usually considered to be the most diverse of any of the deciduous forest types. Dominant trees are tulip poplar, American beech, basswood, white basswood, sweet buckeye, Ohio buckeye, sugar maple, northern red oak, white oak, and eastern hemlock.

The easternmost section of the Wayne National Forest, the Marietta District, is located between Woodsfield to the north and Marietta to the south, and parallels the Ohio River for several miles. By driving State Route 26, the Covered Bridges National Scenic Byway, between Marietta and Woodsfield, you will be able to see and walk across several of these marvelous bridges from yesteryear.

Upon entering the Wayne National Forest from Marietta, near the community of Hills, is the southernmost of the covered bridges (pl. 25) along the scenic byway. When you walk across the bridge, you can join the North Country National Recreation Trail that heads off to the north, staying near the Little Muskingum River. Nearby is Lane Farm Campground with canoe access to the river. Beyond the community of Moss Run, the scenic byway climbs to a ridge featuring a commanding vista of the countryside.

A few miles east of Dart is another trailhead where you may join the North Country National Recreation Trail. If you hike south for a little way, you will come to the fascinating Irish Run Natural Bridge, carved naturally in the sandstone. The bridge is 51 feet long, 16 feet thick, and 39 feet high. You may walk across the bridge or under it. North of Steel Run is another covered bridge at the Hune Bridge Recreation Area. Here is another place to put your canoe in the Little Muskingum River. You are permitted to drive across this covered bridge. A short distance north is Haught Run and the Rinard Covered Bridge. Near the community of Cline is the Knowlton Covered Bridge. From this covered bridge you may either hike or drive to Ring Mill picnic area and tiny campground. At Ring Mill is the site of the Walter Ring house, a sandstone dwelling built in 1846 and occupied by four generations of millers. The first mill at the site was constructed in 1817, and eventually the property contained a sawmill and grist mill. Three miles east of Ring Mill is Rockcamp Run where there are scenic stands of white pine and eastern hemlock.

North of Cline, a side road to the west off State Route 26 will bring you to the Lamping Homestead Recreation Area, which includes picnic facilities and a 5-mile loop hiking trail. The recreation area also has a fishing pond where anglers try their luck at landing bass, bluegill, and catfish.

Southeast of Dart along a creek known as Reas Run is a unique natural area that has been designated as the Reas Run Natural Area. The topography of the natural area consists of steep ridges and hills and V-shaped valleys. The significance of the natural area are the 35 acres of Virginia pine trees with a mixture of shortleaf pine on one of the ridgetops. Some of the Virginia pines are up to 70 feet tall. Intermingled among the pines are sassafras, mapleleaf viburnum, and serviceberry, with an entanglement of grape and greenbrier vines. One wildflower of significance is pipsissewa. Other forest types in the natural area include oak–hickory forests on the upper slopes and sugar maple–American beech forests in the valleys.

By driving State Route 7 that parallels the Ohio River from Marietta northward, you will be able to access that part of the Wayne National Forest that lies near the Ohio River. Near the community of Wade is a pull-out for a scenic view of the river, and at Leith Run is a campground with another superb view of the Ohio River. At Leith Run, boaters have a launch area to enter the Ohio River. One mile north of Leith Run is the Capitol Christmas Tree Site and picnic area. In 1987, the Christmas tree for the Capitol grounds in Washington, D.C., was selected from this site. The tree was a huge Norway spruce, planted along the banks of the Ohio River around 1900 by an early homesteader. Only the stump of this tree remains now, and the homesteader's farmhouse has been reduced to only a few stone foundation blocks.

If you were to drive State Route 800, you would cross Dismal Creek about 2 miles north of the community of Antioch. The gorge that has been carved by Dismal Creek contains beautiful specimens of white pine and eastern hemlock.

Where Ohio drops down to its southernmost part between the towns of Portsmouth and Gallipolis is the Ironton District of the Wayne National Forest that extends all the way to the Ohio River. State Route 93 bisects this region north to south, with U.S. Highway 52 and State Route 233 along the western and eastern boundaries, respectively.

Most of the activities in this part of the Wayne National Forest center around the Lake Vesuvius Recreation Area. Long and narrow Lake Vesuvius bisects the recreation area with swimming beaches at Bald Knob and Big Bend. There are campgrounds and picnic areas in the recreation area, as well as a number of hiking trails. Near the western end of Lake Vesuvius is the historic Vesuvius Iron Furnace, a blast furnace built in 1833. For 3 years, this was a cold-blast furnace using charcoal and air. By 1836, the more efficient hot-blast method was employed at the Vesuvius Iron Furnace. The furnace ceased operation in 1906 and was restored by the Civilian Conservation Corps during the 1930s. It is said that during the Civil War, the Vesuvius Iron Furnace was one of only three in the country capable of providing high-quality iron for use by cannons.

Several hiking trails are in the Lake Vesuvius Recreation Area. The 16-mile loop Vesuvius Trail is the longest, beginning at the iron furnace and taking a wide circular course around the lake. The trail crosses Storms Creek and Aldridge Creek, climbs the ridge overlooking Kanady Hollow, and eventually returns to the iron furnace. Those opting for a shorter hike have three other trails to choose from. The 0.75-mile Beach Trail stays near the shore of Lake Vesuvius, passing through a mixed forest of hardwoods and pines. Whiskey Run Trail is 0.5 mile long, forming a loop from the northern end of the Ridge Campground. This trail passes long-abandoned charcoal pits and an old whiskey still. Rock House Trail originates at the Rock House Picnic Area and meanders down to a natural rock shelter known as the Rock House. The entire loop is about 0.75-mile long.

Cambria Creek is located a couple of miles south of the northern edge of the Ironton Ranger District, about 3 miles south of Oak Hill. An old furnace stack may be seen in one area along the creek, and a nice wetland complex lines the creek in places.

South of Cambria Creek on State Route 93 is a junction with Forest Road 193. By taking this forest road to the east, you will come to Forest Road 46 in about 3 miles. This road crosses and then follows the north side of Caulley Creek. The woods you are driving through are excellent examples of the mixed mesophytic forest association. Two miles west of Bockhorn and State Route 93 is Youngs Branch where another fine mixed mesophytic forest may be enjoyed.

Southwest of the Vesuvius Iron Furnace is Little Storms Creek, which flows from Ellisonville south past LaGrange, eventually flowing out of the Wayne National Forest. The more moist bottomlands and mesic slopes along the creek have good examples of the mixed mesophytic forest association, while the uplands are fine examples of the oak–hickory association.

The third district of the Wayne National Forest located a few miles north and west of Athens is the Athens Ranger District. U.S. Highway 33 crosses the southern end of this region, while State Routes 13, 78, and 93 provide easy access to the rest of the area.

At the eastern end of the Athens Ranger District, on the west side of Burr Oak Lake, is a Forest Service primitive campground. Three miles north is the trailhead for the 15-mile Wildcat Hollow Trail. Where the trail follows Eels Run is one of the best mixed mesophytic forests in the area.

East of State Route 78 and north of State Route 685 is a tiny remnant of native prairie, one of the farthest eastern extensions of prairie in the country. Known as Buffalo Beats, this unique area is home to plants that are the same as those in the great prairies of the Great Plains.

A small isolated part of the Athens Ranger District centers around 1,200-acre Timbre Ridge Lake, where there is a boat ramp and campground. Fishing is good here for bluegill, largemouth bass, channel catfish, and rainbow trout. An off-road vehicle trail is available at Hanging Rock at the southwestern corner of the district.

South of the community of New Straitsville and west of State Route 216 is the historic Shawnee Lookout Tower. The tower is the only lookout left of four that were constructed in the district by the Civilian Conservation Corps in 1939. In use until the 1970s, the steel tower is 100 feet tall, with a 7-foot square cab and wooden steps. At one time near the base of the tower were a log guard station and log garage. After falling into disrepair, the tower was restored in 1989 and is open to visitors today for a breathtaking view from the top.

West of the Shawnee Lookout Tower and a short distance south of New Straitsville is the historic Payne Cemetery, which was refurbished in 1995. The cemetery is a military one where Civil War soldiers of the U.S. Colored Troops are buried. Nearby is the Sand Run Pond, where there is a secluded picnic area.

The 70-mile Monkey Creek Horse Trail begins a short distance north of Nelsonville. By following State Route 33 north from Athens and then taking a forest road to the west, you will come to pleasant Utah Ridge Pond, which has a 0.25-mile nature trail.

The Hocking River enters the Wayne National Forest at Haydenville and is paralleled by U.S. Highway 33 until it leaves the national forest near the headquarters of the Wayne National Forest. The river with its rocky outcrops is exceptionally scenic.

Glen Ebon, immediately west of the forest supervisor's office, has been an extensive coal-mining center in southern Ohio. Several different types of coal-mining activities are explained in this area.

Buffalo Beats

In 1935, botanist Edward N. Transeau described patches of prairie found east of the Great Plains, in Michigan, Illinois, Indiana, and Ohio. He believed these isolated patches were remnants of what had once been an arm of prairieland, which he called the *prairie peninsula*. One of the easternmost of these patches, Buffalo Beats, lies near Athens in southeastern Ohio, on land now part of Wayne National Forest. It is named for the American bison, which may well have grazed there when it roamed this part of Ohio.

I visited Buffalo Beats one warm May afternoon in the company of forest ranger Joe Newcomb and botanist Marilyn Ortt of the Ohio Heritage Program, an organization that assists the Forest Service in managing the prairie. After a short but steep climb through a dense forest of black, scarlet, and white oaks, we reached the top of a ridge where the forest ended abruptly and about 0.5 acre of prairie vegetation was exposed to sunlight. The clearing was pink with hairy phlox (fig. 13) in full bloom, punctuated by an occasional splash of orange from false dandelions, yellow from cinquefoils, and gold from golden Alexanders. Leaves of many other species were evident, some just beginning to unfold: blazing star, yellow gentian, rattlesnake master, stiff goldenrod, and whorled rosinweed. The dominant grasses (big bluestem and Indian grass) would be flowering vigorously by autumn.

Figure 13. Hairy phlox, Wayne National Forest (Ohio).

Lifelong residents of the area recall that this forest opening existed during their childhood, and several botanists have traced its condition for more than half a century. Howard Wistendahl described Buffalo Beats in 1962, when it covered about 2 acres. To help explain the presence of prairie plants in the heart of the predominantly oak forest, Wistendahl examined the soil. He noted that the dense forest, with its oaks and an understory of mapleleaf viburnum, grew on a light-colored, shaly soil, while the prairie had developed on a layer of dark red clay that was deposited several inches deep over the shaly soil. The clay held more water, provided more nutrients, and was less acidic than the adjacent forest soil.

The clay was deepest at the center of the prairie and thinned out at its border. Wistendahl defined three zones of vegetation. The prairie, devoid of all trees, occupied the deeper red clay. A forest with an understory of mixed forest and prairie species grew where only a thin layer of red clay covered the shaly soil. Beyond this "transition zone" was the regular forest.

Two decades after Wistendahl's work, Dennis Hardin restudied Buffalo Beats and discovered that the forest had encroached so much that only about 0.5 acre of reasonably intact prairie remained. Even there, woodland trees, shrubs, and herbaceous plants were invading. Two prairie species, the whorled rosinweed and the milk spurge, had become more common, and several others, including the hairy sunflower, continued to flourish in the transition zone. But rare prairie herbs had become still rarer, and big bluestem, the dominant prairie grass, had decreased from 50 percent of the cover to only 16 percent.

To halt the advance of the forest, Hardin recommended a management plan, which the Forest Service and the Ohio Heritage Program are now implementing. In early spring of 1986, 50 trees with a diameter of 6 inches or more were cut down within the prairie and removed, along with some woody plants in the transition zone. One huge branch of an oak tree that extended halfway across the prairie was also pruned because the shade it cast encouraged the growth of forest species. These steps aided the growth of some prairie plants during the spring and summer, but they also brought an unwelcome influx of poison ivy, which, along with woody seedlings and stump sprouts, is now being weeded out periodically by hand.

Despite the increased exposure to sunlight following the tree cutting, many prairie plants in Buffalo Beats still did not flower or set seeds. As a further experiment, during the spring of 1987, the southern half of the prairie was burned. This triggered rattlesnake masters, yellow gentians, hairy sunflowers, and stiff goldenrods to bloom profusely that summer and autumn, and they have continued to do so in the years since. The management team is reluctant to burn the entire prairie, however, for fear of destroying the insect population. Entomologists have found that Buffalo Beats is a haven for insects, including some very uncommon ones. Two species unique to the area seem to depend completely on the subulate phlox that grows there.

The questions of the origin and age of Buffalo Beats have intrigued botanists for years. In 1942, ecologist Paul Sears suggested that the warming trend that ended the last Ice Age some 12,000 years ago may have eventually rendered the climate in the eastern United States hotter and drier than it is today, allowing the development of Transeau's prairie peninsula. But how can we tell whether Buffalo Beats is a remnant of this ancient prairie or a more recent colonization by prairie plants of an artificial forest clearing?

A recently developed technique has helped to resolve this question. Plants take up silicon from the soil and eventually deposit it in their leaves as granules of silicon dioxide known as *opaline phytoliths*. These microscopic phytoliths, which differ in shape among groups and species of plants, may persist in the soil for as many as 13,000 years. Forest soil scientists P. J. Kalisz and S. E. Boettcher took soil samples in and around Buffalo Beats and discovered that the phytoliths found in the prairie soil were almost exclusively from nonwoody, prairie-type species, whereas those in the transition zone were from a mixture of prairie and forest plants, and those in the forest soil were almost exclusively produced by trees. They concluded that the prairie had occupied the ridgetop for at least 13,000 years and probably once extended over what is now called the transition zone.

The existence of a virgin patch of prairie in this region is remarkable, considering that all other tillable tracts have at least been tried for agriculture. Marilyn Ortt has studied the ownership of the land as far back as the records go and found that except for an inconsequential 3-day period, the land has belonged to absentee owners who were not around to farm this small parcel. Recently, in a fine gesture of conservation, the Quaker State Corporation, which owns the mineral rights beneath Buffalo Beats, gave up its claim in an agreement signed with the Forest Service. As a result, there is renewed hope that the tiny prairie will be preserved for future generations.

NATIONAL FOREST
IN PENNSYLVANIA

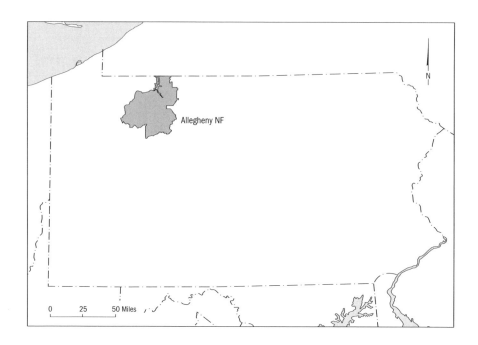

Allegheny NF

0   25   50 Miles

Located in northwestern Pennsylvania is the Allegheny National Forest, which extends north to the New York state line. It is in Region 9 of the United States Forest Service.

Allegheny National Forest

SIZE AND LOCATION: Approximately 470,000 acres in northwestern Pennsylvania, extending from Bradford near the New York border south for nearly 50 miles. Major access routes are U.S. Highways 6, 62, and 219 and State Routes 59, 66, 127, 321, 346, 666, 770, 948, 949, and 3002. District Ranger Stations: Bradford and Marienville. Forest Supervisor's Office: 222 Liberty Street, Warren, PA 16365, www.fs.fed.us/r9/forests/allegheny.

SPECIAL FACILITIES: Swimming beaches; boat ramps; all-terrain vehicle (ATV)/motorcycle areas.

SPECIAL ATTRACTIONS: Tionesta Scenic Area; Heart's Content Scenic Area; Longhouse National Scenic Byway; Allegheny Reservoir Scenic Drive; Highland Hills Scenic Drive; Allegheny National Recreation Area; Allegheny National Wild and Scenic River; Clarion Wild and Scenic River.

WILDERNESS AREAS: Allegheny Islands (370 acres); Hickory Creek (8,560 acres).

The Allegheny National Forest, mostly perched on the Allegheny Plateau, consists of broad-topped ridges and deep, narrow valleys. The land supported primeval forests when the Seneca Indians lived here, but when the first European settlers began to arrive in the late 1700s, cutting the old-growth timber was a major priority. During the 1800s, agriculture was a major venture. As a result, by the beginning of the 1900s, most of the forest had been harvested at least once, and considerable agricultural land had become wasteland.

Majestic white pines were the first trees to be harvested. The logs were floated downriver on rafts to Pittsburgh and New Orleans. Hemlocks were cut for the tanning industry since the bark of the hemlock is rich in tannin. When the railroad arrived in the latter part of the 1800s, most of the hardwood species were cut for shipping. The federal government began to acquire land, and in 1923, President Calvin Coolidge designated the region as the Allegheny National Forest. During the time that the area has been managed as a national forest, the forests have recovered nicely and are now beautiful areas teeming with wildlife and recreational opportunities.

The Allegheny River forms most of the western boundary of the Allegheny National Forest, and the huge Allegheny Reservoir (pl. 26) lies at the northern end of the forest. A few old-growth tracts of timber that has never been harvested still remain. Elevations in the national forest range between 1,200 and 2,360 feet.

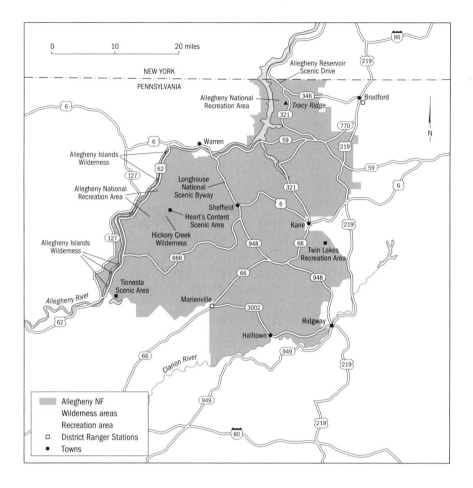

0   10   20 miles

NEW YORK

Allegheny Reservoir
Scenic Drive

PENNSYLVANIA

Allegheny National
Recreation Area

Tracy Ridge

Bradford

Warren

Allegheny Islands
Wilderness

Allegheny National
Recreation Area

Longhouse
National
Scenic Byway

Sheffield

Heart's Content
Scenic Area

Kane

Hickory Creek
Wilderness

Allegheny Islands
Wilderness

Twin Lakes
Recreation Area

Tionesta
Scenic Area

Allegheny River

Marienville

Ridgway

Halltown

Clarion River

Allegheny NF
Wilderness areas
Recreation area
□   District Ranger Stations
●   Towns

The hardwood forests of the Allegheny National Forest contain sugar maple, black maple, red maple, striped maple, mountain maple, yellow birch, sweet birch, bitternut hickory, American beech, white ash, tulip poplar, cucumber magnolia, quaking and bigtooth aspens, wild black cherry, pin cherry, choke cherry, northern red oak, basswood, American elm, and slippery elm. Coniferous trees are primarily white pine and hemlock. There is an abundance of shrubs found throughout the national forest, including wild bush honeysuckle, mapleleaf viburnum, hobblebush, wild raisin tree, arrowwood viburnum, white rhododendron, mountain laurel, mountain winterberry, and two species of elderberry.

An abundance of wildflowers grow in the hardwood forests. Blooming in the spring are Jack in the pulpit, wild leek, bluebead lily, Solomon's-seal, false Solomon's-seal, mayflower, rosy twisted stalk, cucumber root, several species of trillium, pink lady's-slipper orchid, wild ginger, doll's eyes, blue cohosh, toothwort, bishop's-cap, foamflower, both blue and yellow violets, sweet cicely, starflower, Virginia waterleaf, wood betony, and scores of others. During summer and autumn, woodland plants in flower include white hellebore, roundleaf orchid, small purple-fringed orchid, spotted coral-root orchid, tall agrimony, spotted touch-me-not, dotted St. John's wort, whorled yellow loosestrife, closed gentian, blue vervain, skullcap, Oswego tea, partridge berry, goldenrods, asters, fleabanes, golden-glow, and rattlesnake root.

The Allegheny Reservoir and adjacent Kinzua Bay are the focal points of the northern area of the Allegheny National Forest. To get an overview of this beautiful area, drive the two scenic highways that pass through the region.

The Allegheny Reservoir Scenic Drive is in two parts. One part begins in the town of Warren and circles around the northern part of the reservoir. The second segment of the scenic drive is on the eastern side of the reservoir. The first part of the drive from Warren is not in the Allegheny National Forest until the highway comes to Webb's Ferry on the western edge of the reservoir. This huge body of clear water has 12,080 acres, all surrounded by unspoiled scenery. At Webb's Ferry are boat ramps and courtesy docks, a fishing pier, and a hiking trail. Before reaching Webb's Ferry, if you turn east off the scenic drive at Macks Corner, you will come to Roper Hollow on the west bank of the reservoir where there is a boat ramp. Between Webb's Ferry and Roper Hollow is Hooks Brook Campground, reached only by hiking or by boat.

The second segment of the Allegheny Reservoir Scenic Drive goes along the eastern side of the reservoir as it heads south off State Route 346 just east of the Willow Bay Campground. The scenic drive here, which is also State Route 321, passes along the eastern side of Tracy Ridge. This ridge is mostly heavy undisturbed forest of nearly 9,000 acres. On the ridge are rock outcrops, boulders, and very steep slopes. From the Tracy Ridge Campground located along the scenic drive is the Tracy Ridge Hiking Trail System, 33.69 miles of interconnecting loops offering hikes from 1 hour to several days. Camping is permitted along the trail with some exceptions. It is regulated, and a fee is collected at the Tracy Ridge Campground and the Handsome Lake and Hopewell Boat-To/Hike-To Camping area. The Tracy Ridge Hiking Trail also connects with 10 miles of the North Country National Recreation Trail. The North Country National Recreation Trail passes the Willow Bay Campground and proceeds southward for 96 miles to the southern end of the national forest southwest of Marienville. There are boat ramps, cabins, and shower facilities at the Willow Bay Campground. The scenic drive ends at the junction with Forest Road 271. Also at the Tracy Ridge Campground is the 2-mile Land of Many Uses Interpretive Nature Trail.

The Longhouse National Scenic Byway is a loop drive that encircles Kinzua Bay at the southern end of the Allegheny Reservoir. From the Bradford

Ranger Station to the south, the scenic byway is State Route 321. For a while, the road follows North Fork, eventually coming to the east side of Kinzua Bay. The scenic byway then follows the eastern edge of the bay very closely. Red Bridge Campground and Red Bridge Bank Fishing Area provide reservoir access along the way. At the southern tip of Kinzua Bay, the scenic byway leaves State Route 321 and becomes Forest Road 262 as it follows the western side of the bay north to the Allegheny Reservoir. This crooked stretch of the byway provides access to Kiasutha Recreation Area, Elijah Run Boat Launch, Dewdrop Recreation Area, and Kinzua Beach. Kiasutha has a boat launch, courtesy dock, picnic area, swimming beach, playground, and the mile-long Longhouse Hiking Trail. Elijah Run, accessible to visitors with disabilities, has several boat ramps, a fishing pier, and a hardened hiking trail along the shoreline. Continuing north on Forest Road 262, Dewdrop Campground is located in the heart of a hardwood forest along the banks of Kinzua Bay. There is a boat ramp (for Dewdrop campers only) here as well as Campbell Mill Trail, an interpretive nature trail 1.5 miles long. Picnic tables and swimming facilities are available at Kinzua Beach.

The Longhouse National Scenic Byway joins State Route 59 and crosses Morrison Bridge that separates the Allegheny Reservoir from Kinzua Bay. The byway completes the loop back to the Bradford Ranger Station, passing the access road to Rimrock Picnic Area. From the picnic area as well as from a parking area for the Morrison Hiking Trail along State Route 59, you may join the 11-mile Rimrock–Morrison Loop Trail. There is a magnificent view of Kinzua Bay from the overlook at Rimrock. The forests along Morrison Run are in a pristine natural state, with interesting large rock formations and small waterfalls. This is an unimproved trail with no bridges at stream crossings.

The Allegheny Reservoir was formed in 1965 when the Kinzua Dam blocked the flow of the Allegheny River. The concrete Kinzua Dam is 1,897 feet long and rises 179 feet above the streambed. The reservoir was built by the United States Army Corps of Engineers for flood control, low flow regulation, and recreation. The dam is about 3 miles west of the Morrison Bridge on State Route 59. Between Morrison Bridge and Kinzua Dam is Jakes Rocks, an area of rugged natural beauty overlooking the reservoir and the dam. It is perched on sheer cliffs sometimes 600 feet above the water below. Extreme care should be taken while at Jakes Rocks. There are three overlooks along the cliff's edge. The southern overlook, called Picnic Rock, and the middle overlook, known as South Rock, are near the parking area and provide unexcelled views of the dam. It is necessary to hike to North Rock, where you have a beautiful view upriver. Picnic facilities are available near the south parking area.

By continuing to drive west on State Route 59 past Kinzua Dam, you will drive along the south side of the Allegheny River to the bridge to Warren that will take you out of the Allegheny National Forest. About 5 miles west of Warren, U.S. Highway 62 crosses the Allegheny River and reenters the national forest. However, there is a small strip of land on the western side of the river before one comes to the bridge that is a part of the Allegheny National Forest. In this small parcel of land is Buckaloons Recreation Area. The campground has been located on the site of an Indian trading post. Buckaloons extends from the bridge south to the mouth of Brokenstraw Creek. At the campground is a 1-mile interpretive nature trail and a boat ramp into the Allegheny River. This part of the river has been designated as a National Wild and Scenic River. Along the trail you will see white pines, sycamores, shagbark hickories, white oaks, white ashes, wild black cherries, black walnuts, hemlocks, and tulip poplars.

U.S. Highway 62 follows the east side of the Allegheny River closely as well as forming the western edge of the Allegheny National Forest. South from Buckaloons are several islands in the Allegheny River. Seven of these, together comprising nearly 370 acres, are managed by the Forest Service and make up the Allegheny Islands Wilderness. The islands are alluvial in origin, and several of the trees on them may reach diameters greater than 4 feet. Huge royal and cinnamon ferns are often present, some of them with fronds up to 7 feet long.

The northernmost of these islands is Crulls Island, whose 96 acres consist of two-thirds old-growth riparian forest and one-third old-field vegetation. A few miles south of Crulls Island is Thompson Island. This 67-acre island, which has a good riparian forest, was the site of the only Revolutionary War battle in northwestern Pennsylvania. A few more miles farther south is 30-acre R. Thompson Island. The old-growth forest on this island was severely damaged by a violent windstorm in 1985, and slow recovery of the trees is underway. The next island south is Courson Island across from the community of Tidioute. This island has 62 acres of old-growth riparian forest and old-field communities. Across from Courson Island to the east is Tidioute Overlook, offering superb views of the Allegheny River both upriver and down. About 10 miles south of Courson Island is 36-acre King Island. This island has several extremely large riparian trees. Baker Island, with 67 acres, stood in the path of one of the two tornadoes that crossed the forest in 1985 and blew over many of the trees. Immediately south of Baker Island and near the southwestern edge of the Allegheny National Forest is 10-acre No Name Island. The large riparian trees on this island have a very dense undergrowth of vegetation.

From Tidioute south to the community of Tionesta, the Allegheny National Forest does manage some parcels of land on the west side of the Allegheny

River, including Babylon Hill and several streams, called *runs*, that empty into the Allegheny River.

Traveling south along U.S. Highway 62, you will find several recreational opportunities on the eastern side of the highway. A large area across from Buckaloons has been set aside as the Rocky Gap All-Terrain Vehicle (ATV) area, with 20.8 miles of trails. Immediately opposite R. Thompson Island is the Tanbark Trail. This 8.8-mile-long trail passes through undisturbed forest, ending at Dunham Siding just beyond Heart's Content Scenic Area.

Heart's Content Scenic Area is one of two tracts of virgin old-growth forest still in its primeval condition. This majestic area consists of 200 acres, of which 122 acres are original forest. An interpretive trail winds among towering white pines, hemlocks, and American beeches. Some of the trees are estimated to be at least 400 years old. Where the two loops of the interpretive trail cross is Wheeler Spring, whose clear water gives rise to a small stream that eventually empties into the West Branch of Tionesta Creek. The shorter of the two loops of the trail is accessible to people with disabilities. Just across the forest road from the scenic area is the Heart's Content Campground.

Adjacent to Heart's Content Campground to the west is the Hickory Creek Wilderness. The wilderness consists primarily of gently rolling terrain where there has been only moderate disturbance of the forest cover. Middle Hickory Creek flows through pastoral meadows, interrupted by bogs and beaver ponds. The Hickory Creek Trail is an 11-mile loop that begins at the Heart's Content Campground. The northern part of the loop stays on a plateau separating East Hickory and Middle Hickory creeks. After crossing Jacks Run, the loop returns to the campground by climbing in and out of small valleys. In the area between the Hickory Creek Wilderness to the southern end of the Allegheny National Forest are several historic railroad grades.

State Route 666 crosses the lower part of the Allegheny National Forest from East Hickory at U.S. Highway 62 traveling west until it ends at State Route 948, 2 miles south of U.S. Highway 6 near Sheffield. Just east of the village of Mayburg is Minister Valley, a deep valley drained by Minister Creek, a natural trout stream. All around the 1,375-acre valley are conglomerate rock formations. A small campground stands at the junction of State Route 666 and a forest road to the north. Both the North Country National Recreation Trail and the 6-mile-loop Minister Valley Trail may be accessed from the campground.

South of Minister Creek Campground and about 5 miles north of Marienville is Beaver Meadows Campground. The campground is located in a pine woods on the northern bank of Beaver Lake. From the campground is a 6-mile hiking trail. Fishing in Beaver Lake is good for bluegill, pumpkinseed, yellow perch, bullhead, and bass. Salmon Creek, just below the lake's dam, is stocked with trout. Five interconnecting hiking trails around Beaver

Meadows Lake total 6 miles. Beaver Meadows Trail, a 2.5-mile loop, passes south of the lake through grass-covered savannas penetrated occasionally by spruces and pines. The trail eventually passes over the lake on a floating boardwalk. As the trail goes around the northern edge of the lake, it passes through a forest of wild black cherry and red maple. Seldom Seen Trail, 1 mile long, winds through a beautiful forest of American beech, red pine, wild black cherry, and red maple., with a 0.5-acre blueberry patch along the way. Salmon Creek Loop, 1.4 miles in length, enters a forest of red pine and spruce. The 0.8-mile Penoke Path follows Penoke Run on an old railroad grade and goes through a grassy savanna. Lakeside Loop is only 0.2 mile long as it stays near the lake.

West of Marienville and northeast of Muzette is the Muzette Natural Area where a small virgin forest that apparently has never been cut grows along Bear Run. Another scenic area a short distance north of Muzette Natural Area is Lamentation Run, with a pristine natural area along the creek and a nice observation area on Stony Point.

State Route 666 terminates at State Route 948 a short distance south of Sheffield. Continue on State Route 948 north toward Sheffield. West of the highway is lovely Pell Run with nice forests on either side. West of Sheffield is Bull Hill Historic Area. A Civilian Conservation Corps camp was built here in the 1930s; in the early 1940s, this area was used as a prisoner of war camp. U.S. Route 6 north of Sheffield and traveling west goes to Warren. About midway, Farnsworth Run to the west has nice natural areas along it.

State Route 66 winds across part of the southern region of the Allegheny National Forest from Marienville to Kane. From this highway south to the Clarion River is the southernmost part of the national forest. There are several interesting areas in this region. About 4 miles southeast of Marienville via Forest Road 157 is an area known as Buzzard Swamp. It is a series of man-made ponds. To become familiar with some of the wetland vegetation and animal life, hike the Songbird Sojourn Interpretive Trail, a 1.5-mile loop just north of the forest road. If you are interested in ponds, walk around the area and then follow Muddy Fork for a mile or two southeast.

Four miles south of Buzzard Swamp is Loleta Campground along Forest Road 131. There is a loop trail of 3.2 miles from the campground. Loleta was built by the Civilian Conservation Corps in the 1930s on the site of an old logging town.

The area between Kane and Ridgway makes up the southeastern corner of the Allegheny National Forest. The Clarion River forms the southern boundary of the national forest. The river in places is up to 300 feet wide and cuts through deep gorges. A part of it has been designated as a National Wild and Scenic River. East of Halltown is the Irwin Run Canoe Launch into the

Clarion River. Visitors in canoes and rubber rafts will find the river beautiful and exciting. Bear Run joins Spring Creek north of Halltown, and Spring Creek empties into the Clarion River. These bodies of water are home to the rare mountain lamprey.

One mile north of the Clarion River and east of the settlement known as Beuhler Corner is the Little Drummer Historical Pathway. The pathway is named for the breeding grouse called drummers. Two loops comprise the pathway, one of them 1.3 miles long and the other 3.1 miles long. The trails pass historic railroads, pipelines, and wildlife habitat improvement projects. The area is managed for wildlife.

To explore the Allegheny National Forest immediately west of Ridgway, drive the Highland Hills Scenic Drive. This route, approximately 35 miles long, is on back forest roads and encircles the Bear Creek Recreation Area. The drive is pretty, if not spectacular, and you will see evidence of the past history of this part of the national forest. From the Ridgway Office of the Marienville Ranger District, drive west on Forest Road 135. After several miles, including some extremely sharp curves, you will come to Forest Road 136. As you proceed north on this road, you will see, near the settlement of Owl's Nest, evidence of a major fire known as the Bear Creek Fire that burned 13,000 acres of forest on May 17, 1926. Between 1957 and 1964, 4,000 acres of spruce, pine, and larch were planted in the Bear Creek area.

As you continue north on Forest Road 136, keep alert for white-tailed deer and wild turkey. Turn off Forest Road 136 onto Forest Road 124 and head northeastward to the hamlets of Highland Corner and Lamont. From here you may access the Twin Lakes Recreation Area to the east. The lakes were formed in 1936 when the Civilian Conservation Corps constructed the dam on Hoffman Run. Swimming may be enjoyed at the lakes, and there is a beach and bathhouse, as well as individual campsites accessible to visitors with disabilities, a group camping area, and a fishing pier. Hikers have several options at Twin Lakes. The Black Cherry National Recreation Trail, 1.4 miles in length, is a loop interpretive trail through an attractive hardwood forest. Those wanting a longer hike can take the 16.5-mile Twin Lakes Trail that eventually ends at the Tionesta Scenic Area. The Twin Lakes Trail also connects with the Mill Creek Trail a short distance west of the lakes.

The Highland Hills Scenic Drive heads south on Forest Road 185 past the Kane Experimental Forest. Established in 1930, this research facility attempts to find better ways to harvest trees, to improve the quality of young forests, and to study the environmental effects on forests. Forest Road 185 joins State Route 948, and this latter highway returns to the starting point at the Ridgway Office of the Marienville Ranger District.

South of the Twin Lakes Recreation Area is a region of fantastic rock formations amid a forest of oaks and rhododendrons. The Bogus Rocks Scenic Trail winds through the area. Bogus Rocks may be reached off either State Route 66 or State Route 948. (there is no developed or maintained trail by the Forest Service in this area).

From the town of Kane, Forest Road 133 winds through forest and private lands on its way to the Tionesta Scenic Area and Tionesta Research Natural Area. These areas, adjacent to each other, preserve one of the few remaining examples of virgin forest that once covered 6 million acres of the Allegheny Plateau in Pennsylvania and New York. In 1936, the federal government purchased the last remaining uncut hemlock–American beech forest in the area. The Tionesta Scenic Area, on the northern half, consists of 2,018 acres and is for the enjoyment of the general public. Two trails wander through the area. The 2,113-acre Tionesta Research Natural Area, on the southern half, has been set aside for scientific study of the ecology of a virgin hemlock–American beech forest. In 1985, a devastating tornado cut a swath through the heart of the scenic area, causing remarkable damage. Hikers in the area can assess the damage and see how the forest is responding to such a catastrophe.

Of interest to persons wishing to see wetlands and their associated plants and wildlife are Marsh Pond and Duck Pond on either side of Forest Road 150 northeast of Kane. A parking area is provided for those interested. The Willow Creek ATV Trail between Tracy Ridge and Bradford is a 10.6-mile, one-way trail divided into two loops that are designed for different skill levels. This trail is closed from the last Sunday in September until the Friday before Memorial Day.

Mountain bikers are permitted to use all open roads in the Allegheny National Forest as well as most of the closed and gated roads. Since there are some exceptions, it is best to contact one of the district ranger stations or the supervisor's office.

Autumn is a beautiful time of year in the Allegheny National Forest since many of the hardwood trees provide colorful foliage. As you drive through the forest, be on the lookout for some of the black bears that live there.

Winter sports activities are available in most of the Allegheny National Forest. Snowmobiles may be used on designated routes between December 20 and April 1, but there is no cross-country riding off the designated trail system. Cross-country skiing may be done on designated cross-country ski trails, on old railroad grades, and on logging roads. There are eight designated cross-country ski trails, from the 1.5-mile trail at Rimrock to the 11.5-mile trail at Laurel Mill 3 miles west of Ridgway on Spring Creek Road.

Tionesta Forest

Shortly after six on the sultry evening of May 31, 1985, tornadic winds roared eastward across the Allegheny River and slammed into the Allegheny National Forest, toppling hundreds of trees and shattering thousands of others. The violent twisters slashed six corridors through the forest, one nearly a mile wide, before coming together for a final fling at the western edge of the community of Kane. In a matter of minutes, a violent side of nature destroyed what another, more tranquil side had taken hundreds of years to build. One path of devastation pierced the heart of Tionesta Forest, one of only two stands of virgin hemlock remaining in Allegheny National Forest.

Before the arrival of European colonists, some 6 million acres of the Allegheny Plateau in Pennsylvania and New York were covered by a forest dominated by eastern hemlock, American beech, and sugar maple. But little uncut forest remained by 1936, when the federal government purchased 4,000 acres to preserve a remnant of this biological community. In 1973, the Tionesta Scenic Area and Tionesta Research Natural Area, into which this tract is officially divided, were placed on the National Register of Natural Landmarks.

Since 1936, biologists have observed the ecology of Tionesta continuously, recording the area's development. Eastern hemlocks grow up to 125 feet high, with diameters of 4 feet. Nearly as big are the American beeches and sugar maples. Black and yellow birches, yellow poplar, cucumber magnolia, basswood, red maple, and white ash are some of the other trees scattered throughout the forest. Beneath the canopy, shrubs, including the mapleleaf viburnum called hobblebush, grow eye-level high. The forest floor is alive with nearly 75 wildflower species and a variety of ferns—in fact, areas of up to 1 acre are a sea of light green because of the abundance of ferns.

Some of the larger animals in Tionesta have a conspicuous role in shaping the composition of the forest. Porcupines feed on the bark and inner tissues of many of the trees, ultimately girdling and killing some of them. The porcupines seem to prefer American beech and hemlock, but sugar maple and yellow birch also rank high on their menu. White-tailed deer (fig. 14) forage on hemlock and maple seedlings, and they apparently relish hobblebush, sometimes nibbling away until it is eliminated from an area. Since the deer spare the beech seedlings, these are abundant in some parts of the forest.

The 1985 tornado, however, abruptly halted normal life in a segment of the forest and sent things back to square one. As a result, biologists now have an ideal opportunity to observe how a virgin hemlock forest recovers.

After 1863, when Anton Kerner published his landmark study of Europe's Danube Basin, biologists came to realize that in most plant communities, plant groups are constantly replacing one another, a phenomenon now

Figure 14. White-tailed deer, as often seen in Alleghyany National Forest (Pennsylvania).

referred to as *plant succession*. Succession occurs in many habitat types. A new lake bottom, for example, is at first barren of plant life, but soon microscopic plants and animals die and fall to the bottom, forming a layer of organic muck. As the lake bottom's oozy layers thicken, aquatic plants, such as pondweeds, take root. These rooted plants also add organic matter to the lake bottom, creating an environment conducive to the growth of cat-tails, bulrushes, and other emergent plants. As the years pass, a shrub stage may develop, followed by such trees as willows. Lakes all over the globe follow the same sequence.

Similarly, during the first year following abandonment of a midwestern farm field, small plants that complete their life cycle in one season usually prevail. In subsequent years, these annuals are joined by certain perennial herbs that eventually crowd out and eliminate the annuals. Grasses, notably broomsedge, arrive a few years later, and compete with the other perennial herbs. A shrub stage then develops, with sumacs, young sassafras and persimmon trees, and perhaps roughleaf dogwood. The shade that these shrubs provide permits seedlings of maple, oak, and hickory to survive. As they grow, these tree seedlings overtop the shrubs and form the canopy of a forest.

In general, succession involves the replacement of limited plant communities, which produce relatively little living matter, by more complex and increasingly stable plant communities, which encompass far more "biomass." Once a community develops that will persist for a long time under natural conditions, succession is said to have reached the climax stage.

In the Tionesta Forest, the successional climax is a community dominated by eastern hemlock, American beech, and, to a lesser extent, sugar maple. Seedlings of these trees have a high tolerance for shade, surviving for years under the dense canopy created by their parents. Seedlings of most other trees perish under these conditions. When one of the giant hemlocks, beeches, or sugar maples dies of old age and falls, seedlings are ready to grow into the opening created in the canopy.

Because Tionesta was virgin forest and designated a natural landmark, Forest Service personnel decided to let succession take its natural course after the tornado. Nothing was done to tidy up the area or promote the growth of particular species. Giant trees 3 feet in diameter lie where they fell. Elsewhere, upright splinters are all that remain of trees shattered where they stood.

Four months after the tornado, thousands of eastern hemlock seedlings had germinated alongside the forest debris. In the meantime, species that previously had been suppressed by the deep shade began growing in the open. Blackberry bushes and young trees of wild black cherry came up with a vengeance in the spring of 1986. For a few years, until the young hemlocks grew tall enough to shade them out, these species formed nearly impenetrable thickets. Once the blackberries and wild black cherries were reduced, however, spinulose wood ferns, thriving in the shade provided by the hemlocks, once again carpeted the forest floor.

On a visit to Tionesta before the tornado, I had driven to the forest's parking area and hiked a narrow trail that meandered under the hemlocks and beeches. When I returned in May 1986, I saw that the tornado, which had skirted the parking area, seemed to have used the hiking path as one of its boundaries. On one side virgin forest still stood, as it must have appeared hundreds of years ago, while on the other side was total destruction. At one place along the trail, I reached out with my right hand to touch the living trunk of a 100-foot-tall hemlock; my left hand extended to a splintered dead beech tree whose top lay across the devastated area.

NATIONAL FOREST
IN PUERTO RICO

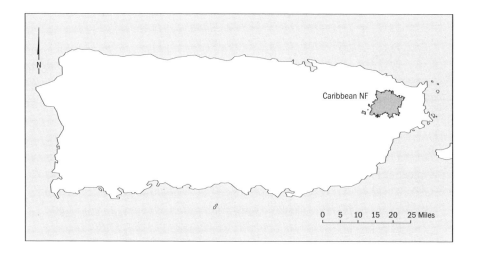

The Caribbean National Forest is the only national forest managed by the United States Forest Service that is not in one of the 50 states. It is also the only national forest with a tropical rainforest. The Caribbean National Forest is in the Southern Region of the U.S. Forest Service.

Caribbean National Forest

SIZE AND LOCATION: Approximately 28,000 acres in northeastern Puerto Rico, between San Juan and Fajardo. Major access routes are Highways 3, 31, 186, and 191. Forest Supervisor's Office mailing address: HC-01, Box 13490, Rio Grande, Puerto Rico 00745, www.southernregion.fs.fed.us/caribbean.

SPECIAL FACILITIES: El Portal; the Rainforest Visitor Center.

SPECIAL ATTRACTIONS: El Toro National Recreation Trail; Baño de Oro Research Natural Area; guided Forest Adventure each hour; Rent-a-Ranger tours by prior arrangement.

Located midway between San Juan and Fajardo, the Caribbean National Forest lies in the northeastern corner of the island of Puerto Rico and is the only national forest that contains tropical vegetation, including rainforests and elfin forests. Although occupying only 28,000 acres, it contains very diverse plant life, with elevations ranging from 600 to more than 3,500 feet above sea level. Much of the national forest has virgin trees, with some trees dating back more than 1,000 years. Only the forests at the lower elevations have been cut for wood and charcoal or cleared for agricultural crops. Most of the Caribbean National Forest receives more than 100 inches of rain annually, with the highest peaks often shrouded with a dense cloud cover. It is on these cloud-covered summits that pygmy or elfin forests develop.

Before the arrival of Christopher Columbus during his second voyage of discovery, followed by groups of European colonizers in the late 1400s, the island was mostly settled by Taino Indians who had migrated by canoe from the headwaters of the Orinoco River in what is now Venezuela more than 1,000 years earlier. Examples of Taino petroglyphs have been discovered throughout the Caribbean National Forest. Most of what comprises today's national forest was set aside as a forest reserve by King Alfonso XII of Spain in 1876. The United States Forest Service acquired the reserve in 1903, calling it the Luquillo Forest Reserve; the area was officially named the Caribbean National Forest in 1935. The Man and Biosphere Program of the United Nations named the Luquillo Experimental Forest (an area of the forest devoted to research) as a part of the International Biosphere Reserve in 1976.

Botanists have recorded 225 species of native trees in the Caribbean National Forest, with 23 of them known from no other place in the world.

The national forest is centered around the Luquillo Mountains, with high peaks including El Yunque and El Toro. It consists of four fairly well-defined vegetation zones. Virgin tracts of all four forest types may be found in the Baño de Oro Research Natural Area, which was established in 1959. Between 600 and 2,000 feet elevation is the Tabanuco Forest, or Lower Montane Forest, a rainforest where annual rainfall is between 70 and 100 inches. Typically, the trees in this forest are large and straight with widely spreading crowns that cast a dense shade over the understory. This is the forest of numerous epiphytes that grow on the branches of many of the trees. Most of the epiphytes are species of orchids and bromeliads. Huge, picturesque tree ferns grow in this forest community.

The Tabanuco Forest consists of three layers of trees. The uppermost layer contains trees reaching a height of 120 feet and a trunk diameter as much as 6 feet. Twenty tree species are common in this layer, half of them

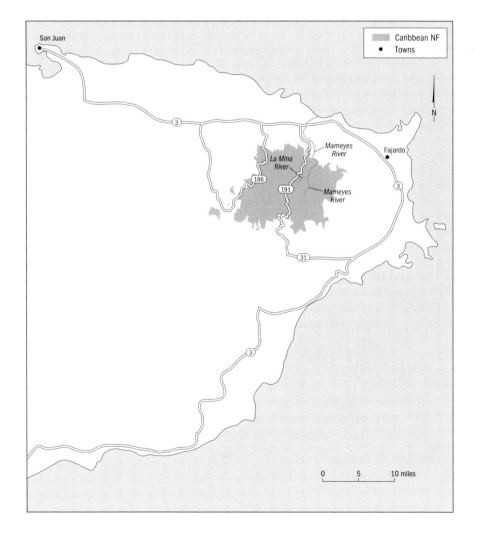

broad-leaved and deciduous and half of them evergreen. The midlayer contains 43 common species of trees that reach heights of 50 to 70 feet. The lower layer consists of 25 common broad-leaved evergreens that reach a maximum height of 20 feet. Since much of the Tabanuco Forest has been cut in the past, the forest that one sees today is called a *secondary forest*. This type of woodland takes its name from a prominent tree called tabanuco. Much of the land that previously had been cleared for agricultural purposes is now reclaimed by forest species, including several nonnative. The tabanuco tree may be recognized by its straight, high trunks that may reach a height of more than 100 feet. The lower half of the tabanuco trunk is devoid of branches. Found with the tabanuco are the motillo (*Sloanea berteriana*) and the asubo (*Manilkara bidentata*). Common birds in the Tabanuco Forest include broad-winged hawk, golden-winged warbler, American redstart, and Antillean euphonia. Endemic species in this forest type are the stripe-headed tanager, Puerto Rican screech owl, Puerto Rican tody, Puerto Rican woodpecker, and Puerto Rican bullfinch.

Between 2,000 and 2,500 feet in the valleys and on gentle slopes is the Palo Colorado Forest, or Upper Montane Forest. Rainfall in this area ranges from 100 to 180 inches annually, and the largest trees in the Caribbean National Forest are found here. The largest of these, the palo colorado, is the same species that grows as a shrub called titi in the swamps of the southeastern United States. The Palo Colorado Forest has two layers of trees. The upper layer has trees about 50 feet tall but with diameters up to 6 feet. All of the trees in this layer are broad-leaved evergreens. The lower layer has trees about 15 feet tall. Huge vines, known as *lianas*, are draped from many of the trees, and epiphytes abound here as well. Pink and white begonias cover the forest floor in many areas. This is the forest in which the federally endangered Puerto Rican parrot lives. Reduced to only 13 individuals in 1972, the Puerto Rican parrot now numbers about 40 birds in the wild, thanks to intensive research conducted on the parrot in the Caribbean National Forest by the U.S. Fish and Wildlife Service, the U.S. Forest Service, and the Puerto Rico Department of Natural Resources. Other birds in the Palo Colorado Forest include green-backed heron, peregrine falcon, ruddy quail-dove, black swift, black-throated blue warbler, Louisiana water thrush, bananaquit, and two endemic species—sharp-shinned hawk and Puerto Rican emerald.

On steep slopes, and along streams, and in valleys above 2,500 feet is the Sierra Palm Forest, where a species of palm tree with silvery leaves grows in great abundance. This palm thrives in areas where annual rainfall is about 90 inches or more. Although the Sierra palms dominate the palm forest, 63 other species of trees and shrubs grow in this zone. The understory is relatively sparse, and lianas are not prominent, so this type of forest is the most open of any in the Caribbean National Forest. Large palmate-lobed leaves of cecropia or trumpet wood tree may often be seen lying on the forest floor. Birds of the palm forest include the spotted sandpiper, scaly-naped pigeon, northern parula, black-and-white warbler, and the endemic Puerto Rican lizard cuckoo.

Above 2,500 feet and on the highest peaks are the pygmy or elfin forests where rainfall may exceed 150 inches annually and sometimes tops 200 inches. Rain often occurs more than 350 days a year. Because of the harsh windy conditions on the peaks, the trees are dwarfed and twisted, and the branches are heavily draped with long-hanging mosses. Thirteen species of broad-leaved evergreen trees occur in this zone. Bromeliads with red flowers grow along the branches of many of the trees. It is in the cloud forest zone where most of the endemic plants and animals occur. Birds seen in the elfin forests are northern parula, black-throated blue warbler, American redstart, and three endemics—Puerto Rican tanager, elfin woods warbler, and Puerto Rican bullfinch.

Contrary to popular belief, there are no poisonous snakes in the Caribbean National Forest. The largest nonpoisonous snake is the Puerto Rican boa, which may reach a length of nearly 8 feet. Small anole lizards are abundant, including a green one that is up to 16 inches long. Any visitor to the forest will be treated to a continuous serenade by coquis (pl. 27), a type of nocturnal tree frog that says its name in its call. From dusk to dawn, particularly after a rain, the voices of the coquis are nonstop.

Most visitors to the Caribbean National Forest will come in from the San Juan area. Highway 3 leads east along the northern edge of the island from San Juan to Fajardo. At Palmer (also known as Mameyes), about 20 miles from San Juan, is Highway 191 that follows Rio Mameyes south to the northern edge of the national forest. Shortly after entering the national forest is El Portal Rainforest Visitor Center (fig. 15), the main visitor center in the national forest, featuring theme pavilions with interactive displays, a theater, and gift shop. There is a short interpretive nature trail here to give the visitor a taste of a tropical rainforest. This is a good place to look for birds in the kadam, mahogany, and Caribbean pine plantations nearby. Look for black-cowled orioles, Puerto Rican bullfinches, red-legged thrushes, and Puerto Rican lizard cuckoos in the kadam trees.

As you continue driving south on Highway 191, you enter the El Yunque Recreation Area, the center for most visitor activities in the national forest. At Kilometer 8.2 is beautiful La Coca Falls to the west of the highway. A parking area and small gift shop are nearby. At Kilometer 8.6 is the La Coca Trail. This gravel trail winds for nearly 2 miles, crossing Quebrada La Coca twice on slippery rocks and small boulders. There are rock steps, usually slippery, along the steeper parts of this rather strenuous trail. You will want to pace yourself as you hike because the return trip over the same trail is mostly uphill.

At Kilometer 9.0 is the Yokahu Observation Tower, built with a Spanish design in 1963. From the tower is an overview of the Mameyes River Valley and the surrounding forest sweeping down to the coastal plain, as well as the offshore islands of Culebra and Vieques. On most mornings, you may spot the native broad-winged hawk overhead. Ornithologists also report that this is a good place to observe Antillean euphonias, Puerto Rican lizard cuckoos, Puerto Rican bullfinches, and stripe-headed tanagers.

After Highway 191 crosses Quebrada Juan Diego at Kilometer 9.9, there is a parking area at Kilometer 10.2 for the Big Tree Trail, one of the favorites in the national forest. After nearly 1 mile, the Big Tree Trail connects with La Mina Trail, which extends for another 0.7 mile to the Palo Colorado Picnic Area at 12.2 on Highway 191. The Big Tree Trail is a self-guided interpretive trail through groves of lush evergreen and deciduous trees of tabanuco, motillo, and *Guarea*, eventually coming to spectacular La Mina Falls. On the way, the trail crosses several streams and goes over small rises by means of stairs. The Big Tree Trail connects with the mostly paved La Mina Trail, which follows the La Mina River to the Palo Colorado Picnic Area. If you look into the clear water of the river, you may see several kinds of freshwater shrimp as well as the goby, the only fish in the national forest. The goby spends its juvenile days in marine estuaries before making its way up the river to complete its life cycle. Inch-wide nesting burrows of the Puerto Rican tody may be observed along the river banks. The Palo Colorado Picnic Area has a visitor center nearby with a well-stocked bookstore and gift shop. If you don't want to take the combined Big Tree–La Mina Trail, you may opt for the La Mina Trail from the picnic area down to the falls, a distance of 0.7 mile.

At Kilometer 10 on Highway 191 is the Sierra Palm Recreation Area with a new, full-service food court, a large picnic area with barbecues, restrooms, and drinking water.

Across the highway from the Palo Colorado Visitor Center at Kilometer 12 is the 0.25-mile Baño de Oro Trail that passes the scenic Baño de Oro pool built by the Civilian Conservation Corps but no longer in use. Where the trail meets the El Yunque Trail, you may either return to the parking area or begin hiking the El Yunque Trail. The El Yunque Trail may also be hiked from the trailhead at Kilometer 10.9 on Highway 191. If you stay on the trail for 2.6 miles, you will ascend to the top of 3,496-foot El Yunque in the heart of an elfin cloud forest. Although the trail is either paved or graveled, it is usually slippery and rather steep in places. An observation deck awaits you at the El Yunque summit where on a clear day you will have spectacular views of the forest, the Atlantic Ocean, and the city of San Juan. The trail passes from the Palo Colorado Forest zone through the Sierra Palm Forest zone to the elfin cloud forest zone. As you hike through the Palo Colorado Forest, you will see some of the big trees, several tree ferns (including one that creeps), a climbing bamboo, and huge lianas known as Marcgravias. The trail climbs into a Sierra palm forest where the trees with the silvery leaves are covered with the Puerto Rican bromeliad (*Guzmania berteroiana*). The elfin forests must be seen to be believed, with the gnarly stunted trees clothed with mosses, lichens, liverworts, and epiphytic flowering plants. Here is the place to see the endemic elfin woods warbler, often flying with flocks of Puerto Rican tanagers.

From the El Yunque Tower you may hike a short but difficult trail to El Yunque Rock or a 0.17-mile trail to Los Picachos Lookout Tower built by the Civilian Conservation Corps in the 1930s. This is a narrow and often muddy, strenuous trail.

At Kilometer 12.6 on Highway 191, where the road ends at a gate, a right turn on the secondary road (#9938) leads to the west. This road will take you to the Mount Britton Trailhead. This 0.8-mile trail climbs to the Mount Britton Lookout Tower in the heart of an elfin cloud forest after passing through a very fine Sierra palm forest. A steep circular staircase permits visitors to reach the top of the tower for unsurpassed views on a clear day.

A short distance south of where Road #9938 rejoins Highway 191, the main highway is closed because of landslides. If you hike south past the gate for 0.2 mile, you will come to the trailhead for the Trade Winds National Recreation Trail. This trail climbs steadily for 4 miles to El Toro Peak, the highest elevation in the forest. The trail is not well maintained and is usually muddy and "wolfish" in spots. At El Toro Peak, the trail becomes the El Toro National Recreation Trail, which continues for another 2.2 miles to Highway 186. These two national recreation trails offer hiking in a tropical rainforest at its primitive level. The trail passes through tabanuco, Sierra palm, and elfin forests.

Highway 186 provides another northern access to the Caribbean National Forest from Highway 3, entering the national forest at the El Verde Work Center. At Kilometer 15.8, where the highway crosses Quebrada Grande, is a picnic facility with restrooms. The highway continues southward and comes to the western trailhead of the El Toro National Recreation Trail at Kilometer 10.8. Beyond this, Highway 186 leaves the national forest.

Back at the El Portal Entrance Station to the Caribbean National Forest on Highway 191, you may take Road 988 southeast into the Caribbean National Forest instead of staying on the main route. If you take this lesser-used route, you will come to the 0.8-mile Angelito Loop Trail that provides access

Figure 15. El Portal Rainforest Visitor Center, Caribbean National Forest (Puerto Rico).

to Las Damas, a serene natural pool in the Mameyes River. A few miles farther along Road 988 is the longer Carrillo Trail that eventually makes its way to La Mina Falls. This remote trail through the Mameyes River Valley passes through nice stands of tabanucos.

Elfin Woodlands

Climbing through the rainforest that cloaks Puerto Rico's Luquillo Mountains and up past the thickets of palo colorado trees and occasional stands of Sierra palms that ring the mountains between 2,000 and 2,500 feet, visitors may suddenly come upon a windswept, misty landscape crowded with dwarfed, twisted trees. This elfin forest grows on the few ridges that rise above 2,500 feet: on Pico del Este, Pico del Oeste, Mount Britton, and, at 3,281 feet, the highest peak, El Yunque—all within the Caribbean National Forest.

Draped with flowering air plants, mostly of the pineapple and orchid families, and veiled (as is the ground) with a layer of mosses, liverworts, and occasional shiny masses of terrestrial algae, 46 species of trees and shrubs live in the elfin forest. Entanglements of morning glories and other vines climb over the treetops, distorting and often breaking the branches in the crown. Ropelike clusters of aerial roots help to anchor some of the trees pulled over by this heavy growth.

Among the trees adapted almost exclusively to this upland habitat are *Tabebuia rigida*, a species related to the North American catalpa; *Ocotea spathulata*, a tropical relative of the sassafras; *Weinmannia pinnata*, a tropical cunonia; and *Eugenia borinquensis*, a tropical myrtle. Other elfin species grow at lower elevations as well, but there they may look quite different— taller, with straight, spreading branches. Several environmental factors account for the stunted and gnarled forms of the elfin forest trees.

The soil, developed from fine-grained volcanic rock by the weathering action of rain, is relatively poor because it is both highly acidic and boggy. Throughout the year, including the dry season, the Luquillo mountain peaks are enshrouded in fog and clouds, and rain falls nearly every day, often at night but most heavily in daytime. The densely packed trees act as a filter, capturing or delaying the water as it makes its way to the ground. For example, during one 11-minute period of precipitation, when 0.54 inch of rain fell above the forest, only 0.38 inch of rain reached the ground, some of it dripping down 2 minutes after the rain had stopped.

As a result of the abundant rainfall and constant dripping of water from leaves, the soil is saturated for nearly a foot beneath the covering mosses. These conditions would severely inhibit plant growth if it were not for the giant, olive-colored earthworms, some up to 2 feet long and 0.5 inch in diameter, that tunnel into the muck, ingesting organic and mineral matter and returning them to the soil in a churned-up condition. The large earthworm tunnels aerate the soil and promote water circulation.

Trade winds, constantly blowing over the peaks at up to 32 miles per hour, also limit tree growth, leveling the tops to a uniform 8 to 12 feet. At similar elevations where there is shelter from the wind, the elfin forest is replaced by a normal-looking forest dominated by Sierra palms.

Perhaps what most stunts the trees, however, are the overcast conditions on the fog- and cloud-bound peaks. Botanist David Gates has noted that the primary influence of climate on a plant is through the transfer of energy, which is consumed in many biological processes. Plants absorb light and heat radiation from direct sunlight, scattered sky light, and reflected light, as well as additional heat radiation from the ground, surrounding vegetation, and air. The biochemical reactions, including photosynthesis, that take place in the leaves and elsewhere usually depend on the level of heat and light. If conditions are less than optimal for a particular species—too warm or too cool, too bright or too dim—vital processes slow down.

Gates found that in the elfin forest, leaf temperatures drop as low as 59 degrees Fahrenheit and never rise above 77 degrees—temperatures slightly below optimum for the same species growing at lower elevations. The rate of photosynthesis is sometimes only half the maximum rate for the species, consequently reducing growth rates and resulting in stunted trees.

El Yunque Rainforest

"Coqui! Coqui!" The high piping of the inch-long tree frog that sings its name lulled me to sleep each night during my stay near El Yunque, a mountain peak of the Caribbean National Forest. There are 12 species of coquis. Sounding more like birds than frogs, they pipe their loudest on rainy evenings, which are frequent. El Yunque is one of the high summits in the Luquillo Mountains, the only area in the U.S. national forest system to harbor a tropical rainforest. The Caribbean National Forest is less than 10 miles from the ocean, and the steep terrain rises from 330 feet near the coast to 3,496 feet. As the moisture that the northeastern trade winds take from the sea strikes the Luquillo Mountains, it condenses into rain. In some places, rain falls an average of 350 days each year, with a yearly accumulation of 200 inches.

Climbing the slopes of El Yunque is like moving rapidly north from the equator.

Every 350 feet ascended in the tropics is roughly equivalent to moving 100 miles toward the pole. Distinct plant communities are encountered on the way. The forest that extends above the hot agricultural lowlands up to about 2,000 feet is the rainforest proper, also called the Tabanuco Forest because of the dominance of the white-trunked tabanuco tree. Tabanuco belongs to the tropical Burseraceae family, a group of plants that includes the gumbo-limbo of southern Florida, the elephant tree of southern Arizona and Mexico, and the trees of biblical fame that produce frankincense and myrrh.

Between 2,000 and 2,500 feet, where evaporation is often inhibited by a heavy, foglike cloud cover and a cooler climate, a forest known as a *montane thicket* generally prevails, characterized by palo colorado trees. In the same altitude range, where the drainage is especially poor, nearly pure stands of Sierra palms attain heights of up to 50 feet. Finally, above 2,500 feet, on the highest ridges, is an elfin forest, where the thick growth of vegetation is dominated by gnarled trees no more than 12 feet tall.

The El Yunque rainforest is all that is left of a rainforest that covered much of Puerto Rico's mountains more than a century ago. Agricultural practices and heavy timbering have reduced it to the 10,000-acre preserve that makes up one-third of the Caribbean National Forest.

My preconceived idea of a tropical rainforest was of an impenetrable, vine-entangled jungle that could be traversed only with machete, with venomous reptiles watching and waiting at every step. In the El Yunque rainforest, nothing could be further from the truth. The trees form such a dense, closed canopy that scarcely any sunlight penetrates to the forest floor. As a result, there is little brushy undergrowth, and the vines grow quickly and vertically to the tops of the trees, seeking the sunlight they need for optimum leaf, flower, and fruit production. Only near streams or where a tree has fallen, opening the canopy to sunlight, does a junglelike growth of plants develop. As for poisonous snakes, there are none in the El Yunque rainforest, although the Puerto Rican boa, which may grow 8 feet long, lives here. Since this snake is rare enough to be considered endangered, the casual visitor is not apt to encounter it.

The Tabanuco Forest provides a new experience for persons accustomed to temperate regions. The trees form a continuous canopy up to 120 feet above the forest floor, with an occasional tree projecting even higher. The first side branches appear about halfway up the trunks, contributing to the openness of the understory. I was surprised to learn that growth of the tall rainforest trees is not like that of Jack's beanstalk but is relatively slow for some species. Foresters from Puerto Rico's Institute of Tropical Forestry have shown that in the mature rainforest, tabanuco stems may increase in diameter by only 0.125 inch per year.

The leaves of most rainforest trees and of the shade-loving plants in the understory are remarkably thin, facilitating photosynthesis in the diffused light that filters through the canopy. The bark of most rainforest trees is mottled by the growth of lichens, while their branches are crowded with epiphytes (air plants) of all sizes and shapes. Giant "bird's-nest" bromeliads and small, delicately flowering orchids share the upper branches.

Although the trunks of the tall trees are usually relatively slender, most of them flare out dramatically at the base to form buttresses. Much of the root system creeps over the ground, which is usually covered by fallen leaves that are rapidly broken down by microorganisms and fungi. These nutrients, along with those released by decomposition of the bedrock material, are absorbed by the feeder roots lying on or near the surface of the soil. Eventually these nutrients are replenished when leaves and other plant parts drop. Thus, the tropical rainforest plants set up a cycle of nutrients that enables them to thrive even when soil fertility is low.

To a botanist, perhaps the most interesting, as well as the most overwhelming, feature of the tabanuco rainforest is the great diversity of trees present. Although the tabanuco is the dominant species, 167 other kinds of trees have been recorded in the community, with 33 different types of trees appearing in any given acre. No forest anywhere in the mainland United States approaches this diversity, and only seven of the Puerto Rican species are also native to the U.S. mainland (all but one confined to southern Florida).

Above the Tabanuco Forest, in the cooler realm of the palo colorado tree, I observed one behemoth that stood 60 feet tall and had a trunk circumference of nearly 21 feet. The palo colorado's red-barked trunk is often hollow, serving as a favorite nesting site for the nearly extinct Puerto Rican parrot. Curiously, the tree is the same species (*Cyrilla racemiflora*) that grows as a native shrub from Florida to Virginia. Called the titi bush in the United States, it typically grows in Coastal Plain swamps. The reason for its very different stature in Puerto Rico is not known.

Epiphytes

In 1991, the United States Forest Service commemorated the 100th anniversary of the birth of the national forest system, for on March 30, 1891, President Benjamin Harrison created the Yellowstone Timber Land Reserve, later divided into the Shoshone and Teton national forests. Today there are 155 national forests, spread over 43 states and Puerto Rico. Their 275,000 square miles include nearly every North American forest type, as well as prairies, savannas, bogs, sandy beaches that front both the Atlantic and Pacific oceans, and even a tidal estuary. But only one, the Caribbean National Forest in eastern Puerto Rico, contains a tropical rainforest, El Yunque. And although it became a U.S. national forest in 1903, known as the Luquillo Forest Reserve, the Caribbean National Forest region was actually set aside as a preserve by the Spanish Crown well before, in 1860.

El Yunque rainforest occupies about 47 square miles in the rugged Luquillo Mountains, noted for their lush plant life and fast-flowing streams. Rain falls nearly every day, particularly on the highest mountains. At the lowest elevation is the lower montane rainforest, a dense growth of tabanucos and other 100-foot-tall trees. Above this is the upper montane rainforest, dominated by huge palo colorado trees. On the steepest slopes is the palm forest, consisting almost exclusively of Sierra palm trees. Finally, on the summits of the highest mountains, enshrouded by mist and clouds, is the elfin forest, where stunted trees grow in dense, nearly impenetrable stands.

Because the trees in the rainforest grow so thickly, permitting little sunlight to reach the forest floor, ferns and flowering plants are usually unable to survive beneath them. Instead, some of these plants find refuge in the upper canopy, closer to the source of light, where they live as epiphytes— plants that grow on other plants while producing their own food by photosynthesis. Epiphytes are a diverse category that also includes various algae, lichens, liverworts, and mosses, plants that lack specialized vascular tissues. But of the nearly 250,000 kinds of vascular plants in the world, roughly 1 in 10 are epiphytes, including both ferns and flowering plants. Only one large group of vascular plants, the gymnosperms (cone-bearing plants), are not represented among the epiphytes.

Epiphytic ferns and flowering plants grow most vigorously in the tropics, where rain, spray from waterfalls, or dew provides continuous moisture (only one, the resurrection fern, is commonly found in temperate North America). They differ from vines and lianas, which remain rooted in the soil as they climb trees. They also differ from parasites, such as dodder and mistletoe, which penetrate the host to obtain needed water and nutrients. The only burdens an epiphyte places on its host are competition for sunlight and possibly mechanical stress (in some tropical countries, epiphytes are routinely weeded off their hosts in parks and gardens to keep tree branches from breaking).

Some epiphytic ferns and flowering plants cannot live without a host plant in nature; others are able to live either on hosts or in the soil. The bark of a prospective host tree plays a major role in determining whether epiphytes will use the tree as a perch. Rough, spongy bark that retains water provides the spores of ferns and seeds of flowering plants a better opportunity to germinate and survive. Smooth bark, on the other hand, dries out quickly and discourages germination. Some trees even produce chemicals that keep the bark free of epiphytes.

Since epiphytic ferns and flowering plants do not have roots in the soil, they are always in danger of desiccation, especially when exposed to the tropical sun. To counter this, in some epiphytes the stomata (the tiny openings in the leaves) open only in the evening to admit the carbon dioxide needed for photosynthesis. The carbon dioxide is stored chemically and then used in the daytime. This same metabolic trick is found in certain desert plants.

Three major groups of vascular plants—ferns, orchids, and bromeliads—include epiphytes. Many epiphytic ferns, such as the birdnest fern, form a nest of broad leaves clustered together on a very recognizable stem. Dead leaves and other organic material fall into the nest and decompose, while hundreds of roots grow from the stem into the accumulated organic matter. The mass of humus and fern roots acts as a sponge that soaks up and stores rainwater. The rotting material also provides nutrients. In some epiphytic ferns that do not form leaf nests, the leaves curl up and become almost brittle during a prolonged drought, only to revive quickly when moisture once again becomes available. As its name suggests, the resurrection fern has this capability.

Most orchids that live in the tropics are epiphytes, in contrast to temperate orchids, which are all terrestrial. Epiphytic orchids have water-storing tissues in their succulent leaves or bulbous stems, while their aerial roots are surrounded by a spongy tissue of dead cells that keep them from losing water to the atmosphere. When drought becomes severe, a cell layer at the base of each leaf allows most orchids to shed their leaves to prevent further water loss. Both the water storage tissues and the root tips contain chlorophyll, enhancing the orchid's ability to carry out photosynthesis.

A third major group of epiphytes are bromeliads, or members of the pineapple family. The aerial roots of epiphytic bromeliads do not absorb water; they merely secure the plant to its host. The leaves, with expanded bases, are arranged in a dense, circular cluster, reminiscent of the nest of fern leaves but called a *tank*. The tank collects rainwater and fallen organic matter. Special scales or hairs that cover all or part of each leaf surface take up water and nutrients from the tank for distribution throughout the plant.

Epiphytic bromeliads are particularly abundant in the elfin forest on the highest peaks of the Caribbean National Forest. Among them are the shiny red guzmania and the showy vriesia, species frequently imported by florists in the temperate north to tempt their clients.

NATIONAL FORESTS
IN SOUTH CAROLINA

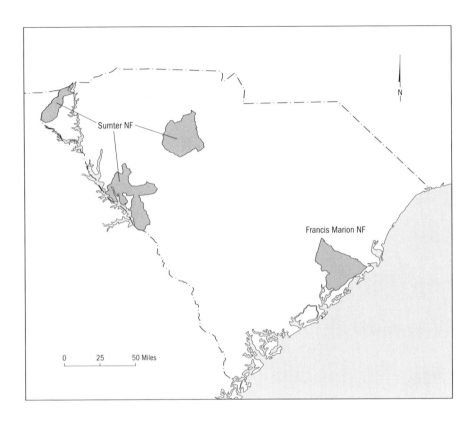

South Carolina has two national forests that cover all three physiographic provinces of the state. The Francis Marion National Forest is in the Coastal Plain Province, and the Sumter has ranger districts in the Piedmont and Mountains provinces. Both forests are administered by a single forest supervisor whose office is at 4931 Broad River Road, Columbia, SC 29210. They are in Region 8 of the United States Forest Service.

Francis Marion National Forest

SIZE AND LOCATION: Approximately 250,000 acres in southeastern South Carolina north of Charleston. Major access roads are U.S. Highways 17, 17A, and 701 and State Routes 41, 45, 48S, 133S, 171S, and 377S. District Ranger Stations: Cordesville and McClellanville. Forest Supervisor's Office: 4931 Broad River Road, Columbia, SC 29210, www.fs.fed.us/r8/fms.

SPECIAL FACILITIES: Boat ramp; swimming beach; motorcycle trail; horseback trail.

SPECIAL ATTRACTIONS: Sewee Shell Mound; Guilliard Lake Scenic Area; Santee Experimental Forest; Battery Warren.

WILDERNESS AREAS: Hellhole Bay (2,150 acres); Little Wambaw Swamp (5,154 acres); Wambaw Creek (1,937 acres); Wambaw Swamp (4,767 acres).

The area now occupied by the Francis Marion National Forest was once the region where Sewee Indians lived before the coming of European settlers; where the earliest settlers built plantations and grew indigo, rice, and cotton for 150 years; and where Hurricane Hugo in 1989 swept through, destroying many trees more than 9 inches in diameter. The live oak–lined King's Highway, in use when South Carolina was an English colony, still exists west of McClellanville. Historical churches and cemeteries are found along the road, including the handsome St. James Church.

The Sewee Indians took advantage of the rich estuaries of Bull's Bay for oysters and other seafood. One may visit the historic Sewee Shell Mound southeast of Awendaw, estimated to be more than 4,000 years old. While in the vicinity of Awendaw, look for lean-tos along U.S. Highway 17 where descendants of early slaves sell beautiful baskets woven from sedge leaves and pine needles.

North of the Elmwood Campground and situated between Wambaw Creek and the Santee River is the Watahan Historical Area. This is the site where an early French Huguenot, Daniel Huger, established a plantation in the 1690s. Just upriver from Watahan Plantation is Cedar Hill Plantation where a Mrs. Tydiman lived. In February 1782, a battle between General Francis Marion's men and British soldiers took place. General Gilbert Lafayette is said to have stayed at Watahan for several days while awaiting word on his request for asylum in the United States.

Less than 100 years later, during the Civil War, an area of earthen embankments and trenches known as Battery Warren was constructed near the Santee River about midway between the Watahan Plantation and Guilliard Lake. The Battery was built in 1863 by Confederate general Pierre Beauregard to protect the Santee River from the Union. The six guns Beauregard ordered from the Charleston Gun Factory were put in place at The Battery and fired upon a northern gunboat proceeding up the Santee River late in 1863. Many years after the Civil War had ended, the guns of The Battery were rediscovered, in place and loaded.

Scattered in and around the Francis Marion National Forest are several plantations, plantation sites, and historic churches worth visiting. Along the King's Highway north of McClellanville stands St. James Church, known locally as the "Old Brick Church." Erected in 1768, this venerable building is noted for its circular brick pillars, high-backed cypress pews, massive pulpit, and a communion rail constructed out of Santo Domingo mahogany. The church was formed when a number of French Huguenots merged with the Anglican Church at the fork of two roads, one fork leading to the English community and the other to the French community. The church has two front doors, one nearest the English road used by the Anglicans, and one nearest the French road used by the Huguenots.

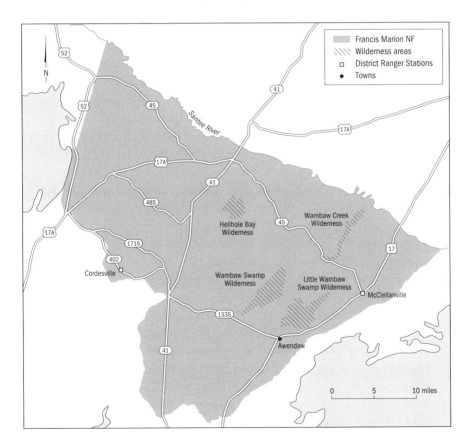

Plant communities of the Francis Marion National Forest range from coastal tidelands to black swamps, to pine forests on sandy ridges. Much of the interior of the national forest consists of swamps, where fine bald cypresses and other large swamp-inhabiting trees used to occur. The devastating Hurricane Hugo, which hit the area on September 21, 1989, destroyed many of these large trees in a 5-hour time span with winds as high as 135 miles per hour. In addition to the destruction of the trees, more than 500 miles of roads were blocked, trails were obliterated, and wildlife populations were depleted.

The swamp forests have now begun to recover, and four of the more pristine swamps are now in wilderness areas. The Wambaw Creek Wilderness is on either side of the creek, and there are two boat ramps for entrance into the creek. Wambaw Swamp Wilderness preserves a typical black swamp.

Longleaf, shortleaf, and loblolly pines are abundant on the sandy ridges that make up much of the forest, while pond pine is present in low, flat woods. Slash pine, somewhat removed from the remainder of its range, occurs in one isolated stand in a low damp woods in the vicinity of Shulerville.

More than 150 miles of streams and rivers traverse the forest, with much of the area adjacent to them occupied by swamps and bogs. The deep swamps, where quiet, black water prevails, are dominated by bald cypress and tupelo gum, although swamp cottonwood, water locust, water hickory, willow oak, and planer tree often occur as well.

In boglike pocosins, which are formed from undrained, shallow depressions in savannas, evergreen shrubs dominate. Many of the plants of the swamps are also in the pocosins, along with large gallberry, titi, southern bayberry, and sweet bay magnolia.

Rich, low woods adjacent to some of the streams exhibit dense, lush vegetation. Loblolly pine is common, along with tulip poplar, sweet gum, American beech, cherrybark oak, bitternut hickory, and green ash. These trees tower above a shrub zone of Virginia sweetspire, red bay, red buckeye, and inkberry, and an herbaceous layer of atamasco lily, crane-fly orchid, and Walter's blue violet. If you are lucky, you might encounter the beautiful-flowering silky camellia and loblolly bay in the low woods.

Of unusual biological and scientific value is the Guilliard Lake Scenic Area adjacent to the Santee River. The area includes slender, finger-shaped Guilliard Lake, an old oxbow. An old growth bottomland forest surrounds the lake on the east, south, and west. Huge cypress knees, some of them nearly 9 feet tall and 3 feet in diameter, are outstanding. A natural levee separating the northern border of the lake from the Santee River is characterized by a mixed hardwood stand of loblolly pine, American holly, sweet gum, overcup oak, water oak, and Carolina willow. West of the lake is Dutart Creek, lined in several places by limestone outcrops.

The Guilliard Lake Scenic Area has three trails. The 0.75-mile Lake Trail completely surrounds the lake and originates at the campground. The River Trail parallels the Santee River for nearly 1 mile. The Dutart Trail follows Dutart Creek, crossing it twice. Boating is permitted in Guilliard Lake.

Canal Recreation Area, adjacent to Lake Moultrie in the western side of the Francis Marion National Forest, features a swimming beach. A boat ramp at the Buck Hall Campground makes boating in the Intercoastal Waterway accessible, and there is a boat ramp at the Huger Recreation Area for access to the Cooper River from Huger Creek.

Edmund Campground just outside the northern end of the Wambaw Creek Wilderness is particularly attractive as it is situated beneath huge, moss-draped trees. An old forest fire lookout tower at Honey Hill, built during the Great Depression by the Civilian Conservation Corps, is the center of attention at the Honey Hill Campground.

The 20-mile-long Swamp Fox National Historic Trail begins at the Huger Campground, passes through the Halfway Campground about midway, and ends at Buck Hall Recreation Area near the community of Awendaw. The trail goes through most of the national forest's plant communities, with boardwalks and bridges where necessary, and is part of the Palmetto Trail that links the mountains to the sea.

Ocean Bay Research Natural Area contains several Carolina bays where Venus flytrap and the rare Savanna milkweed grow and where the endangered red-cockaded woodpecker lives.

Little Wambaw Swamp

A visit to the Francis Marion National Forest is like a trip back through history. Before the Civil War, nearly 100 plantations occupied this territory, their owners taking advantage of the low-lying land and cheap slave labor to become wealthy from rice farming. When slavery was abolished, the rice industry collapsed. The remains of the earthen dikes and floodgates that once controlled the flow of irrigation water are now barely detectable. Piles of rubble overgrown by dense vegetation are all that's left of Watahan Plantation; the only remnant of what was once the Mepken Plantation is a drive lined with live oaks draped with Spanish moss.

The forest also contains several sites and ruins from the Revolutionary War, for this is where General Francis Marion and his band of partisans developed the art of guerrilla warfare after the British drove the Continental Army out of South Carolina. Marion and his men harassed the British supply lines leading inland from Charleston to North Carolina; they would then disappear into nearby coastal swamplands for protection. Marion was so adept at these stealthy maneuvers that he became known as the Swamp Fox. One of his swampy havens may have been Little Wambaw Swamp, today part of a 5,223-acre area managed as a wilderness area by the Forest Service.

The swamp is a coastal plain depression that serves as a collecting basin. The outskirts of the swamp are covered part of each year with shallow water. Here the archetypal swamp trees—bald cypress and water tupelo—begin to appear, but the forest contains a variety of trees: water oak, laurel oak, red maple, sweet gum, and a scattering of cottonwood, hackberry, and loblolly pine. The undergrowth is often so thick that it impedes passage. Shrubby hollies, wax myrtle, and saw palmetto grow above a colorful understory of pickerelweed, lizard's tail, and wild iris.

Boggy depressions called *pocosins* dot the periphery of the swamp and support several kinds of evergreen shrubs. At Little Wambaw, the pocosins are dominated by sweet bay and large gallberry. Pitcher plants and wild orchids commonly grow in the acidic soil.

If you are not preoccupied with watching out for cottonmouths and several other snake species, you may catch a glimpse of parula, Kentucky, prothonotary, black-throated green, and hooded warblers (pl. 28) and red- and white-eyed vireos, as well as hear the distinctive call of the pileated woodpecker.

In the center of this wilderness, 60 swampy acres are dominated by bald cypress (pl. 29) and water tupelo trees. In places, the water stands up to 3 feet deep throughout the year. Where sunlight penetrates the dense canopy, duckweeds may form green patches on the water, and here and there the purplish frond of a mosquito fern floats on the surface. Fallen logs may harbor their own rich flora of mosses, pink St. John's wort, swamp beggarticks, and water horehound.

The bald cypress and the water tupelo are the trees best adapted to the continuously flooded conditions of the swamp's deepest recesses. Physiological experiments conducted on trees that grow in water show that these two varieties are the most capable of absorbing adequate amounts of nitrogen, phosphorus, potassium, and calcium from the saturated soil.

Bald cypresses, water tupelos, and other trees that grow in standing water often develop the familiar characteristic of swollen, seemingly "buttressed" trunks. These buttresses do not give the trees additional anchorage, however; the trunks taper to normal diameters below the water level. The swelling is greatest in the narrow zone where the trunk is nearly always in contact with both water and air; this may be a special circumstance that favors extra growth.

Another well-known but unexplained feature of the bald cypress is its tendency to form the blunt-tipped, conical woody structures known as *knees* that usually project above the water level in the swamp a short distance from the main trunk. The water tupelo may also form kneelike projections, particularly in the southernmost part of its range. The knees are actually roots that begin as small swellings on the upper surface of a horizontal root and grow very rapidly. Susceptible to a wood-rotting fungus, the knees often become hollow as they age.

A theory that has been around for more than a century is that the knees carry out an active exchange of gases, enabling the trees that live in standing water to "breathe." Microscopic examination of the texture of cypress knees, however, reveals no openings like those found in leaves or stems (called *stomatas* and *lenticels*, respectively). Botanist Paul Kramer and his colleagues at Duke University, seeking to test the theory of gas exchange a few years ago, covered cypress knees with inverted oil cans and sealed the bottoms of the cans with wax. After periodically testing the gases within the cans, they concluded that the exchange of gases by cypress knees was negligible.

Since bald cypresses planted outside standing water rarely produce knees, some botanists have suggested that the knees help support the trees in their watery habitat. These trees do grow more than 100 feet tall, with their roots embedded in unstable muck, so the idea that the knees serve as anchors is at least plausible.

Botanist Clair Brown of Louisiana State University has noted that the knees contain large quantities of starch. Apparently, glucose manufactured in the leaves of the bald cypress is transported to the knees, where it is converted into starch until it is needed. Perhaps, therefore, these enigmatic structures are really energy storehouses.

Sewee Shell Mound

A gently sloping zone that reaches from Long Island to Florida, the Atlantic Coastal Plain consists of sands, clays, and marls that were laid down over a very long period, from 140 million to 2 million years ago, as the ocean shoreline shifted back and forth. Located in a part of this low country immediately north of Charleston, South Carolina, is the Francis Marion National Forest, whose nearly 250,000 acres include bays, tidal marshes, upland ridges, and bottomland swamps. Among the forest's unusual attractions is the Sewee Shell Mound Historical Area, which includes the remains of two Indian shell middens, or refuse heaps, that lie close together in tidal marshland adjacent to the Cape Romain National Wildlife Refuge.

Human occupation of the region dates back to about 11,000 years ago, near the end of the last Ice Age, when Paleo-Indians living in small, mobile groups often camped along the coastal riverways. The animals they hunted included tapirs, mammoths, mastodons, and giant sloths. After the glacial ice retreated, the native population adapted to the warming climate, a rising sea level, and the disappearance of some of the large Ice Age mammals. Eventually, a pine forest came to dominate much of the Coastal Plain. From about 4,500 to 3,000 years ago, in the Late Archaic period, the inhabitants became increasingly sedentary, leaving behind shell middens, which also contained fragments of pottery.

The two middens are named for the Sewee Indians, who did not create them but lived in the region when Charles Towne was settled by the English in 1670. According to W. J. Rivers's 1856 history of South Carolina, a small group of Sewees remained until 1715 but were wiped out when the Yamasee War was fought to control rebelling Indians.

First studied in 1965 by archaeologist William Edwards, the older of the two middens at Sewee Shell Mound (pl. 30) consists of tons of oyster shells deposited within and on top of a low, sandy ridge. This midden is about 415 feet across and is ring- or doughnut-shaped. Edwards estimated the highest part of the ring to be about 15 feet tall, but souvenir hunters and perhaps others seeking fill for roadways may have reduced it. In addition to oyster shells, Edwards found the remains of snails, clams, scallops, crabs, rays, several kinds of fish, turtles, alligators, birds, deer, raccoons, and opossums.

Edwards speculated that the site was used as a base camp where Indians dug fire pits, storage pits, and refuse pits and built dwellings on the accumulating mound. Based on the amount of refuse and the type of pottery shards excavated, he estimated that the site was occupied for 150 to 400 years. A few pockets of charcoal in the midden yielded radiocarbon dates of about 3,300 years ago.

According to Robert Morgan, former national forest heritage program manager, current thinking is that the ring was a long-term habitation site or base camp and may have served as a gathering place where scattered groups came together to exploit the abundant coastal resources, share in communal and seasonal activities, and celebrate annual rituals. The ring shape may be the result of maintaining a cleared, central area for common use.

The second midden, which consists primarily of clam shells, dates to about 1,000 years ago.

I first visited the oyster shell ring in 1979, walking a short distance through forest to the tidal marsh. The marshland looked as though it would not support the weight of a hiker, but the water-logged sand was surprisingly springy. During high tide, however, some of the mound is covered by several inches of water.

The plant life that I saw in the tidal marsh consisted of small species, many of them succulents, adapted to life in brackish water. The lowest-growing plants, usually not more than 6 inches tall, included glassworts, whose jointed, cylindrical, leafless stems turn crimson in autumn; sea blite; sea purslane (a type of portulaca); a fleshy-leaved, pink-flowered member of the mustard family called sea rocket; a round-leaved creeping plant known as the smooth water hyssop; and a couple of salt-tolerant grasses—saltgrass and coastal dropseed. Growing up to a foot higher than these diminutive species were seaside goldenrod, the stiffly branched sea lavender, oxeye sunflower, narrowleaf loosestrife, and two kinds of asters. A trailing morning glory (fig. 16) is also present.

Figure 16. Morning glory, Sewee Shell Mound, Francis Marion National Forest (South Carolina).

The maritime forest adjacent to the mound is unusual, owing to the more alkaline soil created by calcium from the oyster shells. Species rare or absent elsewhere in Francis Marion National Forest flourish under these conditions, notably small-leaved buckthorn and basswood. Other trees that are somewhat unusual for coastal South Carolina are bitternut hickory and Shumard oak. The handsome climbing hydrangea crawls over many of the branches.

When I returned to the site in 2003, however, this maritime forest had been leveled by the winds of Hurricane Hugo, which roared through Francis Marion National Forest on the night of September 21–22, 1989. In addition, the site suffered a devastating wildfire in 1991. Since then, the U.S. Forest Service has constructed an interpretive trail leading to the ancient middens, which have been included in the National Register of Historic

Places since 1970. The oyster shell ring can now be viewed from a 120-foot-long boardwalk.

Sumter National Forest

SIZE AND LOCATION: Approximately 360,000 acres in central and western South Carolina. Major access routes are Interstate Highway 26; U.S. Highways 76, 176, 221, and 378; and State Routes 18, 23, 28, 33, 37, 53, 72, 81, 82, 86, 107, 121, 215, 283, and 389. District Ranger Stations: Edgefield, Mountain Rest, and Whitmire. Forest Supervisor's Office: 4931 Broad River Road, Columbia, SC 29210, www.fs.fed.us/r8/fms.

SPECIAL FACILITIES: Boat ramps; swimming beaches.

SPECIAL ATTRACTIONS: Chattooga National Wild and Scenic River; Long Cane Scenic Area; Broad River–Henderson Island Scenic Area.

WILDERNESS AREA: Ellicott Rock (8,274 acres, partly in the Chattahoochee and Nantahala national forests).

Two of South Carolina's physiographic regions, the Piedmont and the Appalachian Mountains, are represented in the Sumter National Forest. The third region, the Coastal Plain, is in South Carolina's Francis Marion National Forest. The Appalachian Mountains comprise the Andrew Pickens Ranger District, while the Piedmont has a northern district, the Enoree, and a southern district, the Long Cane.

From Ellicott Rock, southwest to the Tugaloo Lake Dam, the Chattooga River is in the beautiful mountains of the Appalachian system. A popular river for whitewater rafting, it has rapids ranging from rather mild ones to Class III that should be attempted only by experienced rafters. Several scenes from the film *Deliverance* were filmed along the river. On both sides of the river near its upper end is the rugged and scenic Ellicott Rock Wilderness. Ellicott Rock marks the spot where Georgia, North Carolina, and South Carolina come together. Scientist and astronomer Andrew Ellicott was commissioned to determine the exact location among the states, and in 1811 he chiseled "NC SC LAT 35 AD 1811" on the face of the rock. The rock may be reached by hiking the Ellicott Rock Trail from the Chattooga Picnic Area near the Walhalla Fish Hatchery on State Route 107; by hiking the Chattooga River Trail, which stays near the river from Burrell's Ford to the Rock; or by coming in from the Nantahala National Forest in the north from Bull Pen Road.

The Chattooga River may be accessed by taking Burrell's Ford Road (Forest Road 708), from a short distance south of Burrell's Ford Wayside Picnic Area on State Route 107, or by taking State Route 28, which crosses the river at the Russell Bridge. Burrell's Ford Road crosses the river and continues in the Chattahoochee National Forest in Georgia. There is a walk-in campground at Burrell's Ford. You may also access the Foothills National Recreation Trail at the ford. This 25-mile-long trail begins on Tamassee Knob in the Sumter National Forest, crosses Ocoee State Park, and then reenters the national forest where the trail crosses State Route 107. The trail then heads to the Chattooga River, which it parallels to Burrell's Ford before taking an easterly direction and following the southeastern boundary of Ellicott Rock Wilderness. It then crosses State Route 107 again at the Sloan Bridge Picnic Area and continues in the Nantahala National Forest in North Carolina, terminating at Whitewater Falls.

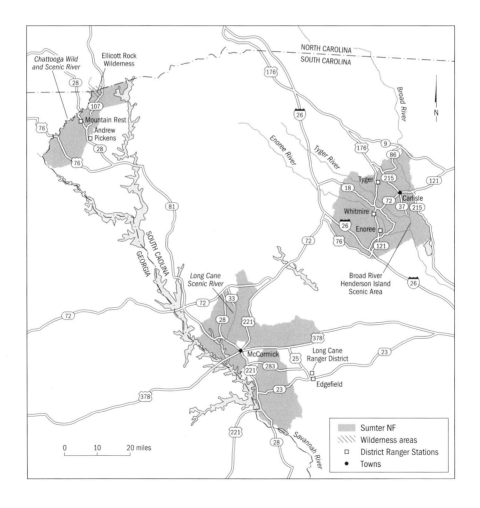

Tamassee Road (Forest Road 710) is an interesting back road that starts at a parking area along State Route 107 a few miles south of Cherry Hill Campground. The road twists and climbs all the way to Lake Cherokee (not in the Sumter National Forest) and eventually to State Route 95. Where Tamassee Road crosses Crane Creek, there is the southern terminus of Winding Stairs Trail, a rugged trail full of switchbacks that crosses State Route 107 at Cherry Hill Campground and continues to the Chattooga River where it connects with the Foothills National Recreation Trail. Several waterfalls are in this part of the Sumter National Forest, including Big Bend Falls south of Burrell's Ford and Lee Falls.

An 0.8-mile nature trail at Cherry Hill Campground passes through a forest of tulip poplar, red maple, flowering dogwood, mountain laurel, American holly, chestnut oak, white pine, hemlock, shadbush, witch hazel, rhododendron, and azalea. On the forest floor are lady fern, cinnamon fern, partridge berry, rattlesnake plantain orchid, and cucumber root. If you stray off the trail, you are apt to get entangled in thickets of greenbriers.

West of the Andrew Pickens Ranger Station is Yellow Branch Picnic Area with a shelter, picnic area, and a small waterfall nearby. A short nature trail will permit one to see white pine, shortleaf pine, black gum, red maple, white oak, pale hickory, and the shrubby farkleberry, or highbush blueberry.

U.S. Highway 76 crosses the southern end of the Andrew Pickens Ranger District. Toxaway Campground is situated along crystal-clear Toxaway Creek. From the campground you can explore Rockhouse Mountain, Longnose Mountain, and Grassy Mountain. At one place Toxaway Creek has carved a beautiful rocky gorge.

The Enoree Ranger District is the northern district of the two Sumter National Forest districts in the Piedmont. This district lies between the towns of Union to the north and Newberry to the south, a distance spanning about 30 miles. Before the area was acquired by the Sumter National Forest in 1936, its timber had been cut and the land well depleted. Through good forestry management, the forests have made substantial recovery. There are still many private parcels of land among the holdings of the Sumter National Forest. Several historical sites are included within the area.

The Tyger and Enoree rivers cross the forest west to east, both emptying into the Broad River near the eastern side of the district. Just south of the confluence of the Tyger and Broad rivers is the Broad River–Henderson Island Scenic Area. This 210-acre area preserves a nice section of the rolling Piedmont, ranging from alluvial river bottoms to steep bluffs. The bottomlands contain cottonwoods, sugarberries (fig. 17), bitternut hickories, black willows, musclewoods, and river birches, with brookside alders along the streams. In the uplands are loblolly pines, red oaks, southern red oaks, blackjack oaks, post oaks, slippery elms, and red mulberries. A nice forested area occurs on the west side of the Broad River under a steep bluff. Common in this forest are post oak, shellbark hickory, sweet gum, cherrybark oak, swamp white oak, tulip poplar, green ash, and sugarberry.

Figure 17. Sugarberry tree, Sumter National Forest (South Carolina).

The Broad River Recreation Area is near the northern border of the Enoree District near the junction of State Routes 86 and 389. In the recreation area are huge cottonwood trees on the Neal Shoals impoundment of the Broad River. Fishing is usually good in the Broad River. Across the river from the recreation area is Woods Ferry Campground with boat ramps.

State Route 72/121 crosses the Broad River 2 miles east of Carlisle. In the river is a chain of rocks that Indians placed as fish traps. During the Revolutionary War, the Americans defeated the British at this site, and General Thomas Sumter was severely wounded.

Where State Route 45 crosses the Enoree River at Keitts Bridge, there is a canoe access to the river, and where State Route 81 crosses the river at Brazzlemans Bridge, there is a Forest Service boat ramp.

Molly's Rock Picnic Area is at the end of a short spur road from U.S. Highway 176. This favorite picnic spot is set under a forest of sweet gum, flowering dogwood, shagbark hickory, black walnut, American holly, red mulberry, hackberry, and tulip poplar. Trumpet vines, with their large, orange, trumpet-shaped flowers, climb on some of the trees. Molly's Rock is in an area where a local girl is said to have preached to and converted 90 wild animals to Christianity.

Equestrians will find the 30-mile Buncombe Trail pleasant as it winds through typical Piedmont hills and valleys, crossing streams and passing old cemeteries, plantation sites, and abandoned wagon roads. A good starting point for this long trail is the Brick House Campground. Just off the northwestern corner of the horse trail is the Enoree off-road vehicle area.

John's Creek Lake, Sedalia Lake, Wildcat Lake, and Macedonia Lake are four small lakes in the northwestern corner of the Enoree District where fishing is said to be good. East of John's Creek Lake where State Route 18 leaves the Sumter National Forest is the hamlet of Cross Keys. Barrum Bobo, an English sailor and purser, erected a square, two-story brick building here between 1812 and 1814 with the purser's insignia of cross keys on its great chimney. The building is at the intersection of the old Piedmont Stage Road and Old Buncombe Road. The house also served as a stagecoach stop during the Civil War.

Just north of the junction of State Routes 34 and 72 is the Duncan Creek Presbyterian Church, built in 1763 by Scots-Irish emigrants from Pennsylvania. Numerous Revolutionary War soldiers are buried in the nearby graveyard.

The Long Cane Ranger District is along the western edge of South Carolina, separated from Georgia by the huge Clark Hill Reservoir. The district receives its name from a tall, native bamboo grass called giant cane that grows along rivers and streams in the area. With intensive agriculture, much of the cane has disappeared, but a fine stand occurs in the Long Cane Scenic Area a few miles southeast of Abbeville and reached by taking Forest Road 505E east off State Route 33. Giant cane grows in dense stands, known as canebrakes, which are difficult to hike through. The only native woody grass in the eastern United States north of Florida, giant cane is rarely seen in bloom or in fruit because the spikelets that bear the flowers and fruits are on short stalks from the main stem a foot or two above the ground. Some cane plants may reach a height of 15 feet, and their stems are popular for old-fashioned cane fishing poles. Like the century plant of the West, once a cane stem bears flowers, it then dies.

The 655-acre scenic area is in rolling to moderately steep terrain, with high bluffs along the main drainage of Long Cane Creek. In addition to the canebrakes along the creek, the scenic area also contains bottomland hardwood forests of black gum, sweet gum, water oak, red maple, cottonwood, and sycamore. There are also forested Appalachian coves, as well as stands of loblolly and shortleaf pines on the ridge. The Old Charleston Road, which at one time passed through the area, was once the main route from Abbeville to Augusta and the Port of Charleston. Now, the 26-mile Long Cane Loop Trail, originally constructed for horseback riding, goes through the scenic area. Within this area, too, is a huge shagbark hickory that measures more than 11 feet in circumference and more than 60 feet tall, with a crown spread of nearly 150 feet.

Less than 3 miles southwest of the Long Cane Scenic Area is 30-acre Parsons Mountain Lake, which features boat ramps, swimming beaches, and a campground. A hiking trail to the south climbs to the top of 832-foot Parsons

Mountain where there is an 80-foot-tall lookout tower. The Parsons Mountain Motorcycle Trail loops south of the lookout tower. The Living on the Land Trail near the lake passes under white oaks, American beeches, shagbark hickories, hop hornbeams, shortleaf pines, redbuds, red mulberries, flowering dogwoods, red maples, sugar maples, sweet gums, tulip poplars, black gums, northern red oaks, winged elms, blackjack oaks, black willows, water oaks, post oaks, white ashes, wild black cherries, persimmons, and Hercules' clubs. Christmas ferns are common on the forest floor, and wildflowers such as phlox, green dragon, violets, wild forget-me-nots, white bluets, and Virginia snakeroots bloom in season. If you venture off the trail, you may get tangled up in poison ivy, Virginia creeper, trumpet vine, and greenbriers.

Along State Route 38 south of Parsons Mountain Lake is the Long Cane Church near which the Cherokee Indians on February 1, 1760, overtook and massacred a party of settlers fleeing to Augusta. Catherine Calhoun, grandmother of John Calhoun, was killed. Two small girls, Mary and Ann, were captured. After 12 years, Ann was returned. A third girl, 15-year-old Rebecca, hid in the forest for 3 days. She later became the wife of General Andrew Pickens and mother of Colonel Andrew Pickens and Francis W. Pickens, both later governors of South Carolina.

Northwest of the village of McCormick is the John de la Howe School. De la Howe, a French medical doctor, founded the school in the late 1700s on 2,600 acres for 12 poor boys and 12 poor girls, most of whom were orphans. Among subjects taught were reading, writing, arithmetic, geography, geometry, matting, brewing, fixing colors, and making vinegar, soap, and cheese. He is said to be responsible for naming Abbeville. In his will, de la Howe stipulated the construction, dimensions, and epitaph for his tomb. After he died on January 2, 1797, his tomb was erected beneath a nice stand of shortleaf pine a short distance south on land surrounded by the Sumter National Forest.

West of Parsons Mountain and at the edge of the Sumter National Forest along Forest Road 521 is the Calhoun Cemetery and site of the Calhoun Farm. In the cemetery is buried Patrick Calhoun, his wife, and children. Calhoun was leader of the first settlement in the area in 1756 and the father of John C. Calhoun. Near the burial grounds was the site of one of the first log cabins in South Carolina and the birthplace of John C. Calhoun.

Southwest of the community of Edgefield at the edge of the Sumter National Forest is Lick Fork Lake Recreation Area with swimming beaches, boat ramps, hiking trails (including the Horne Creek Trail), and a campground.

The Forest Service also maintains a boat ramp and picnic area along the Savannah River at Furys Ferry at the southern tip of the ranger district. A short distance north on State Route 28 is a Forest Service lookout tower.

Chattooga River

Originating north of Highlands, North Carolina, the Chattooga River twists and turns its way through rugged, mountainous terrain, eventually reaching Ellicott Rock. As the river continues southward, forming the border between South Carolina and Georgia, the water becomes more rapid and full of boulders. Whitewater rafters from all over the country come to test their skills in the Chattooga River, particularly the 7-mile stretch between Russell Bridge and Earles Ford. On either side of these 7 miles, the river is too wild for the average river runner and should be avoided. Even the 7 miles recommended for rafting contain Class III rapids where waves are high, rocks and eddies are plentiful, and very narrow passages exist.

A good place to enter the river is at Russell Bridge where State Route 28 crosses the Chattooga River. Here the water is shallow and slow for about 500 feet until the mouth of West Fork is reached. For the next 1.5 miles, the water remains fairly calm until you come to Long Bottom Ford where you will probably have to pull your canoe over the low-water bridge. Beyond Long Bottom Ford, you will encounter small rapids as you approach Turn Hole. Just past Turn Hole is a series of rapids with narrow passages through rocky outcrops. From here to Big Shoals are a number of swift rapids and small pools. At Big Shoals is a pair of Class III rapids with an abundance of boulders and overhanging trees.

For the next 1.5 miles, you will paddle through continuous moderate rapids until encountering very turbulent water just before Earles Ford. Unless you are an expert rafter, you should take your canoe out of the river at Earles Ford because the Chattooga River becomes very boisterous as it proceeds southward.

Long before white settlers came to the area, the land surrounding the Chattooga River was a great hunting ground for game such as elk and bison. The vegetation also provided plentiful nuts, berries, and roots that the early natives survived on. During the middle of the 16th century, a permanent town known as Chattoogee was settled by the Cherokee Indians in the valley where State Route 28 crosses the Chattooga River today. Called Chattooga Old Town by English traders, it was a major town of the Cherokee nation. During the 1600s, it is estimated that as many as 1,500 people lived in and near the town. The small amount of flat land along the river was used to grow corn and squash.

In 1792, Andrew Pickens negotiated a treaty with the Cherokees that left the Indians with little land adjacent to the Chattooga River, and in 1816, the last Cherokee land was given up in the Treaty of Washington. Most of the Cherokees moved to the mountains, but one named Walter Adair wanted to remain in South Carolina and was given 640 acres of land in the valley of the

Chattooga. Unable to make it on his land, Adair sold his 640 acres in 1817 to William Clark, who sold the acreage in 1819 to Soloman Palmer.

During this early time when the land was changing hands frequently, two naturalists visited the area along the Chattooga River and reported their discoveries of plants not seen anywhere before. André Michaux and William Bartram made several significant discoveries, including the rare wildflower known as Oconee bells.

The land along the Chattooga changed hands several times until Ganaway Russell bought the 640 acres in 1867 for $1,200. Russell built a large two-story home and several outbuildings near State Route 28 where it crosses the Chattooga River. These structures remained intact and were of historical significance until the large home burned to the ground in 1988. For a while, the Russell home served as a stagecoach stop midway between Highlands, North Carolina, and Walhalla, South Carolina, where it was known locally as the Halfway House. It is alleged that when the stage reached the turnoff for the Halfway House, the driver would blow a trumpet. The number of passengers was indicated by the number of times the trumpet was blown. This signal let the Russell family know the number of beds to provide and the number of places at the dinner table. A Forest Service note said that the children of the family showed great ingenuity and industry. Often, when the parents and older members of the family were away, and travelers would come to the house for food and lodging, the young children, some under the age of 10, would pitch in and cook a complete meal, care for the travelers horses, provide beds, and send the travelers on their way well rested and well fed.

The Halfway House was a center for activities in the Chattooga River valley. It was the site of community meetings, a polling place in elections, and for a while a post office. The house was in operation as a traveler's rest stop until the mid-1920s.

Russell is said to have won many bets from travelers who would not believe the coldness of the spring water behind the house. A standing bet of $5 was that no one could keep his hand submerged in the spring for 5 minutes. Russell never lost a bet. The springs house kept milk and butter chilled and preserved for daily use. For long-term refrigeration and for ice during the summer, Russell built an ice house. Ice was cut from the river during the winter and stored in sawdust in the ice house.

In 1970, the United States Forest Service purchased the Russell home and 200 acres that surrounded it and developed it into a visitor center for the Sumter National Forest. The center had various displays describing the history of the house and the area until the house burned down in 1988. Today you may visit the site of the Russell house along State Route 28 just after crossing the Chattooga River.

NATIONAL FOREST IN TENNESSEE

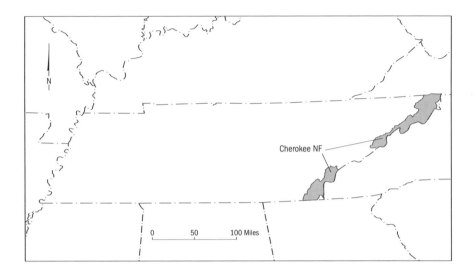

The Cherokee, the only national forest in Tennessee, is located in the central and southeastern parts of the state. This national forest is in Region 8 of the United States Forest Service.

Cherokee National Forest

SIZE AND LOCATION: Approximately 630,000 acres on either side of the Great Smoky Mountains National Park, between Chattanooga and Bristol, Tennessee. Major access roads are Interstate Highways 40 and 181; U.S. Highways 19E, 23, 25, 64, 70, 321, 411, and 421; and State Routes 30, 40, 67, 68, 70, 91, 107, 133, 143, 165, 315, and 395. District Ranger Stations: Benton, Greenville, Tellico Plains, and Unicoi. Forest Supervisor's Office: 2800 N. Ocoee Street, Cleveland, TN 37320, www.southernregion.fs.fed.us/cherokee.

SPECIAL FACILITIES: Boat ramps; swimming beaches; whitewater rafting areas; off-road vehicle (ORV) areas.

SPECIAL ATTRACTIONS: Unaka Mountain Auto Tour; Cherohala Skyway; Ocoee National Scenic Byway; Doe River Gorge Scenic Area; Rogers Ridge Scenic Arch.

WILDERNESS AREAS: Bald River Gorge (3,721 acres); Big Frog (8,082 acres); Big Laurel Branch (6,332 acres); Citico Creek (16,226 acres); Cohutta (36,977 acres, partly in the Chattahoochee National Forest); Gee Creek (2,493 acres); Joyce Kilmer–Slickrock (17,394, partly in the Nantahala National Forest); Little Frog Mountain (4,666 acres); Pond Mountain (6,890 acres); Sampson Mountain (7,991 acres); Unaka Mountains (4,496 acres).

The Cherokee National Forest occupies the Tennessee side of the Blue Ridge Mountains from Chattanooga to Bristol, being broken up in the middle by the Great Smoky Mountains National Park. Several mountain ranges are in the national forest, including Chilhowee Mountain, Bald Mountain, Rich Mountain, the Unaka Mountains, the Iron Mountains, and Holston Mountain. In addition to the mountains, the Cherokee National Forest is traversed by more than 1,000 miles of rivers and streams, including the Ocoee National Wild and Scenic River. Fishermen find plenty of fish to cast their lines for in the streams as well as the larger lakes such as Watauga, South Holston, Parksville, and Ocoee.

You will observe beautiful scenery and have access to the southernmost part of the Cherokee National Forest by driving the Ocoee National Scenic Byway (U.S. Highway 64) that follows the course of the Ocoee River through the Ocoee River Gorge from Ducktown to Parksville. Along the river, which has been dammed to form Parksville and Ocoee lakes, and around the lakes are several Forest Service campgrounds and places to put in the river for whitewater rafting. Where the scenic byway enters the Cherokee National

Forest west of Ducktown, stop at the Boyd Gap Observation Site for scintillating scenery.

The scenic byway enters the Ocoee River Gorge, with Little Frog Mountain Wilderness to the northeast and Big Frog Wilderness to the southwest. The road passes the Ocoee Whitewater Center where the 1996 summer

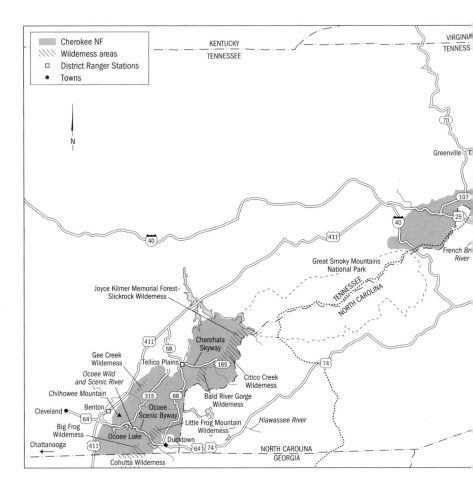

Olympic Games held their slalom canoe/kayak events. The byway twists along the Ocoee River, eventually reaching Parksville Lake. Just before the Parksville Beach Campground, the road turns north off U.S. Highway 64 and climbs into the Chilhowee Mountains on State Route 40.

Eventually the Ocoee National Scenic Byway leaves the Cherokee National Forest just before State Route 40 comes to Benton. A worthwhile side trip is to take the Oswald Dome Road that heads north off the scenic byway before it exits the national forest. The Oswald Dome Road goes to a Forest Service lookout tower. There is a branch road to the Chilhowee Seed Orchard where tree seedlings are grown for distribution into forests of the southeastern United States. A hiking trail south of the seed orchard follows Rock Creek through the Rock Creek Gorge Scenic Area. This area is notable for Benton Falls, a gorgeous cascade over rocky outcrops where Rock Creek falls for 65

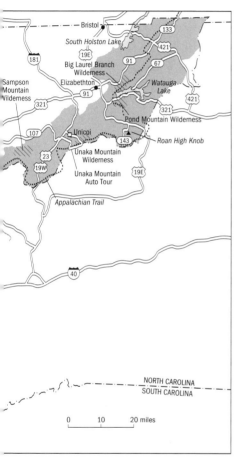

feet. From a parking lot in the Rock Creek Gorge Scenic Area, the Benton Falls Trail passes a small lake before continuing for 1.3 miles to the falls. Several slippery wooden steps lead to the base of the falls.

Big Frog Wilderness lies south of U.S. Highway 64 and extends to the Georgia state line. In fact, a very tiny part of the wilderness (89 acres) is in the Chattahoochee National Forest. Tumbling Creek Campground east of the wilderness and Sylio Campground west of the wilderness are good bases for hikers. The dominant feature of the wilderness is 4,224-foot Big Frog Mountain, and there are trails along Licklog Ridge and Folk Ridge as well.

Little Frog Mountain Wilderness, bordered on its southwest side by the Ocoee National Scenic Byway, has a nice hiking trail. Little Frog Mountain, at the northern edge of the wilderness, is accessible via the Kimsey Highway (Forest Road 68). A lookout is on Little Frog Mountain's 3,322-foot summit.

The Hiawassee River cuts across the Cherokee National Forest less than 10 miles north of the Ocoee River. State Route 30, which is the western segment of the Kimsey Highway, follows the southern side of the Hiawassee River from the Quinn Springs Campground on the west to its junction with State Route 315 and Forest Road 108. About 1 mile east of Quinn Springs, a short hike south along Lowry Branch will bring you to Lowry Falls, a neat little 10-foot waterfall that drops into a shallow catchpool.

From the junction of State Routes 30 and 315 and Forest Road 108, you will eventually come to the eastern border of Gee Creek Wilderness. Chestnut

Mountain Trail follows this boundary of the wilderness, and Starr Mountain Trail follows the western boundary. By continuing north on State Route 315, you will have access to Bullet Creek and Yellow Creek falls. Bullet Creek drops approximately 35 feet in a rather narrow falls, while broader Yellow Creek falls nearly 30 feet into a rocky gorge. You may hike from one falls to the other.

At the junction of State Routes 30 and 315, Forest Road 108 soon comes along the north side of the Hiawassee River where there are four campgrounds within less than 3 miles. Although the forest road ends at the last campground, the John Muir National Recreation Trail continues to follow the river all the way east to State Route 68.

The areas along either side of the Hiawassee River are very scenic, with several interesting trails and roads available for the hiker and driver. South of the river is the Turtletown Creek Falls Scenic Area, while Coker Creek Scenic Area lies to the north. Turtletown Creek Falls Scenic Area is in the historic Copper Basin mining district where copper mining was the major industry of the area in years past. From the tiny community of Farner on State Route 68, take Forest Road 1166 west to a parking area at the end of the road. In a little more than 1 mile, the hiking trail from the parking area comes to a view of the 30-foot Turtletown Falls that drops in two segments. The trail then crosses rapidly to the base of the falls. By hiking another 0.5 mile past Turtletown Falls, you will be able to see Lower Tower Falls; this cascade drops gently for about 50 feet but with a width that is even wider.

By driving 7 miles north of the Hiawassee River bridge on State Route 68, you will come to County Road 628 to Coker Creek Falls. In the vicinity are four small but picturesque waterfalls—Coker Creek Falls, Upper Coker Creek Falls, and Hiding Place Falls, all along Coker Creek—and Towee Falls along Towee Creek, a few miles north of Coker Creek.

The community of Tellico Springs is situated along the Tellico River at the western edge of the Cherokee National Forest. State Route 68 to the south eventually comes to the Coker Creek Scenic Area, but 2.5 miles south of town, Forest Road 341 veers to the right on to Bakers Grave Mountain. A 0.5-mile trail through a fine forest dominated by hemlocks goes to Conasauga Falls where Conasauga Creek drops in three stages for a total of about 50 feet.

State Route 165 east from Tellico Plains is the scenic Cherohala Skyway (pl. 31) that crosses the Cherokee National Forest for 20 miles before entering the Nantahala National Forest in North Carolina. The first side road to the south in 1.5 miles brings you to the northern end of Warrior's Passage National Recreation Trail that proceeds for a few miles.

From near the Tellico Ranger Station, Forest Road 210 twists its way to spectacular Bald River Falls at the north end of Bald River Gorge Wilderness. From the bridge over the river, one gets a face-on view of thunderous Bald

River Falls. This broad falls plummets nearly 100 feet into a rocky gorge. Forest Road 210 parallels the Tellico River (pl. 32), which takes several drops near the falls and is crowded with tourists and whitewater enthusiasts during the summer. The paved road then continues for several miles to the North Carolina state line, passing seven Forest Service campgrounds. Fishing is good in the streams adjacent to each campground, and a pleasant hiking trail is at Pheasant Fields Campground.

Forest Road 126 from near the Pheasant Falls Campground heads west to Holly Flats Campground at the southern edge of Bald River Gorge Wilderness. Before reaching the campground, a 2.5-mile trail south along Brookshire Creek comes to Upper Bald River Falls. To see the nearly 10-foot falls, however, you must ford a small creek as well as the Bald River.

The Cherohala Skyway veers northward to Indian Boundary Lake, where there is a campground and a nice hiking trail around the lake. The lake provides a great opportunity to fish for bream, bass, catfish, rockfish, and muskie. A swimming beach is located at the northeast corner of the lake. The area gets its name from a boundary created by a treaty in the early 1800s between the Cherokee Indians and the U.S. government.

The skyway then turns sharply south and then east, forming the southern boundary of the Citico Creek Wilderness before entering North Carolina. The wilderness is crossed by several picturesque streams and hiking trails, and it contains Falls Branch Falls, one of the prettiest in the national forest. The falls rushes for 60 feet into a rocky splash pool. To view these falls requires a 1.5-mile hike from the trailhead along the north side of the Cherohala Skyway, about 1.5 miles west of the North Carolina state line. The trail passes through a virgin forest of white ash, tulip poplar, wild black cherry, hemlock, and sugar maple.

The eastern edge of Citico Creek Wilderness merges with the western edge of the Joyce Kilmer–Slickrock Wilderness, the latter extending into the Nantahala National Forest. A hiking trail is all that separates the two wilderness areas. At the extreme northeast corner of the Cherokee National Forest's part of the Joyce Kilmer–Slickrock Wilderness are two neat waterfalls. Lower Falls of Slickrock Creek makes a 15-foot narrow plunge into a catchpool below, while 3.5 miles farther southeast along Slickrock Creek Trail is Wildcat Falls, which drops for a total of 20 feet in two separate plunges. Wild columbine occurs on rocks next to the falls. To reach the trailhead to these two falls, take U.S. Highway 129 in North Carolina and cross the Little Tennessee River bridge. The trail begins at the south end of the bridge, a short distance north of the hamlet of Tapoco.

The southern and northern regions of the Cherokee National Forest are separated by the Great Smoky Mountains National Park. The northern region, consisting of the Nolichucky, Unaka, and Watauga ranger districts, extends from the Great Smoky Mountains National Park along the Blue Ridge Mountains to the Virginia border. The westernmost part of the northern division consists of Stone Mountain, located between the Pigeon River to the south and the French Broad River to the north. You may reach this region from Interstate Highway 40 and U.S. Highway 25/70 south and east of Newport, but there are no developed Forest Service areas.

By driving U.S. Highway 25/70 east of Newport and Stone Mountain, you will follow the French Broad River, with the Meadow Creek Mountains to the north and the western edge of the Bald Mountains to the south. This part of the Bald Mountains may be reached by taking State Route 107 where it branches south off U.S. Highway 27/70. There is a Forest Service lookout tower on Round Mountain and the Round Mountain Campground nearby. The Appalachian Trail passes a short distance east of the campground.

To visit the Meadow Creek Mountains and then the main range of the Bald Mountains, take State Route 107 north of the French Broad River, staying in the Cherokee National Forest. Houston Valley Campground is along the south end of Meadow Creek Mountain, including a hiking trail as well. State Route 107 continues into the Bald Mountains. South of the highway are several campgrounds and picnic areas in beautiful mountain settings, along with some pretty waterfalls along Paint Creek. One-half mile south of Moses Turn Picnic Area on Forest Road 41 is Kelly Falls, where Paint Creek drops for about 25 feet. Two miles west near the Dudley Falls Picnic Area is a small falls that drops into a serene pool where swimming is popular. Stephen Falls is just a short distance to the west.

State Route 107 turns north, where it junctions with State Route 70 and the Forest Service's Nolichucky Work Center. The Appalachian Trail crosses State Route 70 after a few miles south of the work center. Just before reaching the trail, a good road follows Paint Creek into the Bald Mountains, eventually reaching the edge of Bald Mountain Ridge Scenic Area. The Appalachian Trail follows along the southern boundary of the scenic area. Along the northern edge of the scenic area are Old Forge and Horse Creek campgrounds, both with hiking trails. Adjacent to the scenic area to the east is Sampson Mountain Wilderness.

Within the Bald Mountain Scenic Area are three waterfalls worth seeking. Petes Branch Falls, side-by-side falls each 45 feet tall, may be reached by trail from the Horse Creek Campground. Davis Creek Falls and Margarette Falls are accessed by trail from the Round Knob Campground. Margarette Falls is particularly beautiful as Dry Creek tumbles over a rocky precipice for nearly 60 feet.

State Route 107 eventually takes an eastward course and parallels the Nolichucky River. Forest Road 25 branches south off State Route 107 and follows Oak Creek to the historic site of the Clarkville Iron Furnace. By continuing on Forest Road 25, you will come to Sill Branch Falls and Buckeye Falls. A 0.5mile hike will bring you to Sill Branch Falls, which drops as much as 30 feet in two side-by-side falls. Buckeye Falls consists of several falls separated by nearly horizontal runs of water. Counting all of the flows, Buckeye Falls drops a total of about 500 feet, but it is not as spectacular as it sounds.

The Unaka Mountains, partly in Tennessee and partly in North Carolina, are a few miles south of Johnson City, Tennessee. Two of the more beautiful parts of this mountain range in Tennessee are in the Unaka Mountains Scenic Area and the Unaka Mountains Wilderness. State Route 395 from Erwin passes the western edge of the scenic area and provides access to Rock Creek Campground. You may hike into the scenic area and the wilderness from the campground. Nearby Rock Creek Falls consists of an upper wide falls and a lower narrow falls, together dropping about 50 feet into a small, rocky gorge. A short distance to the east is 60-foot Red Fork Falls, which drops in a series of stair steps. The Appalachian Trail follows the southern boundary of the scenic area and wilderness. By continuing on State Route 395, you will intersect Forest Road 230, the start of the Unaka Mountain Auto Tour. By turning left, this roadway follows the state line between Tennessee and North Carolina and stays along the southern edge of the scenic area and wilderness, ending at State Route 107.

Several miles east of the Unaka Mountains Wilderness and south of the community of Roan Mountain, Tennessee, is Roan High Knob, one of the prettiest areas in any national forest. Roan Mountain is noted for the bald on its summit and its unusual vegetation, including the incomparable Catawba rhododendron. (The best group of these magnificent shrubs is at Roan Mountain Gardens across the state line into North Carolina in the Pisgah National Forest.)

U.S. Highway 19E through Elizabethton separates the Unaka and Watauga ranger districts. The Watauga Ranger District east of the highway includes Holston Mountain to the north and the Iron Mountains to the south.

Large South Holston Lake is at the northeastern edge of Holston Mountain. The Forest Service maintains two campgrounds around the lake, with boat ramps at Little Oak and a swimming beach at Jacobs Creek. Holston High Knob along the Appalachian Trail provides for outstanding views of the surrounding areas. On the east side of Holston Mountain is the Stony Creek Scenic Area, with the Appalachian Trail passing along its east side. State Route 91 is in the valley between Holston Mountain and the Iron Mountains.

About 10 miles north of Elizabethton, Panhandle Road leads off west from State Route 91. In less than 1 mile is Blue Hole Falls, where Mill Creek drops four times for a total of 50 feet. The trail to the falls is only 0.1 mile long. Where State Route 91 crosses U.S. Highway 421, it becomes State Route 133.

Just before reaching the Virginia state line, State Route 133 goes through an opening in Backbone Rock. The vertical rock through which the highway passes is about 75 feet tall. Just before reaching the tunnel from the south, a very short trail to the east will bring you to Backbone Falls, where a scenic 45-foot waterfall is on a small tributary to Beaverdam Creek.

Lake Watauga, built by the Tennessee Valley Authority, is a few miles east of Elizabethton. Several campgrounds and boat ramps are maintained by the Forest Service along the lake. On the north side of the lake and northeast of the dam is Big Laurel Branch Wilderness in the Iron Mountains. The Appalachian Trail crosses the center of the wilderness from north to south. South of Watauga Lake is Pond Mountain Wilderness. County Road 50 (Dennis Cove Road) is just outside the southwestern edge of the wilderness and provides access to it and to a number of waterfalls in the immediate vicinity. North of Dennis Cove Road are Laurel Falls and Middle Laurel Falls, while south of the road are Coon Den Falls, Dennis Cove Falls, and Upper Laurel Falls. All are impressive, but Laurel Falls is the most spectacular as it drops about 50 feet over a wide rock face. Laurel Falls is near the Dennis Cove Campground and can be reached by a 1-mile hike. The Appalachian Trail passes close to each of the falls. South of Dennis Cove Campground and reached by the Appalachian Trail is the Forest Service's Whiterock Lookout Tower.

Southwest of the Dennis Cove Road is the Doe River Gorge Scenic Area where Doe River has formed a picturesque and rugged gorge between Fork Mountain and Cedar Mountain.

County Road 50 continues for many miles past the Dennis Cove Campground, eventually arriving at the North Carolina state line. Just north of the road and on the Tennessee side is lovely Jones Falls, but it is difficult to get to because of the absence of well-marked trails. A few miles north is Twisting Falls where the Elk River splits into two impressive falls side-by-side for nearly 40 feet. Twisting Falls is accessible from the north by following U.S. Highway 321 to Elk Mills and County Road 2607 south.

The Cherokee National Forest has an isolated tract in the northeastern corner of Tennessee where Tennessee, North Carolina, and Virginia come together. Most of the Forest Service land is in the Rogers Ridge Scenic Area, highlighted by Gentry Creek Falls. The falls are actually two parallel falls of 40 feet each. To reach the falls, you will need to hike more than 2 miles and ford Gentry Creek several times.

NATIONAL FOREST IN VERMONT

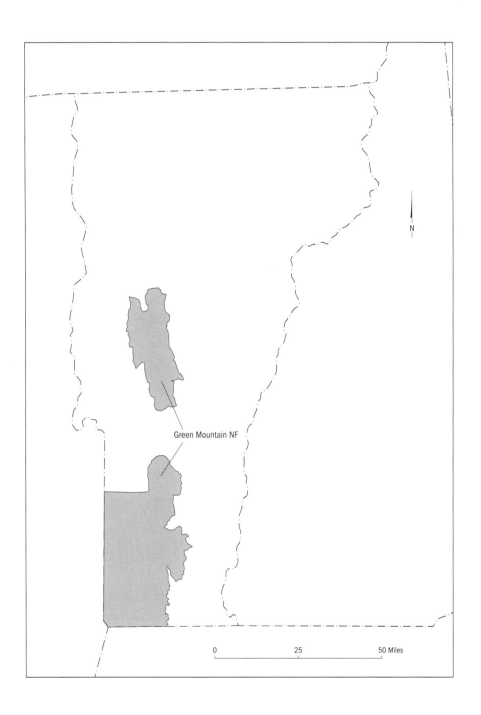

N

Green Mountain NF

0   25   50 Miles

Green Mountain National Forest, the only national forest in Vermont, occurs primarily in the central part of the state. It is in Region 9 of the United States Forest Service.

Green Mountain National Forest

SIZE AND LOCATION: Nearly 400,000 acres, from Bristol, Vermont, on the northern end, to the Massachusetts border on the southern end along the Green Mountain ridge. Major access routes are Interstate Highways 89 and 91; U.S. Highway 7; and State Routes 9, 11, 30, 73, 100, 116, 126, 140, 155, and 165. District Ranger Stations: Manchester Center, Middlebury, and Rochester. Forest Supervisor's Office: 131 N. Main Street, Rutland, VT 05701, www.fs.fed.us/r9/gmfl.

SPECIAL FACILITIES: Swimming beach; boat ramp; winter sports areas; historic fire tower.

SPECIAL ATTRACTIONS: Hapgood Pond; Texas Falls; Robert Frost Wayside, Trail, and National Historic Site; White Rocks National Recreation Area; Glastenbury Mountain; Robert Stafford National Recreation Area.

WILDERNESS AREAS: Big Branch (6,720 acres); Breadloaf (21,480 acres); Bristol Cliffs (3,738 acres); George D. Aiken (5,060 acres); Lye Brook (15,680 acres); Peru Peak (6,920 acres).

As its name indicates, the Green Mountain National Forest consists of a major part of the Green Mountains in south-central Vermont. These mountains form a spine nearly 200 miles long down the center of the state. The Green Mountains are a very ancient range that has been eroded down to rounded summits and gentle to relatively steep slopes. They feature sparkling mountain streams, spectacular rocky gorges, large exposures of white limestone, and pleasant waterfalls. Black bears still roam the national forest, along with white-tailed deer, moose, squirrels, porcupines, and rabbits. Water is plentiful with more than 3,000 acres of ponds and lakes and nearly 450 miles of streams, most of them teeming with trout.

The Green Mountain National Forest is divided into two distinct parts separated in the middle by Killington and Pico ski resorts. Rutland, Vermont, the location of the forest supervisor's office, is also in the separation of the north and south halves of the forest. The northern area consists of two ranger districts, Middlebury and Rochester, while the southern area is in a single ranger district, Manchester.

Following occupation of land in what is now the Green Mountain National Forest by Algonquin tribes and the Iroquois Federation, European settlers began arriving in the 1770s. These early pioneers arrived at an area heavily forested by tall white pines, hemlocks, red spruces, and hardwood trees

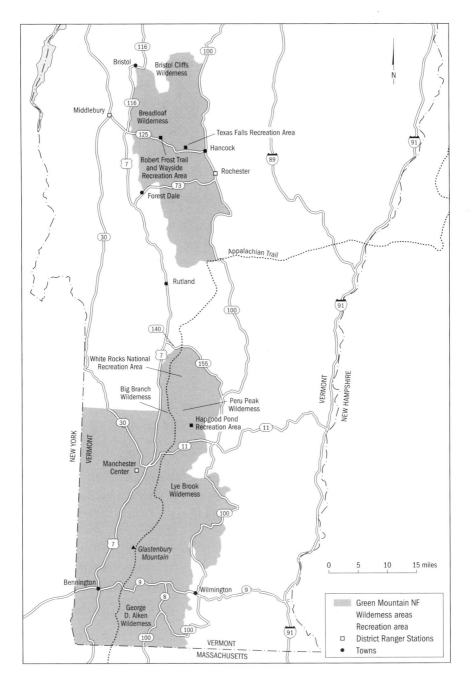

that included sugar maples, American beeches, birches, and oaks. Logging was heavy as the forests were being cleared for farming and sheep grazing. By 1932, the original forest consisted of cut-over wasteland and often abandoned, depleted farmland.

The United States Forest Service, in conjunction with the Green Mountain Club, maintains the 127 miles of the Long Trail, which passes through the entire length of the national forest. The first formal hiking trail in the United States, the Long Trail (fig. 18) was completed in 1931. It is known as the "Footpath in the Wilderness" because it passes through four wilderness areas in the Green Mountain National Forest. The trail actually runs along the crest of the Green Mountains from Canada to Massachusetts, a distance of 255 miles. A part of it is also used by hikers on the Appalachian Trail, which comes into Vermont at the Massachusetts border and splits off just outside Killington, Vermont, toward New Hampshire and Maine.

The Green Mountain National Forest offers three-season recreational experiences. Summer is a great time to visit for backcountry hiking, camping either in formal or primitive campgrounds, fishing, birding, and wildlife watching. Fall offers hiking, camping, and hunting. Wintertime is the most visited, with plenty of downhill and cross-country skiing, snowshoeing, sledding, and maple tapping. Spring is wet and muddy, and the black flies make most outdoor adventures unpleasant.

Although there are no through roads north to south in the Green Mountain National Forest, several paved roads cross the forest from east to west and provide fairly easy access to most parts of the national forest.

At the northern end of the Green Mountain National Forest is Mount Ellen, the highest elevation in the national forest at 4,083 feet. The Long Trail crosses the mountain. A short distance south of Mount Ellen is Mount Abraham, just a few feet less in elevation. Between these two mountains to the east is Sugarbush Valley, one of the forest's popular ski areas. A back road crosses the national forest connecting the villages of Lincoln and Warren, skirting south of Mount Abraham and then following picturesque Lincoln Brook.

South of the Lincoln Brook Road and extending nearly to State Route 125 is the Breadloaf Wilderness, the largest in the Green Mountain National Forest. Breadloaf Mountain, near the middle of the wilderness, is the highest point in the wilderness at 3,835 feet, but other nearby mountains are nearly as high. Mounts Cleveland, Grant, Roosevelt, and Wilson, all with trails to their summit, are also in the wilderness. This area also contains the headwaters of the New Haven and White rivers. As you hike, be aware of black bears and moose.

Outside the eastern edge of the Breadloaf Wilderness is the village of Granville on State Route 100. As you drive through scenic Granville Gulf, a tiny parking area will provide access to Moss Glen Falls, which results from a small stream dropping over large boulders for about 30 feet. You may spot the falls from the highway. From the parking area, a fenced boardwalk will take you to a closer view without having to step over slippery rock surfaces.

At the northwest corner of the Green Mountain National Forest, immediately south of Bristol, is the Bristol Cliffs Wilderness. Consisting of 3,740 acres, the wilderness allows visitors spectacular views of the Champlain Valley over the steep talus slopes on the west face of South Mountain. North Pond and Gilmore Pond are picturesque bodies of water on the mountain. This scenic area is seldom visited, however, because it has no developed trails.

Figure 18. Hiking on the Long Trail, Green Mountain National Forest (Vermont).

The northernmost state highway to cross the Green Mountain National Forest is State Route 125, from Middlebury to Hancock. A few miles east of East Middlebury is the Robert Frost Wayside, Trail, and National Historic Site. The site has a red pine forest that the poet, who lived in a nearby cabin, planted in the mid-1940s. In the village of Breadloaf, south of State Route 125, are the Sugar Hill Snowmobile Trails, and a short distance to the east is the Middlebury College Snow Bowl. The highway veers slightly south, dropping into Middlebury Gap and paralleling Robins Branch. State Route 125 next comes to impressive Texas Falls, where Hancock Branch plummets over a rock ledge. There are several picnic tables along the east side of Hancock Branch, and a short interpretive nature trail goes from the picnic area to the falls. The trail passes through a forest of sugar maple, hemlock, white pine, birch, and American beech. A common shrub with maplelike leaves is known as witch hobble, a species of *Viburnum*.

State Route 53 forms the western border of the Green Mountain National Forest south of East Middlebury, with Chandler Ridge to the east in the national forest and Lake Dunmore to the west. Paralleling State Route 53 about 3 miles to the east is a back road from Ripton to Goshen that passes the Falls of Lana. This impressive falls drops over a ledge and cascades between two solid rock walls that flank either side of it. The falls drops in stages, each with a catchpool at the base. A trail leads to the top of the falls and provides a splendid view of not only the falls but also the surrounding countryside, including Lake Dunmore. The Moosalamoo Campground southeast of Mount Moosalamoo is along the way.

State Route 73 crosses the Green Mountain National Forest from Forest Dale to Talcville. For much of the way, the highway follows scenic Brandon Brook, dropping into Brandon Gap with Goshen Mountain rising to the south and White Rock Mountain to the north. East of Brandon Gap is the Brandon Branch Picnic Area. Forest Road 45 follows Chittenden Brook south for about 4 miles, ending at Chittenden Brook Campground. Nearby is the Chittenden Winter Sports Area.

Between State Route 73 and U.S. Highway 4 at the south edge of the northern unit of the Green Mountain National Forest, there are no roads that cross the forest and few that even penetrate it. Within this primitive area are Farr Peak, Bloodroot Mountain, and Mount Carmel, all reached by the Long Trail. Chittenden Reservoir and the smaller Lefferts Pond are south of Mount Carmel. Blue Ridge Mountains, a small range southeast of the village of Chittenden, are in the Green Mountain National Forest.

State Route 40 curves around the northern part of the south district of the Green Mountain National Forest about 10 miles south of Rutland. In this area is perhaps the most striking scenery to be found in the national forest: the White Rocks National Recreation Area. From a considerable distance, the white limestone rock faces of the cliffs may be seen. Several rocky trails, often difficult because of the rock-strewn surface, wind through the recreation area. The Long Trail, which crosses the area, is the best way to get a firsthand look at this impressive region. At the north end of the White Rocks National Recreation Area is a short trail from the White Rocks Picnic Area near State Route 140 to the White Rocks, passing the interesting Ice Beds.

To explore the uppermost part of the southern district, drive along Forest Road 10, which begins at the town of Darby on Interstate Highway 89. This highly scenic road follows Big Branch in a northerly direction, coming to Big Branch Campground and Black Branch Campground. South of the Big Branch Campground is the northern edge of Big Branch Wilderness. In addition to Big Branch Stream, lovely Lake Brook is also in the wilderness. The steep, heavily wooded mountain slopes make climbing to the summit difficult. A mixed hardwood forest is the main community type, including fine specimens of American beech, sugar maple, birches, red spruce, balsam fir, and hemlock. Black bears are not scarce in this wilderness. It also contains wetlands where moose are often seen, particularly in the Elbow Swamp region. Five miles of the Long/Appalachian Trail system pass through the wilderness.

East of Big Branch Wilderness and extending south and nearly to Hapgood Pond is Peru Peak Wilderness, one of the most remote regions in the Green Mountain National Forest. This wilderness is similar to Big Branch Wilderness in that it has steep-sloped mountains, including Pete Parent Peak, Peru Peak, and Styles Peak, and wetlands around Big Mud Pond and Little Mud Pond and along Utley Branch. Most of these areas may be reached via marked hiking trails except Pete Parent Peak, which has no developed trails to its summit.

Forest Road 10 eventually crosses Kilns Brook before coming to a formation known as Devil's Den Cave, which you may explore carefully. Forest Road 10 then follows Mount Tabor Brook south, skirting the eastern edge of Peru Peak Wilderness and coming to a junction with Forest Highway 3. To the west is the Hapgood Pond Recreation Area, the most developed site for activities in the Green Mountain National Forest. This 7-acre impoundment offers a beach for swimming, a ramp for boating, and many fish for fishing, plus a campground as well as ample picnicking facilities. A forest trail, nearly 1 mile long, circles around the north side of Hapgood Pond, connecting the campground with the picnic area. Originally used as a grist mill, Hapgood Pond used to feature a water wheel. South and west is the Bromley Winter Sports Area on the north side of State Route 11. Across the highway is the Snow Valley Winter Sports Area.

South of Manchester is the 14,300-acre Lye Brook Wilderness. Most of the wilderness is in primitive woodland that is crossed by the Long/Appalachian Trail (pl. 33). Bourne Pond, near the southeastern corner of the wilderness, is not only scenic but popular with fishermen. The trail between Bourne and Stratton ponds has been closed because of a severe windstorm in the area on July 20, 2003, but may very well be open by the time you visit. Steep cliffs are on the northern and western sides of the wilderness, while peaceful-looking Lye Brook Meadows is near the southern border. Lye Brook Falls, in the heart of the wilderness, is one of the most beautiful in the national forest, dropping nearly 100 feet in stair-step fashion. To get to the falls requires a one-way, 2.3-mile hike, much of it following an old railroad grade through picturesque beech–maple forests.

South of Lye Brook Wilderness, a pretty part of the Green Mountain National Forest is crossed by Forest Highway 6. The forest highway follows Roaring Branch, passing south of Whetstone Bluff, and then proceeds along the northern side of Beebe Pond. Farther east is lovely Grout Pond. Beebe Pond is at the northern end of a vast wild area that is being considered for wilderness status as the Glastenbury Mountain Wilderness. This 35,000-acre region extends south nearly to State Route 9. Near the center of the wilderness is Glastenbury Mountain, whose summit rises to 3,748 feet and has an old, historic lookout tower. The Long/Appalachian Trail goes through the entire length of the wilderness through forests of American beech, sugar maple, yellow birch, white birch, mountain ash, balsam fir, and red spruce. The forest floor is covered with typical north woods plants such as club-mosses, bluebead lily, bunchberry, cucumber root, starflower, and many others. Ferns are extremely abundant. Many of the smooth-barked trunks of the American beech exhibit claw marks from black bears. In addition to Beebe Pond, other interesting wet areas in the wilderness are Little Pond, Boles Brook, and Hell Hollow Brook. The Glastenbury Mountain area attracts bird-watchers because the rare Bicknell's thrush, Swainson's thrush, yellow-rumped warbler, Cape May warbler, winter wren, dark-eyed junco, and white-throated sparrow may all be observed here.

Southeast of the Glastenbury Mountain area is another large region that includes the headwaters of Deerfield River. This region is under consideration for designation as the Robert Stafford National Recreation Area. Along the Deerfield River are several bogs, while the adjacent mountains are heavily wooded except for rocky barrens where wild blueberry bushes abound.

On the eastern side of the Green Mountain National Forest, adjacent to State Route 100, are three important winter sports areas: Mount Snow, Carinthia, and Haystack Mountain.

The southernmost road to cross the Green Mountain National Forest from west to east is State Route 9, between Bennington and Wilmington. This scenic highway follows several bubbling brooks, with the Red Mill Brook Campground along the way.

Forest Road 74 goes for a short distance south from State Route 9 and the Red Mill Brook Campground. The primitive area to the west of this forest road is in the George D. Aiken Wilderness. Although this 5,060-acre wilderness contains nice forests today, it was logged heavily in the past. Several ponds, many of them full of water through beaver activity, are in the area, and the Long/Appalachian Trail traverses it.

East of State Route 8 and south of State Route 9 are 5,000 acres of wild, mountainous land that is being proposed as the Lamb Brook Wilderness. Several mountain peaks in this area reach an elevation of nearly 3,000 feet, and they are some of the most rugged in the Green Mountain National Forest. One may still make out part of the stagecoach road that was between Albany and Boston about 200 years ago. By hiking along the route, you will pass through neglected apple orchards, along crumbling stone walls and old cellar holes where trees are now growing. Bird-watchers will find Lamb Brook an excellent place to see such neotropical birds as the scarlet tanager, veery, and black-throated blue warbler. The brook empties into the Deerfield River.

NATIONAL FORESTS IN VIRGINIA

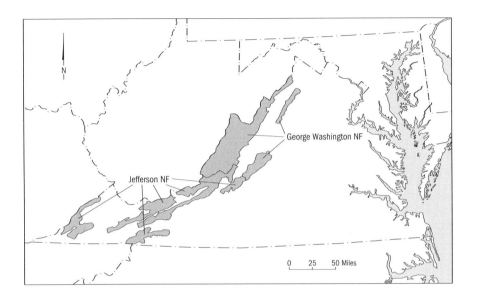

Virginia is home to two national forests. The George Washington National Forest is the more northern of the two, and a small part of it is in West Virginia. The Jefferson National Forest lies below the George Washington National Forest and extends south to the Tennessee and North Carolina borders. Both of Virginia's national forests are in Region 8 of the United States Forest Service and are jointly administered by the Forest Supervisor's Office at 5162 Valleypoint Parkway, Roanoke, VA 24109, www.southernregion.fs .fed.us/gwj.

George Washington National Forest

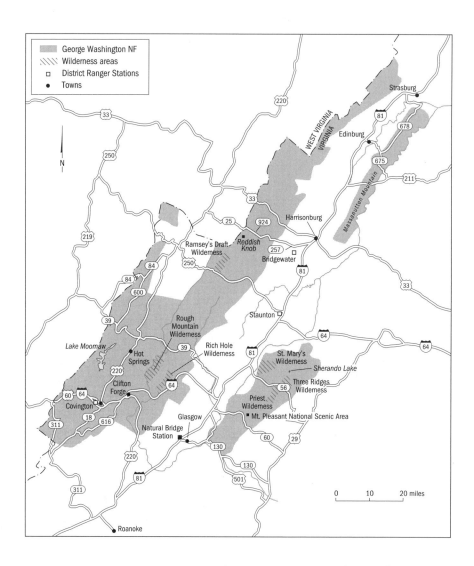

SIZE AND LOCATION: 1,055,000 acres in western Virginia, with a small part in extreme eastern West Virginia. Major access routes are Interstate Highways 64 and 81; U.S. Highways 33, 60, 211, 220, 250, and 501; Virginia State Routes 18, 39, 56, 60, 84, 130, 311, 600, 616, 675, 678, 818, and 924; and West Virginia State Routes 20, 25, and 84. District Ranger Stations: Bridgewater, Covington, Edinburg, Hot Springs, Natural Bridge Station, and Staunton. Forest Supervisor's Office: 5162 Valleypointe Parkway, Roanoke, VA 24019, www.southernregion.fed.fs.us/gwj.

SPECIAL FACILITIES: Swimming beaches; boat ramps; all-terrain vehicle (ATV) trails; horseback trails.

SPECIAL ATTRACTIONS: Mount Pleasant National Scenic Area; Sherando Lake; Lake Moomaw; Highland Scenic Tour; several iron furnaces; shale barrens; old-growth forests.

WILDERNESS AREAS: Priest (5,963 acres); Ramsey's Draft (6,500 acres); Rich Hole (6,450 acres); Rough Mountain (9,300 acres); St. Mary's (9,835 acres); Three Ridges (4,608 acres).

The George Washington National Forest, located in the Appalachian and Allegheny mountains, is the more northern of the two national forests in Virginia, extending from Wardensville and Strasburg in the north to Covington and Glasgow in the south. A small part of the national forest is in West Virginia, stretching along the West Virginia–Virginia border for nearly 35 miles. The national forest contains several mountains, nearly all of them trending northeast to southwest and more or less paralleling each other. The westernmost mountains in the national forest, such as Little Mountain and Back Creek Mountain, are part of the Allegheny range, while the numerous mountains to the east are part of the Appalachian range. These mountains are often separated from each other by low-lying valleys that are generally not in the national forest but are good agricultural lands.

The southwestern-most section of the George Washington National Forest is west of Covington and south of Interstate Highway 64. Although this area contains such geological features as Brushy Mountain, Peters Mountain, and Jingling Rock, there are no major national forest recreation areas. East of the settlement of Jordan Mines, however, is the National Children's Forest, one of three such forests in the United States sponsored in part by Hunt-Wesson Foods, Inc. In 1972, nearly 1,000 young people replanted this burned-out area of the national forest with thousands of tree seedlings.

Scattered throughout the national forest is a unique habitat and plant community known as a *shale barren*. Shale barrens are found only in Virginia, West Virginia, Pennsylvania, and Maryland, and they possess a number of rare endemic plant species. The majority of the shale barrens are in extreme eastern West Virginia and western Virginia. The habitat has a substrate of small acidic rock particles of shale, and their position on west- or south-facing slopes gives them a very harsh environment because of high solar radiation, searing heat, and hot soil temperatures. Because of these conditions, trees are usually sparse and widely spaced, and the rather open ground beneath them harbors the rare plants. One of these shale barrens occurs on the western side of State Route 18 less than 5 miles southwest of

Covington. Known as Gauging Station Shale Barren, this particular one occurs on a very steep, xeric, south-facing slope above crooked Potts Creek. Although only 47 acres in size, this shale barren has, in addition to the usual plants found in such a habitat, two of the rare endemics: the shale barren rock cress and the white-hairy clematis.

At the very southwestern edge of the George Washington National Forest and actually extending into the adjacent Jefferson National Forest is another unique area known as Potts Pond. This is a natural pond situated in a saddle on the crest of Potts Mountain. The pond is surrounded by exposed sandstone. Several species of emergent aquatic plants and one or two carnivorous bladderworts live in this small pond, including an extremely rare and federally endangered sedge known as the northeastern bulrush, one of about 60 occurrences for this aquatic plant in the world. The sedge forms a ring around the outer edge of the pond. Also known from the pond area is the very rare mountain damsel fly (*Lestes eurinus*). Sphagnum moss forms thick patches around the pond, and at the southern end, a quaking mat is present. Surrounding the pond is a shrub zone of wetland shrubs including red chokeberry and Minnie-bush, mixed with large cinnamon ferns. No streams flow into Potts Pond, but one stream flows from the pond, forming the headwaters of Toms Branch.

From Covington to Warm Springs and north, U.S. Highway 220 follows a valley that separates the Allegheny Mountains to the west and the Appalachian Mountains to the east. A few miles west of the spa community of Healing Springs is Lake Moomaw, the largest lake in the national forest. Located at the northern end of Oliver Mountain, the lake is a popular destination for campers, picnickers, fishermen, boaters, and hikers. The Forest Service maintains five campgrounds around the lake. There are boat ramps at Bolar Flat, Coles Point, and Fortney Branch, and swimming beaches at Bolar Mountain Beach and Coles Point.

A few miles farther north is the resort community of Warm Springs. West of town are two campgrounds in picturesque settings, Blowing Springs and Hidden Valley, both associated with Back Creek Mountain. Blowing Springs is a mountain spring producing strong gusts of underground air.

In Hidden Valley is the Warwick Mansion. In 1788, Jacob Warwick bought land in Hidden Valley and named it Warwickton. The mansion was inherited by his son, Judge James Warwick, who lived there with his family until their deaths in the 1890s. After renovation, this imposing structure became a bed and breakfast operated by permit from the United States Forest Service. The Jackson River, a fine trout stream, flows through Hidden Valley, and several hiking trails wind through the area.

Across from House Hollow is the historic Duncan Knob Lookout Tower on a rise at the north end of Wilson Mountain. A scenic part of the George Washington National Forest also occurs east of U.S. Highway 220 between Covington and Warm Springs and south of State Route 39. The mountains here are in the Appalachian range. Forest roads run the entire length of this region. Less than 5 miles north of Covington on a sandstone ridge above Dolly Ann Hollow is a dry oak forest typical of the Appalachian Mountains except that it has the southernmost population in the world of the variable sedge (*Carex polymorpha*). Forest Road 125 climbs onto Middle Mountain past the western side of Douthat State Park. West of the forest road is Bald Knob, which contains a high-mountain heath bald and a scrub forest dominated by dwarf pitch pines and scrub oaks—probably the best examples of these two vegetation types in all of Virginia. On the summit of Bald Knob, the shrubs grow less than 2 feet tall in the harsh conditions and highly acidic soils. There are several species of the heath family, including a number of kinds of blueberries. Much of the vegetation is growing on large boulders of sandstone in which are prominent fissures. Away from the summit, the vegetation increases in size, eventually becoming a forest of northern red oak, sweet birch, striped maple, and witch hazel. You may see the rare smooth green snake on the bald as well.

Interstate Highway 64 east of Clifton Forge loops south around Mill Mountain and penetrates the George Washington National Forest. Within 1 mile south of Clifton Forge are beautiful rock formations along the James River known as Rainbow Rocks. Before the interstate highway reaches the community of Longdale Furnace that grew up around an old iron furnace, there is the Longdale Recreation Area situated by a lake with a sandy beach where swimming is permitted.

State Route 42 exits off Interstate Highway 64 and follows the western edge of the Rough Mountain Wilderness where steep ridges and dry drainages are the rule. The moister sites of the wilderness support forests of upland hardwoods, while the dry, south-facing slopes are dominated by shortleaf pine. Griffin Knob at the south end of the area tops out at 2,842 feet, the highest in the wilderness. On the knob is the beautiful and uncommon yellow western wallflower of the mustard family.

Forest Road 129 goes along the east side of Rough Mountain Wilderness, separating it from Rich Hole Wilderness, which is 4 miles to the east on Brushy Mountain. It takes its name from the rich soils that occur in the head of drainages, or "holes," on the mountain. The rich soil supports old-growth hardwood species. Just outside the northern edge of the wilderness is Bubbling Spring Campground. Two miles east of Bubbling Spring is one of the best shale barrens in the national forest. Brattons Run Shale Barren occurs on a steep shale slope above the creek. The significant plant in this shale barren is the rock cress, but the white-hairy *Clematis* and the Millboro leatherflower are also here. Higher up on the steep shale slope is a shale woodland dominated by oaks and pines.

Between Brattons Run and the community of Millboro Springs are two tunnels for the Chesapeake and Ohio Railroad. On the dry shale slopes above these tunnels is the Millboro Tunnel Shale Barren where you may find the rock cress, Millboro leatherflower, Kates Mountain clover, and the shrubby Allegheny plum.

Interstate Highway 64 temporarily leaves the George Washington National Forest just beyond Exit 45. The gravel road south of the exit is the 19-mile Highland Scenic Tour. This often rough and narrow road curves and climbs through diverse vegetation, scenic vistas, and rock formations. The road climbs to the crest of North Mountain, where there is an upland mixed hardwood forest, and then drops into the valleys of Brattons Run and Simpson Creek where there are dense hemlock forests. Nine numbered stops appear along the route. At Stop 3 is a short, accessible trail through rhododendrons estimated to be 100 years old. At the end of the trail is a fine overlook. Stop 4 allows one to hike the short Knoll Wayside Trail through a mixed hardwood forest. Cocks Comb Trail at Stop 5 is steep and rocky and leads to an overlook where you can see the Rich Hole Wilderness to the west. The easy North Mountain Overlook Trail is at Stop 6. By looking east from this overlook you can see Lake Robertson and, in the distance, the Blue Ridge Mountains.

North of State Routes 39 and 42 and west of Interstate Highway 81 is an extensive part of the George Washington National Forest that includes Shenandoah Mountain and Great North Mountain. Cowpasture River below the western edge of Shenandoah Mountain is also roughly the boundary of the national forest. On a xeric ridge above a ravine along the Cowpasture River known as Blackies Hollow is one of the nicest shale barren communities in the national forest.

The southern ends of Shenandoah and Great North mountains are separated by Deerfield Valley, and State Route 62 and Calfpasture River follow the entire length of the valley. At the north end of the valley is Walker Mountain, which connects the two major mountain ranges.

Shenandoah Mountain stretches for many miles, and the Shenandoah Mountain Trail follows the crest of the mountain. U.S. Highway 250 also crosses Shenandoah Mountain, climbing to the crest by a very crooked mountain road. At the crest crossing are the Confederate Breastworks where Confederate soldiers constructed a trenchlike fortification to protect this part of Shenandoah Mountain during the early part of the Civil War. There is a short loop trail around the remains of the breastworks. South of the Confederate Breastworks is Reubens Draft Shale Barrren on an extremely dry, open, steep shaly slope. Seven miles north of Reubens Draft are Browns Pond and Winterberry Pond, two natural sinkholes on the west side of Cowpasture River. Near the center of Winterberry Pond is a nice clump of winterberry shrubs, a type of deciduous holly. Typical wetland plants in both ponds are three-way sedge and buttonbush. The northern inflated sedge is here near the southern limit of its range.

Two miles east of the Confederate Breastworks is the Mountain House Picnic Area on the site of a tollhouse on the old Parkersburg Pike. Six miles farther east is Braley Pond, where you may picnic beneath a stand of pines and fish for rainbow trout, largemouth bass, bluegill, and channel catfish. The history of the area is described on a panel at the picnic area.

Two miles west of the Confederate Breastworks is the settlement of Headwaters and a steep shaly slope north of town that harbors the Headwaters Shale Barren.

Directly north of Mountain House Picnic Area is Ramsey's Draft Wilderness with stands of virgin forests. Unfortunately, the virgin hemlocks are being devastated by the woolly adelgid. A 14-mile loop trail is in the wilderness, including a difficult trek to the top of 4,282-foot Hardscrabble Knob.

Northeast of Ramsey's Draft Wilderness around Little Bald Knob is a significant group of forest communities on a sandstone ridge. There is a mixed hardwood forest composed primarily of oaks, interspersed with smaller stands of hemlocks. On drier sites are well-developed communities of bear oak and pitch pine. The rare Cow Knob salamander lives on the knob as well as a rare plant known as drooping bluegrass. A forest road passes through the Little Bald Knob area.

All along the Shenandoah Mountain crest in the northern part of the mountain are several rocky knobs, many of them with significant natural features. Most of the knobs are shale capped by sandstone and have very steep slopes. The Cow Knob salamander and the smooth green snake are in these areas. On upper slopes above the headwaters of Skidmore Fork are magnificent old-growth stands of hemlock. Scenic knobs include Reddish Knob, Bother Knob, Flagpole Knob, and High Knob, and Cow Knob far to the north. A lookout tower stands on High Knob.

North of the High Knob Lookout is Middle Mountain, which consists of east-facing, flat-topped ridges of the Shenandoah Mountains. Old-growth northern red oaks are the dominant features on these ridges. South-facing slopes have extensive stands of Table Mountain pines, which serve as nesting sites for the northern red cross-bill.

Cow Knob is near the northern end of the George Washington National Forest. This is the knob where the rare Cow Knob salamander was first discovered on a ridgetop dominated by oaks and hemlocks.

South of Cow Knob is a beautiful stream known as Laurel Run. The areas adjacent to the stream are some of the most significant in the George Washington National Forest. Near Hall Spring is a majestic forest of red spruces beneath which grow the rare three-seeded sedge and bunchberry at its southernmost station. There is an old-growth northern red oak forest on Tomahawk Mountain.

The Great North Mountains is a parallel range east of the Shenandoah Mountains. The highest elevation in the national forest is Elliott Knob, at the southern end of the range. This is one of five knobs in the area above 4,000 feet, and all consist of exposed sandstone that is home to heath thickets, northern hardwoods, and dry woods dominated by bear oak and pitch pine. The extremely rare mountain variety of the Virginia trillium has one of its two world populations on Elliott Knob.

The Great North Mountain Trail runs for 15 miles, merging with the 8-mile-long Crawford Mountain Trail. Much farther north, the crest of the Great North Mountains forms the boundary between Virginia and West Virginia, with the George Washington National Forest in both states. Campgrounds in this vast area are at Tomahawk Pond in Virginia and Hawk, Wolf Gap, and Trout Pond in West Virginia. A fine scenic drive over back roads connects Wolf Gap Campground with Trout Pond Campground. Trails wind through the area.

Capon Furnace is an old iron furnace that was in operation from 1832 to 1880. Southeast of Capon Furnace, and in Virginia, is the Van Buren Furnace. The original furnace was in operation between 1837 and the 1850s, replaced in 1873, and then abandoned entirely in 1884.

North of the Wolf Gap Campground and also straddling the state line on the crest of the Great North Mountains are unusual geological areas known as Big Schloss and Little Schloss. Massive rocky outcrops are here, and the sandstone cliffs of Big Schloss have been used as reintroduction sites for the peregrine falcon. Cold-water streams in the region, such as Little Stony Creek, Waites Run, and Cedar Creek, support native trout.

Two long and relatively narrow sections of the George Washington National Forest are located east of Interstate Highway 81. The southernmost Pedlar District is in the Blue Ridge Mountains; the Blue Ridge Parkway passes through most of the area, as does the Appalachian Trail. Near the southern end of the district, the Appalachian Trail crosses 3,376-foot Bluff Mountain. This district is very scenic, with several waterfalls.

East of U.S. Highway 60 is the 7,580-acre Mount Pleasant National Scenic Area. Three mountains are highlights of the area: Mount Pleasant, Pompey Mountain, and Cole Mountain. Along Little Cove Creek is a small virgin stand of mixed hardwood trees. Although the scenic area is popular with hikers, much of it is steep and extremely rugged. It also features numerous springs, bubbling streams, and interesting rock formations. The 5-mile Henry Lanum Memorial Trail goes to the summit of Pompey Mountain.

East of Buena Vista is Panther Falls, where the Pedlar River falls about 8 feet between two large rock shelves. A 1-mile loop trail leads to and from the falls. Staton's Falls is a few miles north, where Staton Creek drops several times, the longest for about 50 feet. Catchpools are at the base of each of the falls. Still farther north and accessible from the Blue Ridge Parkway is Crabtree Falls, the most spectacular in Virginia. If you add together the five falls of Crabtree Creek, the total drop is 1,200 feet over a distance of 0.5 mile. The largest of the five, called the Grand Cataract, falls approximately 500 feet. To see the entire spectacle requires a 3-mile round trip hike, often steep and slippery in places.

The Blue Ridge Parkway eventually takes an easterly course, and at Milepost 19.9 at the Slacks Overlook, you are near two small falls on White Rock Creek that drop 30 and 15 feet, respectively. The White Rock Creek Falls lie midway between the St. Mary's Wilderness and the Three Ridges Wilderness. Located on the western slopes of the Blue Ridge Mountains, St. Mary's Wilderness is scenic, with a gorge along the St. Mary's River; beautiful wooded slopes above Spy Run, Hogback Creek, and Mineback Creek; and interesting geology at Knob of Rocks. There is also a double waterfall in the wilderness.

Three Ridges Wilderness, adjacent to the south side of the Blue Ridge Parkway, is bisected by the Appalachian Trail. Its V-shaped ravines are steep and rocky. Southwest of Three Ridge Wilderness is Priest Wilderness, another extremely rugged area through which the Appalachian Trail passes. Priest Mountain has the highest elevation in the wilderness at 4,063 feet.

Northeast of St. Mary's Wilderness and adjacent to it is a large region known as Big Levels. (The name *Big Levels* refers to several thousand acres of relatively flat mountaintops, ranging between 3,000 and 3,600 feet in elevation.) Because of the rocks and dense forests in the area, this region was a center for the iron industry during the 1800s. Iron furnaces were constructed, and you can still see the old Mount Torry Furnace along the eastern edge of Big Levels. This furnace was the longest in operation, making iron between 1804 and 1892. Because of the great amount of charcoal needed to fire the furnaces, the surrounding forests were heavily logged several times. With the demise of the iron industry, the forests of Big Levels are making a good recovery. On the ridgetop and upper slopes is a forest dominated by bear oak, with the lower slopes and flat wooded areas supporting a mixed oak community. Smaller forested communities include a rock chestnut oak–type near the top of rugged north slopes, shortleaf pine and hardwood-type on some of the upper slopes, and hemlock forests in moist coves at the head of valleys.

A number of sinkhole ponds are present in the northern part of Big Levels, and those at Maple Flats have significant vegetation. A few miles south of Maple Flats is Sherando Lake Recreation Area with two lakes, a campground, swimming beach, and fishing pier.

The Lee Ranger District lies north of the Pedlar Ranger District and includes Massanutten Mountain. The eastern border of this district is formed by the crooked South Fork of the Shenandoah River.

The Massanutten Visitor Center is at New Market Gap on U.S. Highway 211 and divides the ranger district into a north and south region. At the visitor center is the 0.5-mile Wildflower Trail that follows Stonewall Jackson's steps during the Shenandoah Valley phase of the Civil War. The wildflowers along the trail are at their peak during the spring. Also at the visitor center is the paved 0.2-mile Discovery Way Trail (accessible to visitors with disabilities) that provides an introduction to the forest. For a longer hike from the visitor center, try the 8.5-mile Bird Knob Loop Trail or the 8.5-mile Waterfall Mountain Loop Trail.

In the southern part of the Lee District are Catherine Furnace and Browns Hollow Research Natural Area. Catherine Furnace is a historical structure dating back to the time the furnace was in operation between 1836 and 1885. Browns Hollow Research Natural Area protects several significant shale barren communities. One of the shale barrens is home to the rare shale barren goldenrod.

Immediately north of the Massanutten Visitor Center is the Story Book Trail. This paved 0.4-mile trail interprets the geology of the Massanutten Mountain and includes a scenic overlook into Page Valley.

Forest Road 274 is the Crisman Hollow Road, and it proceeds north from the Story Book Trail area in the mountain for several miles to Camp Roosevelt. Before reaching Camp Roosevelt, the road comes to the Lion's Tale National Recreation Trail, which is accessible to visitors with disabilities. This 0.5-mile loop interprets this part of the forest using the senses of touch, hearing, and smell.

The campground at Camp Roosevelt is on the site of the first Civilian Conservation Corps camp in the country where the first young recruits arrived on April 3, 1933; this date marked the beginning of 1,500 such camps across the United States. You can hike around the area and see the sites where several of the buildings were located.

Near the northern end of Massanutten Mountain is the Elizabeth Furnace with two interpretive trails, a campground, and a picnic area. The 0.6-mile Charcoal Trail describes the process of making charcoal, a necessary component of iron making. The shorter Pig Iron Loop Trail circles the ruins of Elizabeth Furnace, with signs explaining the operation of the furnace.

At the northernmost tip of the Lee Ranger District is Signal Knob, which may be hiked to over a rugged 10.5-mile trail from the west side of State Route 678.

Blackies Hollow

Halfway up the side of a steep hill, Bob Glasgow, a wildlife biologist for the George Washington National Forest, pulled his green truck off the narrow gravel road and stepped out. I followed him through a thick forest of sugar maples and oaks down into Blackies Hollow (pl. 34), a ravine carved by a stream that now flows only intermittently. Pausing briefly to examine a violet or two growing on the forest floor, we started up the other side of the ravine. The steep climb to the top was difficult because the rich soil was permeated with countless fragments of shale, but slender trunks of flowering dogwood and redbud were spaced just right to provide convenient handholds.

The ridgetop was flat for just 2 or 3 feet, and then it dropped at a 60 percent grade for several hundred feet to the meandering Cowpasture River in the adjacent valley. The footing was treacherous, for the ridgetop was covered by a thin (2- to 6-inch) mantle of loose shale flakes, and a few weathered pines and gnarled chinquapin oaks were the only trees to interrupt the slope's rocky surface. This dry habitat, known as a *shale barren*, is one of many that dot the steep hills found on both sides of the Virginia–West Virginia border and as far northeast as central Pennsylvania. They generally face south, at an elevation between 1,000 and 2,000 feet, and are undercut by a stream.

Frederick Pursh and other early botanists explored some of these shaly slopes in the 19th century, but it wasn't until 1911 that naturalist Edward Steele coined the term *shale barren* and began to realize that many flowering plants that lived in the shale were unlike those found anywhere else. On the 50-acre barren off Blackies Hollow, where more than half the ground is devoid of plants, I spotted a buckwheat, a clematis, a rock cress, an evening primrose, a clover, and a groundsel—nearly half of the endemic species botanists have identified. Later, when I visited shale barrens farther north, I saw another kind of clematis and a wild onion, an aster, a goldenrod, and a phlox.

In 1951, ecologist Robert Platt, working from Emory University, analyzed shale barrens and found that although their surface is dry, a 4- to 13-inch-thick soil layer beneath the mantle provides moisture for plant growth as well as a medium for roots. He also noted that during the growing season, the temperature at the shaly surface sometimes reaches 140 degrees Fahrenheit, too intense for plant seedlings from the adjacent forest to survive. The seedlings of the shale barren endemics have apparently been able to adapt to this extreme heat, with the result that they reach maturity with little competition.

There have been several hypotheses to explain the origin of the barrens and their endemics. Geologists know that the shale and associated siltstone were formed about 400 million years ago. The real question is, Did the shale barrens develop after the great glaciers retreated north at the end of the last Ice Age, which lasted from 1.8 million to some 12,000 years ago? Or did the barrens already exist before and during Ice Age times?

Some believe that as the North American glaciers advanced, the existing eastern deciduous forest shifted south, even reaching the Gulf of Mexico. In the hills of Virginia and adjacent areas, conditions became cooler and moister. As a result, forest covered the ancient shale and siltstone, just as today it covers the more shaded northern slopes of the shale barren hills. As the glaciers retreated and the climate warmed, some of the southern slopes became barrens. If this is what happened, the dozen endemic plants were probably not present during maximum glaciation but arose during the last 12,000 years on the newly exposed areas.

Another hypothesis is that the Ice Age caused few major changes in the position or composition of the eastern deciduous forest. Proponents of this view believe that the forest was shaped some 15 million years ago but that a cycle of erosion of the softer shale and siltstone began about 13 million years ago, at which point the shale barrens began to develop. This suggests that the endemic shale barren plants could have emerged over a period of at least 10 million years. Since all the endemics are herbs that overwinter belowground, they could have survived the Ice Age winters.

A look at the endemics themselves seems to provide some resolution of this debate. Botanist Carl Keener of Pennsylvania State University has divided the endemics into two categories. Some shale barren species closely resemble plants that grow in the adjacent deciduous forests. In fact, some botanists do not consider all of them to be distinct species. For example, the shale barren onion differs from the common nodding onion of the forest only in its more round-tipped flower parts; its longer, more slender flower stalks; and its blooming time, which is 2 weeks later.

Similarly, there is a purple-flowered shale barren clematis, whose only obvious difference from the common Appalachian clematis is that the tufts of hairs attached to its fruits are white instead of cinnamon or buff colored. Other examples in this category are the shale barren rock cress, the shale barren goldenrod, and the shale barren aster. Since these endemics have not diverged a great deal from the nearby forest species that probably spawned them, they may have sprung up relatively recently. They also have the most restricted ranges in the shale barrens, possibly because they have not had much time to expand. Accordingly, Keener calls all these plants *neoendemics*.

Some shale barren endemics, however, bear little resemblance to other plants of the region, and their closest relatives grow far away. The erect, non-climbing white-tailed clematis, whose nearest relative is the clematis that lives on dry exposures in the Ozarks of Missouri and Kansas, is an example. At one time, there was probably one species of erect clematis that lived in barren habitats all across the eastern and midwestern United States. This species was then split into two isolated populations, which eventually diverged into two distinct species. The ice sheet that separated parts of the West from the East could have accounted for this split.

Other plants with this kind of distribution are the shale barren evening primrose, whose closest relative grows in the southwestern United States and Mexico; the Kates Mountain shale barren clover, which appears close to the buffalo clover from the Midwest and South; the shale barren buckwheat, related to Correll's buckwheat from Texas; and the shale barren phlox, whose nearest relative may be a species that grows on the Colorado Plateau. Because these plants may have originated before or during the Ice Age, Keener calls them *paleoendemics*. He argues not only that their closest relatives grow far away but that the plants themselves are more widespread on the shale barrens than the neoendemics, probably because they have had much more time to adapt to this specialized habitat.

Flagpole Knob

When Bob Glasgow alerted me to the discovery of a virgin stand of eastern hemlocks on Shenandoah Mountain, he promised me that it looked like the "Redwoods of the East." A few days later I met Bob and forest botanist Steve Croy at the forest headquarters in Harrisonburg and with eager anticipation set out on a 22-mile drive, traveling through Mennonite country and passing huge turkey farms. Our first destination was 4,397-foot Reddish Knob in the middle of Shenandoah Mountain, where we could get an overview of the entire region.

Shenandoah Mountain is one of a series of northeast- to southwest-trending mountains that are separated from each other by broad valleys of farmland. Across the valley to the east of Shenandoah Mountain is the northern section of the Blue Ridge Mountains and its famed Shenandoah National Park and Skyline Drive. The mountain to the west of Shenandoah Mountain is West Virginia's Cheat Mountain, with its highest point, Spruce Knob, topping out at 4,860 feet. All of these mountains, with their lush vegetation and rounded ridges, are part of the Appalachian system, formed several hundred million years ago.

As we approached Reddish Knob, we noticed a few quaking aspens and paper birches, uncommon trees for this area, in the dry, rocky woods. At one inviting spot we stopped and walked into a parklike forest whose understory had been kept clear by lightning fires. Pennsylvania sedge formed a grass-like carpet over much of the ground, occasionally punctuated by painted trillium, cucumber root, fly poison, and false lily-of-the-valley, all spring-flowering members of the lily family. We paused to examine a low, craggy, sandstone outcrop and found the succulent blue-leaved sedum and the related white alumroot and Virginia saxifrage. The wood rat, a pack rat, builds its nest along these types of rocky ledges. Lumbering in wood rat territory has reduced the animal's number nationwide.

After driving on to the summit, we viewed the rounded crest of the slightly lower Little Bald Knob 5 miles to the south. As its name implies, Little Bald Knob, like the other higher knobs in the area, is devoid of tall trees. Instead, scraggly bear oaks surround a summit of Pennsylvania sedge. Little Bald Knob has the added significance of being at the southern end of the range of the Cow Knob salamander. Recognized as a species only about 30 years ago, this secretive amphibian lives in a 24-mile-long by 1-mile-wide zone along the crest of the Shenandoah Mountains. The animal has a black-and-white speckled body and large round eyes. Its flat snout helps it to burrow under rocks and stones.

The Cow Knob salamander apparently developed as a distinct species because it was isolated on Shenandoah Mountain for thousands of years. The forests on the mountain summits, inaccessible for timber harvesting, provide its ideal living conditions. Similar but different species of salamanders live in isolation on the adjacent mountains. The Cheat Mountain salamander lives to the west, while the Shenandoah salamander lives on the Blue Ridge range to the east.

North from Reddish Knob we followed the Forest Service road toward Flagpole Knob, stopping again to explore a Pennsylvania sedge opening surrounded by gnarly, lichen-covered hawthorns, Table Mountain pines, and an attractive shrub known as the many-flowered pieris. The pieris, a member of the heath family, is so full of white, bell-shaped blossoms in mid-May that the leaves are nearly hidden. This sedge opening and one other, also on Shenandoah Mountain, are the only places in the world where the dwarf Virginia trillium grows. This plant's nearest relative lives in the Coastal Plain.

From a roadside pullout near Flagpole Knob, we blazed our way down the western side of Shenandoah Mountain in the direction of a ravine formed by Skidmore Creek. Because it had rained hard the evening before, we had to descend slowly over the treacherous jumble of slippery rocks. Soon we found ourselves in a parklike forest that had rarely been disturbed by humans. Above a nearly continuous thicket of fetterbush rose the giant eastern hemlocks. True to Glasgow's description, the grove was like a redwood forest but on a smaller scale; the hemlocks grew straight and tall, with several of the larger ones approaching a girth of 3.5 feet at shoulder height. Several species of birds use the forest for nesting, including the red-breasted nuthatch, the brown creeper, and the red cross-bill.

These trees had escaped the mass lumbering that has destroyed the native forests of the East and soon threatens to consume all of the old-growth forests of the West. To prevent the loss of such natural communities that remain in the George Washington National Forest, Glasgow and his colleagues have joined with the Virginia Heritage Program to identify the best areas and set them aside as reserves. As Glasgow commented, this approach not only protects biodiversity but also enhances the public perception of the George Washington National Forest and relieves loggers from any concern that they might inadvertently wipe out important populations of plants or animals. Unfortunately, the eastern hemlocks are being decimated by the hemlock woolly adelgid.

Maple Flats

Maple Flats, a nearly level forested area on the western fringe of the Blue Ridge Mountains, is shot through with sinkhole ponds, each distinctive in size, shape, and associated plant and animal life. The sinkholes, 21 in all, developed when cavities dissolved out of the bedrock—dolomitic limestone—and the overlying soil collapsed into them. All are fairly shallow. Some have standing water, some are merely muddy, and some are dry, but what their water levels will be at any given time is not predictable. Some receive water from underground sources; all are fed by rainwater runoff. The sinkholes at lower elevations are often the first to fill up after it rains, but this varies depending on the amount of groundwater and the porosity of the soil and bedrock.

To give me a tour of some of the sinkholes, wildlife biologist Bob Glasgow led me through a forest composed predominantly of red maples. The drier ridges, however, are covered with a mixture of Table Mountain and shortleaf pines, oak, hickory, and black gum. On the day of our visit, we even passed an American chestnut, a tree rarely encountered ever since the chestnut blight fungus nearly extinguished the species at the turn of the century. Lower-growing trees include sassafras, flowering dogwood, sourwood, and mountain laurel. The last two, as well as the shrub deerberry, are members of the heath family and grow well in the acidic soil that has built up over the limestone bedrock.

The first sinkhole we came to was just 20 feet across, a circular, dry depression strewn with fallen leaves. A dense stand of trees, mostly shortleaf pine and red maple, surround it, forming a closed canopy. After heavy rains during the spring, this usually dry sinkhole retains water for a few days.

A thicket of dwarf willow has become established in the shallow water at Oak Pond. This sinkhole is ringed with pin oaks, whose lower limbs hang down close to the water surface. Not far away is the deepest of the sinkholes, Deep Pond. Its lowest depressions hold as much as 4 feet of water and are filled with Robbins' spikerush, a plant that has grasslike stems but only a single head of minute, green flowers. This pond is also ringed by pin oaks; a few feet beyond them, a slightly elevated ridge supports species, such as post oak, that prefer drier conditions.

Cricket frogs, salamanders, and other animals abound at Deep Pond. Like frogs, salamanders are amphibians. In early life the wetland species possess gills and lead an aquatic existence; later they metamorphose into an adult form more adapted to terrestrial life, usually losing their gills as their lungs develop. Several kinds of salamanders live at Maple Flats, one of the more common and beautifully patterned ones being the marbled salamander (pl. 35). The rarest is the tiger salamander, a species normally found on the flat Coastal Plain adjacent to the Atlantic Ocean. It is densely speckled and may grow up to 13 inches long.

Marbled salamander eggs are laid and hatch in autumn, when temporary sinkhole ponds begin to accumulate water. In winter the hatchlings can be found floating in the lower levels of the water. The young start growing quickly during the spring, and by summer, when the ponds dry up, they are able to survive as terrestrial animals.

Tiger salamanders, on the other hand, don't lay their eggs until spring. When the young hatch, they are often eaten by the young marbled salamanders, which have about a 6-month head start. Tiger salamanders are also more vulnerable in dry years. In drought years, many egg masses that the tiger salamanders lay inside the shallow ponds dry up.

Painted turtles were common at the next sinkhole I visited that day, Twin Pond—so named because at low water levels it is divided in two by a low ridge. Two plants that grow in Twin Pond are particularly unusual. One is the seven-angled pipewort, a species rare in Virginia; the other is the Virginia sneezeweed, a yellow-flowered plant found in wet areas in and near Maple Flats and nowhere else, except, unexplainably, the Missouri Ozarks.

The largest pond is Spring Pond, which has an underground water supply and fluctuates little from its average 2-foot depth. Surrounded by a profuse mat of sphagnum moss, the pond is home to many of the plant species recorded at Maple Flats. The rarities include swamp pink, one of the prettiest pink wild lilies in the world; maidencane, a sedge normally found in coastal areas; Virginia chain fern; and Oakes' pondweed, which lives in standing water. More common wetland plants include skunk cabbage, cucumber root, wild cranberry, roundleaf sundew, three-way sedge, and swamp azalea.

Some of the bog-loving plant species at Maple Flats, such as wild cranberry and skunk cabbage, are boreal forest plants that apparently migrated south because of Ice Age glaciation and remained behind even after the climate warmed. Evidence for this includes core samples of mud taken from Spring Pond that contain pollen of red spruce, the dominant boreal forest tree. But according to Julia Gorey of The Nature Conservancy, Coastal Plain plants at Maple Flats are relicts from before the last glaciation, when the Coastal Plain extended farther westward.

Glasgow and I took an alternate route back so we could examine Football Field Pond and Mosaic Pond. The first of these is rectangular and, during low-water periods, covered by low-growing, mat-forming grasses that resemble the playing surface of a football field. Mosaic Pond, named for its diverse patterns of vegetation, is home to the spotted turtle, a Coastal Plain species not found elsewhere on the western side of the Blue Ridge Mountains.

Ramsey's Draft

The timber in the western, mountainous part of Virginia was cut extensively near the turn of the 20th century, and by 1920 very little of the original forest was left. Among the harvested trees were Canadian hemlocks, but they were not particularly valued for their wood. Instead, lumbermen stripped the tannin-rich bark from the trunks to sell to the tanning industry. Because the tissues that conduct water and dissolved nutrients lie in the bark, the mutilated trees soon perished. Today only a single stand of virgin hemlock forest survives in Virginia—Ramsey's Draft, in the George Washington National Forest. Just a few hours' drive from Washington, D.C., and Richmond, Virginia, Ramsey's Draft is nestled below the summit of Shenandoah Mountain. Unfortunately, the hemlock woolly adelgid has decimated many of the Canadian hemlocks in the area.

The name *Ramsey's Draft* attracted me long before I was able to visit the place itself. One of the definitions of *draft* is "gorge," and I had little trouble envisioning a deep ravine filled with stately trees and colorful mountain flowers. Looking at a geologic map, I noted that the creek that has carved the draft through centuries of cutting and washing is named Ramsey's Draft Run. *Run* has an air of excitement about it that is not usually conveyed to me by the more common word *creek*.

Ramsey's Draft is a lush ravine nearly 1,000 feet below Shenandoah Mountain to the west and Bald Ridge to the east. Through the ravine, Ramsey's Draft Run tumbles and sparkles over a bed of smooth, polished rocks. The waters that feed the run come from temporary rain-fed streams, mountainside springs, and a major tributary known as Jerry's Run, which begins as a trickle high up on Shenandoah Mountain. A primitive, dead-end road once followed Ramsey's Draft Run for nearly 5 miles, but in 1972, Hurricane Agnes washed away some of its stone fords. Today, this route is the main basis for a hiking trail that begins at a parking area off U.S. Highway 250.

The trail wends its way upstream, crossing the run 16 times. For the first 2 miles, the trail climbs imperceptibly from 2,252 to 2,914 feet, where Ramsey's Draft Run forks into the Right Prong and the Left Prong (the Left Prong is Jerry's Run). The forest in the ravine bottom is deeply shaded by thick stands of coniferous Canadian hemlocks and white pines, occasionally punctuated by deciduous trees—cucumber magnolia, American beech, shagbark hickory, and red maple. Mountain laurel adds a dash of color during June, while witch hazel, whose woody seed capsules explode with a sharp cracking sound in autumn, hangs gracefully over the run in several places.

Records indicate that the forest along the first 2 miles of the run has not been timbered since the early 1900s. But the part of the forest that has never been cut lies beyond the fork of Ramsey's Draft Run, stretching for 3 miles along the Right Prong. It was probably spared because of its remoteness and because its previous owners appreciated its magnificence. This virgin forest contains Canadian hemlocks nearly 150 feet tall and up to 4 feet in diameter. Occasional specimens of broad-leaved yellow and gray birches provide a contrast to the dark green needles of the hemlocks.

Canadian hemlock is common in the Maritime Provinces of Canada, as well as in much of New England, northern New York, and western Pennsylvania, but it also occurs in isolated areas as far away as southern Indiana and northern Alabama. During its early years, the Canadian hemlock is shaped like a broad-based pyramid. The long, slender branches on the lower part of the trunk are usually horizontal or even pendulous, whereas those higher up are ascending. In the dense shade of Ramsey's Draft, however, the lower branches die off after a few years, leaving the trunks nearly naked for up to two-thirds their length. Except for a dense covering of the evergreen spinulose woodfern on the forest floor, there is little underbrush, so an extensive view through the forest is possible, interrupted only by the large tree trunks.

The hemlock's bark, often cinnamon red, contains high concentrations of tannin, which discourages insects. The tree's flat, narrow leaves, each less than 1 inch long, are arranged on the branches in flat sprays, giving each leaf maximum exposure to the limited sunlight. Individual leaves persist on the hemlock for 3 years. They emerge from tiny buds a light yellow-green, but during their first year they become deep green and shiny.

In early May, millions of yellow pollen grains develop in small, spherical clusters at the base of the year-old leaves. While most of the pollen falls uselessly to the ground, some lands on the hemlock's erect, green, 0.125-inch-long female "flowers." Self-pollination is unlikely because a tree's female parts do not mature until a few days after pollen is released. Following pollination, tiny cones begin to develop. By the end of the year, they are 0.75 inch long and hang gracefully on slender stalks. Most of their seeds fall during the winter and are dispersed on transparent wings by the wind.

Although each mature hemlock produces thousands of seeds annually, only those that land in dense shade are likely to germinate. If there is an opening in the canopy left where a giant tree fell, the forest floor is more conducive to the germination of maples and other tree species. After a few years, however, as the canopy closes overhead, the seedlings of these other trees die. As long as the hemlocks of Ramsey's Draft are protected from lumbering and fire, other species will not gain a significant foothold.

While the spinulose woodfern dominates the forest floor, several mountain wildflowers occasionally emerge through the fronds. The delicate bloodroot and several kinds of white and purple violets are the first to bloom in the spring, followed by Jack in the pulpit, cucumber root, Solomon's-seal, and clintonia. In summer, the dainty enchanter's nightshade puts forth its tiny white flowers. As the trail continues ever higher along the Right Prong, occasional patches of flame azalea and rose azalea fill in the shrub layer of the forest.

After passing through the virgin hemlock forest, the trail climbs 1,400 feet in less than 2 miles to Hardscrabble Knob. The knob is very steep and covered with loosely packed gray and green fossil-bearing sandstone rocks. Hiner Spring, which feeds Ramsey's Draft Run, seeps out a few feet below the summit. The huge shade-forming trees found at lower elevations are absent on Hardscrabble Knob. With more sunshine, red maples and black, white, and red oaks become common; and thickets of black huckleberry and an abundance of May apples, yellow stargrass, wild geranium, and other wildflowers grow beneath the trees.

Jefferson National Forest

SIZE AND LOCATION: Approximately 723,000 acres in southwestern Virginia, and a small area in West Virginia. Major access routes are Interstate Highway 81; U.S. Highways 21, 23, 52, 58, 421, 460, 580, and the Blue Ridge Parkway; and State Routes 16, 42, 43, 68, 80, 83, 311, 600, 601, 603, 606, 611, 615, 617, 619, 621, 623, 625, 706, 708, and 822. District Ranger Stations: Blacksburg, Marion (Mount Rogers National Recreation Area), Natural Bridge Station, New Castle, and Wise. Forest Supervisor's Office: 5162 Valleypointe Parkway, Roanoke, VA 24019, www.southernregion.fs.fed.us/gwj.

SPECIAL FACILITIES: Swimming beaches; boat ramps; horseback trails; all-terrain vehicle (ATV) trails.

SPECIAL ATTRACTIONS: Mount Rogers National Recreation Area.

WILDERNESS AREAS: Barbours Creek (5,700 acres); Beartown (5,609 acres); James River Face (8,886 acres); Kimberling Creek (5,542 acres); Lewis Fork (5,700 acres); Little Dry Run (2,858 acres); Little Wilson Creek (3,900 acres); Mountain Lake (11,113 acres); Peters Mountain (3,328 acres); Shawvers Run (3,467 acres); Thunder Ridge (2,344 acres).

The Jefferson National Forest, lying south of and adjacent to the George Washington National Forest, occupies much of the land mass of southwestern Virginia. Extending all the way to the Tennessee and North Carolina state lines to the south, to within a few miles of Kentucky to the west, and including a small part of West Virginia to the northwest, most of the national forest is in the Valley and Ridge provinces of the Blue Ridge Mountains.

The far western part of the Jefferson National Forest is in the Clinch Ranger District, several miles distant from the remainder of the national forest. This district itself is in two contiguous areas. The northern part is just across the Kentucky state line and extends from Breaks Interstate Park at the northern tip to the North Fork of the Pound Reservoir at the southern end. The reservoir is the main attraction in this region, and the Forest Service maintains campgrounds at Cane Patch and Laurel Falls. There are boat ramps at Cane Patch, Laurel Falls, and Phillips Creek.

The larger part of the Clinch Ranger District lies south of Norton. U.S. Highway 23 forms part of the northern border before turning south and bisecting the district. This very scenic, mountainous region features clear, babbling streams, waterfalls, and rocky gorges, as well as Lake Keokee. Stone Mountain forms the northern edge of the district, while Powell Mountain is along the south end and the central part of the district. At the western end of

Stone Mountain is Cave Springs Campground, nestled in a forested setting near a clear natural spring. A hiking trail is here, as well as a swimming beach. Just off State Route 606 is Lake Keokee, which is fed by a number of mountain streams and surrounded by wetland vegetation and several dead trees that provide great habitat for fish and wildlife. A loop trail encircles the lake and also joins the Olinger Gap Trail that passes through dense rhododendron thickets and forested areas. Within Stone Mountain is a steep-walled, rocky gorge that has been carved by Devil's Fork. This clear-running trout stream has scoured out rock basins and pools. On the dry ridges above the gorge are forests dominated by oaks and hickories; on the lower slopes are occasional coves with beautiful hardwood trees. Here and there along Devil's Fork are nice stands of hemlock.

South of Norton, near the junction of State Route 619 and Forest Road 238 in the Powell Mountains, is High Knob. From the summit of the knob, at 4,223 feet, you may be able to see five states on a clear day. Nearby is a hiking trail and a campground at a spring-fed lake suitable for swimming.

Big Stony Creek and Little Stony Creek have carved deep, rocky gorges in the Powell Mountains at the eastern end of the Clinch Ranger District. Big Stony has carved Chimney Rock Gorge and Mountain Fork Gorge, the two separated by a rocky plateau. There are waterfalls in both gorges. Mixed hardwood forests are found on the upper slopes of both gorges, while hardwood coves sprinkled with hemlocks are on the lower slopes. Dense thickets of rhododendrons occur along the streams at the bottom of the gorges.

Little Stony Creek has formed a 440-foot-deep gorge that is 1.75 miles long and up to 2,500 feet wide. Two 24-foot waterfalls and one 8-foot falls are in the gorge, and whitewater rafting is popular for the strong-hearted. Although most of the upper slopes are forested with oaks and hickories, the driest areas have pitch pine and Table Mountain pine. There are huge entanglements of rhododendrons and mountain laurels.

Along State Route 806 in Powell Mountain is Bark Camp Lake, with a campground, boat ramp, swimming beach, and hiking trail.

The southernmost part of the Jefferson National Forest is so unique and scenic that it has been designated as the Mount Rogers National Recreation Area. This area has everything that the outdoors person would want, including unsurpassed scenery, rock formations, mountain streams, camping areas, hundreds of miles of hiking trails, and three designated wilderness areas. All of these features are grouped around Mount Rogers, the highest elevation in Virginia at 5,729 feet. The 154,000-acre recreation area has rugged mountainous terrain, steep ridges, and broad valleys, the latter used in the past for grazing animals. On the crests of the mountains are spruce–fir forests, while oaks and hickories dominate the upland woods. More mesic

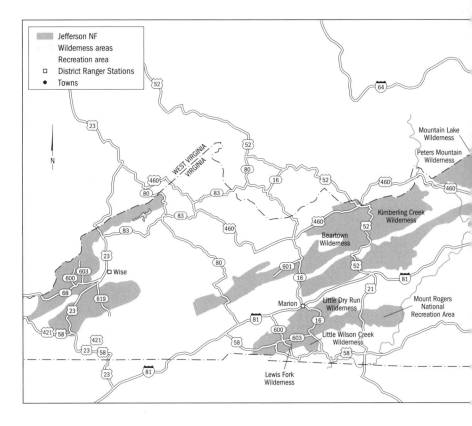

forested areas include American beech, yellow birch, and sugar maples. Elevations within the area range from 2,000 to 5,729 feet.

The Pat Jennings Visitor Center, 7 miles south of Marion on State Route 16 in Brushy Mountain, will introduce you to what the Mount Rogers National Recreation Area has to offer. Trails also originate at the visitor center. If you continue south, you will come to a nice campground along Raccoon Creek below Dickey Knob.

Toward the western end of the Mount Rogers National Recreation Area is the Beartree Recreation Area that lies northwest of Konnarock off U.S. Highway 58. The area, located in the Straight Branch Valley, has campsites and a lake for swimming and fishing. A spur trail connects the recreation area with the Appalachian Trail. A high-mountain bog is in the area.

Southeast of Beartree is Whitetop Mountain, the second highest mountain in the national recreation area. Beautiful open meadows stretch on the slopes just below the mountain's summit. The Appalachian Trail meanders between Beech Mountain and Buzzard Rock before climbing over Whitetop Mountain. East of Whitetop Mountain is pleasant Elk Garden, where the trail to the summit of Mount Rogers and the Lewis Fork Wilderness begins. Trails

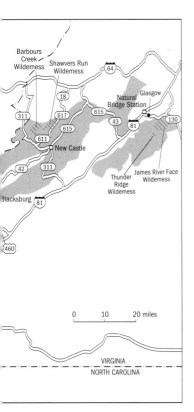

in the wilderness include the popular and therefore much-used Appalachian, Virginia Highlands, and Mount Rogers trails and the less frequently used Grassy Branch, Helton Creek, Cliffside, Pine Mountain, and Sugar Maple trails.

Just outside the northeastern edge of the Lewis Fork Wilderness is Fairwood Valley, a 5-mile-long valley between Troutdale and Konnarock. Here are 4,300 acres of densely wooded hillsides and pastures. Fox Creek, a clear trout stream, flows through the middle of the valley. Grindstone Campground has nature trails and an amphitheater.

Southeast of Fairwood Valley is the Little Wilson Creek Wilderness, also crossed by the Appalachian Trail as well as First Peak, Bearpen, Little Wilson Creek, Kabel, and Hightree Rock trails. Little Wilson Creek flows through the entire wilderness.

Between Lewis Fork Wilderness and Little Wilson Creek Wilderness, in the remaining high country around Mount Rogers, is the large Crest Zone Special Management Area where high-altitude vegetation is carefully being managed.

North of Fairwood Valley is Hurricane Mountain and Hurricane Campground. A 1-mile loop trail leads from the campground. The Appalachian and Iron Mountain trails pass near the area.

Farther east in the Iron Mountains, the Iron Mountain Trail runs across the ridge. At the northern edge of the Mount Rogers National Recreation Area is Little Dry Run Wilderness. This small wilderness features Little Dry Run, a native trout stream. To the south of the wilderness is an unusual rock formation known as Comer's Rock, with a campground nearby.

East of Little Dry Run Wilderness and reached via Forest Road 14 is the Hussy Mountain Horse Camp and the Horse Heaven Trail. Areas to explore here are Falls Branch Waterfall, Jones Knob, and High Knob. The Iron Mountain Trail passes through this area.

At the eastern side of the Mount Rogers National Recreation Area are Raven Cliff Recreation Area and New River Recreation Area. Raven Cliff features the remains of an old iron furnace that was in operation until the early 1900s. The short trail to the furnace first crosses a neat bridge over Cripple Creek. Near the southern end of the Raven Cliff area is a fine example of a hardwood cove called Collins Cove. The New River Recreation Area has boat ramps for small boats, a campground, and an adjacent New River State Park rail-to-trail.

The most easterly part of the Jefferson National Forest lies a short distance northeast of Roanoke and south of Interstate Highway 81. Some of the most scenic areas in the national forest are in this region, and the Blue Ridge Parkway passes through the entire area. At the extreme northeastern corner of this district is the James River Face Wilderness (pl. 36), the oldest designated wilderness in the Jefferson National Forest. The most dramatic part of the wilderness is the James River Gorge, with lush vegetation on both sides of the deeply cut gorge. Elevation rises to 3,073 feet on Highcock Knob. At the western edge of the wilderness is Devil's Marbleyard, 600 acres of rocks in jumbled piles. State Route 721 and the Blue Ridge Parkway form the southern boundary of the wilderness. South of these roads is the Thunder Ridge Wilderness on the northeastern slopes of the Blue Ridge Mountains. Apple Orchard Mountain at the southwestern corner of the wilderness has the highest elevation at 4,200 feet. Nearby is Apple Orchard Falls, one of the best in the eastern United States as it drops 200 feet. The Apple Orchard Falls Trail is 1.4 miles long from the Sunset Fields Overlook at Milepost 78 on the Blue Ridge Parkway. A short distance south of the falls is the Cornelius Creek Trail, which is worth taking.

Between the Blue Ridge Parkway and Interstate Highway 81 is Cave Mountain Lake with a campground and hiking trail. North of here on the east side of State Route 159 is the Glenwood Iron Furnace (fig. 19), in operation between 1849 and 1887.

The remainder of the Jefferson National Forest lies north of Interstate Highway 81. At the western side of this region, Little Brushy Mountain fea-

Figure 19. Glenwood Iron Furnace, Jefferson National Forest (Virginia).

tures Steer Knob and Goat Knob. To the northeast is the Beartown Wilderness, named for Beartown Knob that rises to 4,710 feet at the northeastern edge of the wilderness. This wilderness is in one of the most remote and rugged parts of the national forest. There are fine spruce–fir forests at the higher elevations, and hardwood and hemlock coves at lower and moister areas. The Appalachian Trail crosses the southern edge of the wilderness as it climbs over Chestnut Ridge. Between Beartown Wilderness and Interstate Highway 77, the Appalachian Trail winds over Little Brushy Mountain and drops down to cross scenic Laurel Creek.

Immediately east of Interstate Highway 77 and north of State Route 42 is Kimberling Creek Wilderness. Kimberling Creek and the Appalachian Trail both cross this area. The Appalachian Trail continues eastward, crossing over Brushy Mountain and just south of the Falls of Dismal. This 15-foot drop, located between Brushy and Flat Top mountains, is reached by a 4-mile trail from the Appalachian Trail, although a rough forest road that branches off Forest Road 606 will bring you very close to the falls.

The New River twists its way through Virgina, separating Pearis Mountain from Peters Mountain, both in the Jefferson National Forest. The wildest part of Peters Mountain has been designated as a wilderness area and includes the crest of the mountain at an elevation of 3,956 feet. Pine Swamp is in the wilderness and accessible on the Appalachian Trail. Peters Mountain is also home to the very rare and endangered Peters Mountain mallow. South of the wilderness along Little Stony Creek are the Cascades. A 2-mile hike will take you to these 66-foot falls.

East of the Cascades, State Route 613 goes between Kire Mountain and Rocky Mountain. A spur road leads to scenic White Rocks with an adjacent campground. South of this is the Mountain Lake Wilderness. Nearby Mountain Lake is the largest natural lake in the western part of Virginia. The wilderness consists of a high plateau that contains a mountain bog surrounded by virgin stands of spruces and hemlocks. If you hike to War Spur Overlook, you will get a magnificent view of the surrounding countryside. Within this wilderness is a beautiful forest of red spruce, Mann's Bog, which is a peat moss bog; a hemlock–hardwood forest near the War Spur Overlook; Appalachian deciduous forests that contain white oak, red oak, black oak, rock chestnut oak, basswood, American elm, and sourwood, particularly in Johns Creek Valley; and impenetrable thickets of bear oak and members of the heath family on top of Johns Creek Mountain and Salt Pond Mountain. The wilderness is reached by trails off State Route 700.

Between State Routes 613 and 311, Peters Mountain is in West Virginia, but this is still a part of the Jefferson National Forest and includes Moteshard Mountain. East of State Route 311 and back in Virginia is Hanging Rock

Valley nestled between Middle Mountain to the north and Potts Mountain to the south. Hanging Rock, a part of Potts Mountain, is in the Shawvers Run Wilderness where hemlocks and white pines are common along the streams. Farther east, on the southeastern slope of Potts Mountain, is Barbours Creek Wilderness. In addition to the usual forest species in the national forest, you will find good stands of white pine and shortleaf pine here as well. Lipes Branch has native brook trout. The Pines Campground is just east of the wilderness.

Southeast of Barbours Creek Wilderness is the site of the Fenwick Mines Iron Furnace, which has a picnic area. At the northern-most tip of the national forest is the Roaring Run Iron Furnace, in operation between 1832 and 1864.

Mallow Fellows

In two habitats 800 miles apart, scientists have saved two closely related wildflowers from extinction by suppressing the growth of competing plants and using fire to stimulate the germination of dormant seeds. One of these sites is the Jefferson National Forest; the other is Langham Island, which lies about 40 miles southwest of Chicago in a rocky gorge of the Kankakee River. Built on a bed of dolomite, the northern, downstream end of the island is a flat-topped upland, while the southern, upstream end consists of a low-lying deposit of alluvial soil. Trees cover most of the island's 24 acres.

The island's claim to botanical fame traces back to the morning of June 29, 1872, when the Reverend Ellsworth Jerome Hill of Kankakee, a public school teacher and avid amateur botanist, rode on horseback to the village of Altorf on the banks of the Kankakee River and then crossed by boat to Langham Island. The island had most recently been owned by a Potawatomie who had grown some crops there, but most of it was in a natural state. When Hill climbed ashore on what he called the "gravelly island," he found a number of rare native plants, including leafy prairie clover, buffalo clover, and a white-flowered violet.

But Hill's greatest discovery was a group of 4- to 8-foot-tall plants whose pinkish flowers resembled those of hollyhock but were a little smaller. As it turned out, Hill had found a plant new to botanical science. Like the hollyhock, the new plant belonged to the mallow family, and it most resembled a species in the Rocky Mountains. But its scientific identification remained undecided until 1906, when botanist Edward Lee Greene, of the Smithsonian Institution, decided to distinguish it as a species in the same genus as the Rocky Mountain flower. He named it *Iliamna remota* ("remote" because of its great distance from the Rockies). Today we know this plant as the Kankakee mallow, and it grows in the wild only on Langham Island.

By 1983, the number of Kankakee mallow plants on Langham Island had dwindled to 180. One reason for this decline was the widespread increase of nonnative shrubs that apparently cast too much shade for the mallow to thrive. The U.S. Fish and Wildlife Service listed the plant as a candidate for endangered species designation.

After management practices, including removal of competing nonnative shrubs, enough of the plants now thrive on Langham Island that this species has never had to be designated as endangered.

Coincidentally, 800 miles to the east, a related story has unfolded. On August 3, 1927, several members of the West Virginia University Botanical Expedition, led by professors P. D. Strausbaugh and Earl L. Core, set out to look for plants on Peters Mountain, Virginia, a few miles across the state line. About 500 feet above the New River, at an elevation of approximately 2,000 feet, the group entered a dry, open forest dominated by rather stunted post oaks and rock chestnut oaks. Beneath the trees were several bare patches of sandstone. Where soil had filled pockets in the sandstone, wildflowers grew.

One of the wildflowers that the botanists discovered grew about 6 feet tall and had pinkish flowers that looked like small hollyhock blooms. About 50 of these plants were scattered over a small area, many growing in full sunlight because the trees were so sparse and stunted. Taking a few specimens back to the laboratory for identification, Strausbaugh and Core learned that their plant was nearly identical to the Kankakee mallow. There were a few differences, however: The Illinois plants were a little taller on average, and they had a slight fragrance, which the Virginia plants lacked. The two also differed slightly in the shape of their leaves. To this day botanists cannot agree on the exact status of these two plants, although the Fish and Wildlife Service recognizes them as two distinct species, calling the Virginia plant the Peters Mountain mallow.

In 1984, botanists checking the status of the Peters Mountain mallow found it faring even more poorly than the Kankakee mallow. They were able to locate only five plants and no young seedlings. A hiking trail that had cut through the colony had introduced a coarse, weedy wildflower called Canadian leafcup, which had overtaken the understory and outcompeted the mallow. To try to protect the remaining plants, in May 1986 the Fish and Wildlife Service added the Peters Mountain mallow to the federal list of endangered species and authorized a plan to prevent its extinction. The Canadian leafcup was eliminated, and burns were carried out in 1992 and 1993 under the leadership of the Virginia Department of Conservation and Recreation. As a result, some 500 new seedlings came up.

To ensure protection of the Kankakee mallow, Langham Island—which is part of Kankakee River State Park—has been designated an Illinois Nature Preserve. The Nature Conservancy recently purchased the land on which the Peters Mountain mallow grows (it was previously in private hands even though it falls within the boundaries of Jefferson National Forest). With both species apparently on their way to recovery, botanists may now return to the question of why these two closely related plants grow in the wild in only two widely separated locations.

Mount Rogers

At 5,729 feet, Mount Rogers is the highest mountain in Virginia and the focal point of the Mount Rogers National Recreation Area, a complex of mountains, gorges, and clear, fast-flowing streams in the Jefferson National Forest. I was interested in climbing to the top of this southern Appalachian peak, one of the highest in the eastern United States, to compare its vegetation with that of other eastern peaks of similar altitude and with the higher mountains of the west. For example, the summit of Roan Mountain, North Carolina, is treeless, with gorgeous Catawba rhododendrons and other shrubby members of the heath family making up the dominant vegetation. Such heath-laden summits, treeless despite being below the regional timberline, are referred to as *heath balds*. Brian Head, Utah, on the other hand, rises to 11,307 feet and is well above tree line. Its upper slopes are rocky and support dwarf, alpine vegetation, while mountain meadows crowded with colorful paintbrushes and beardtongues occupy the lower slopes.

My 4.5-mile hike to the summit of Mount Rogers began at the parking area at Elk Garden and followed the Appalachian Trail. Low clouds and bluish haze obscured the summit, but stretching before me at Elk Garden, as far as the clouds would permit me to see, was a steep, grassy-looking slope that was virtually treeless. The trail climbed over the slope, its way marked by cairns of small stones. Even though it was the last day of May, the area was alive with colorful flowers borne on low plants.

Hawkweed and hawksbeard, two species with dandelion-like flowering heads, abounded, mixed with white yarrow, whose crowded flower clusters rose above lacy leaves. These common flowers are mostly European weeds, not native species, a sure sign that Elk Garden has had many years of disturbance of some kind. A few 10-foot shrubs of spiny hawthorns were dispersed throughout the meadow. Small colonies of 20-foot-tall sweet buckeyes, another woody plant, joined the hawthorn as the trail climbed steadily up the meadow. The absence of heaths of any kind dispelled any notion that Elk Garden was a heath bald.

Historical records indicate that some of Elk Garden was forested when the first European settlers arrived during the late 1600s, although some of the area was open and served as a grazing land for native bison and elk. Traces of a buffalo wallow at Elk Garden give testimony that bison were plentiful in the area. After the forested part was cleared and burned, however, grasses invaded the slopes, which became ideal pasture for the settlers' sheep, goats, and cattle. Soon the bison and elk disappeared from the region.

At the upper end of Elk Garden, the disturbed meadow gives way abruptly to a forest that consists of taller sweet buckeyes and a number of other trees. The forest marks the beginning of the Lewis Fork Wilderness and continues to the summit of Mount Rogers, less than 4 miles away. The slopes in the forest are very steep and fall off quickly to one side of the trail. On the other side are wet, dripping rock faces, kept moist by the dense shade provided by the closed canopy of the trees. Vegetation on the forest floor and in the cracks of the rocks is lush.

The trees are all deciduous, consisting predominantly of sugar maple and American beech, but in the depths of the woods, mountain maple and yellow birch are common. The dense shrub layer is composed of such shade-tolerant species as wild gooseberry, hobblebush viburnum, wild hydrangea, and northern elderberry. Ferns and wildflowers flourish beneath the woody layers. Spinulose woodfern, silvery spleenwort, and Christmas fern, displaying 2-foot-long leaves, grow in extensive patches. The smaller common polypody fern finds a home in the cracks that have eroded into the rock faces. The best show of mountain wildflowers is from late May to early June, with the blooming of foamflower, doll's eyes, Appalachian bittercress, and wild ginger. During the summer and autumn, several of these spring-flowering plants spend their energy developing new leaves and expanding older ones. A few plants that wait until summer and autumn to flower are golden-glow, mountain oxalis, and black cohosh. Delicate rock saxifrages grow in unlikely niches in the rocks.

As the trail makes its final ascent, cone-bearing red spruces begin to break up the domination of deciduous trees. Finally, at the summit, there is a fine stand of red spruce and Fraser fir—no heath bald or rocky, treeless, alpine meadow. The red spruce, a northern species, is nearly at its southern limit on Mount Rogers, whereas the Fraser fir, a southern species, is at its northern extreme.

The ancestors of both species of trees originally came from the north. During the last Ice Age, about 12,000 years ago, the lowlands ahead of the advancing glaciers became cool and wet, supporting a forest of northern spruces and firs. As the mountain peaks were frozen less and less each year, the tree limit gradually moved upward, and the summits were covered with a forest of spruce and fir, which require the coolness provided by the higher elevation of the mountain peaks. The warmer lowlands became occupied by deciduous maples, beeches, and buckeyes (which migrated up from the south), so that many of the mountain tops soon became islands of spruce and fir. During the thousands of years of isolation, the firs on the Appalachian peaks developed different characteristics from the northern species of firs and are now recognized by botanists as a distinct species.

NATIONAL FOREST
IN WEST VIRGINIA

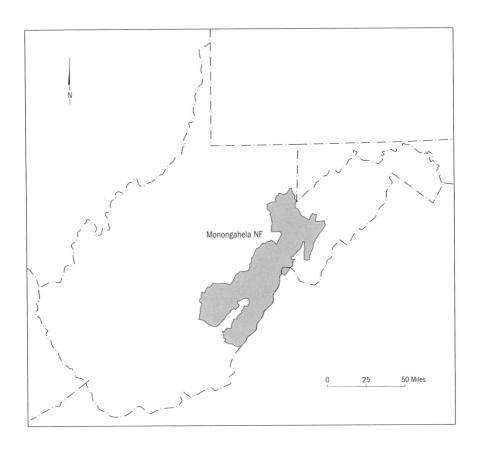

Monongahela NF

0   25   50 Miles

The Monongahela, West Virginia's only national forest, occupies much of the mountainous part of eastern West Virginia. It is in Region 9 of the United States Forest Service.

Monongahela National Forest

SIZE AND LOCATION: Approximately 900,000 acres in the mountains of east-central West Virginia. Major access roads are Interstate Highway 64; U.S. Highways 33, 219, 220, and 250; and State Routes 1, 5, 14, 20, 22, 27, 28, 31, 39, 43, 49, 55, 60, 72, 84, 92, 150, and 600. District Ranger Stations: Bartow, Marlinton, Parsons, Petersburg, Richwood, and White Sulphur Springs. Forest Supervisor's Office: 200 Sycamore Street, Elkins, WV 26241, www.fs.fed.us/r9/mnf.

SPECIAL FACILITIES: Swimming beaches; boat ramps; horseback trails.

SPECIAL ATTRACTIONS: Highland Scenic Highway; Cranberry Glades Botanical Area; Lake Sherwood Recreation Area; Spruce Knob–Seneca Rocks National Recreation Area; Smokehole; Gaudineer Scenic Area; Dolly Sods Scenic Area.

WILDERNESS AREAS: Cranberry (35,864 acres); Dolly Sods (10,215 acres); Laurel Fork North (6,055 acres); Laurel Fork South (5,997 acres); Otter Creek (20,000 acres).

The Monongahela National Forest has more interesting and significant areas in it that you can imagine. Located in the Allegheny Mountains, this national forest includes rugged, scenic mountains, high mountain lakes, unique botanical and geological areas, wild and woolly wilderness areas—all steeped in history. The southern end of the national forest begins a short distance above Interstate Highway 64 and the resort community of White Sulphur Springs, and it extends north to U.S. Highway 50 south of the village of Aurora. Its breadth includes the area between the Gauley River and the town of Beaver on the west and the Virginia state line on the east.

State Route 92 goes for many miles through the southeastern region of the Monongahela National Forest between White Sulphur Springs and Bartow. A few miles north of the southern boundary of the national forest, State Route 16 branches west off State Route 92, taking you to the Blue Bend Campground and Picnic Area nestled in a bend along Anthony Creek. The water in the creek at this point is deep blue, and the forests surrounding the campground consist of hemlock, rhododendron, mountain laurel, and several majestic hardwood trees. Should you decide to follow Forest Road 296 west from the campground, you will climb over Gunpowder Ridge before dropping down to a boat ramp on Anthony Creek that is maintained by the Forest Service. Forest Road 139 goes a short distance north of the campground to the Hopkins Overlook where you may get a fine view over Anthony Valley.

North at the settlement of Neola, State Route 14 to the east of 92 ascends over Meadow Creek Mountain and then heads north just inside the West Virginia–Virginia state line to Lake Sherwood Recreation Area. When Meadow Creek was dammed, it created the 165-acre Lake Sherwood, which has glistening blue water. The recreation area has many campsites and picnic tables, a swimming beach, bathhouse, and boat ramp between the picnic area and West Shore Campground; a nature trail across from this campground; and a

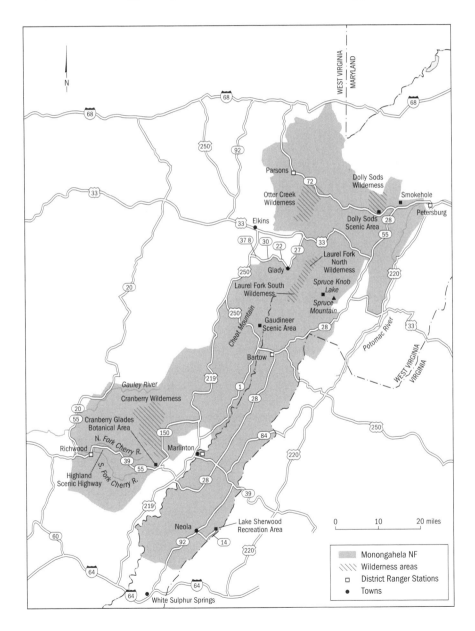

1.5-mile trail that encircles the lake. Fishermen will appreciate the stocking of the lake with bluegill, largemouth bass, and catfish.

At the north end of Meadow Creek Mountain, just before State Route 92 intersects with State Route 39, is the Pocahontas Campground. Nearby to the southeast is the High Top Lookout Tower; to the west are trails ascending Middle Mountain. By following State Route 39 west from the junction with 92, you will come to the communities of Huntersville and Marlinton.

At Marlinton, State Route 39 turns abruptly south and, in a few miles, turns back west again, taking you to the western part of the Monongahela National Forest and the famous and significant Cranberry Glades Botanical Area. This is the southern end of the Highland Scenic Highway. At the entrance to this part of the national forest is the Cranberry Mountain Visitor Center where you may learn all about the Cranberry backcountry and the unique plant communities in the area, see a coal-mining exhibit, and obtain information about forest conservation. There is an interpretive nature trail at the visitor center.

A short distance north of the visitor center is the Cranberry Glades Botanical Area where four high mountain bogs exhibit a unique flora containing plants seen nowhere else in the vicinity. Bird-watchers will also enjoy the opportunity of seeing a wide variety of bird life, including the northern water thrush, hermit thrush, olive-backed thrush, and mourning warbler. A boardwalk has been provided so that visitors may admire the area without damaging the fragile environment. The botanical area is at the southern end of the huge Cranberry Wilderness on the Allegheny Plateau. This is mountain country with deep, V-shaped valleys and drainages of the Middle Fork of Williams River and the North Fork of Cranberry River. The wilderness is also a black bear sanctuary, so hikers on the 50 miles of trails should be alert for these animals.

A few miles west of Cranberry Glades is Falls of Hill Creek Scenic Area. After a 0.75-mile hike down 250 feet into a steep, narrow ravine, you will be treated to one of the most spectacular features of the Monongahela National Forest. Three waterfalls comprise the scenic area. The upper falls drops 25 feet over a 25-foot layer of sandstone and shale. The trail then goes down a stairway to a level area where the stream crosses a hard sandstone layer. A footbridge allows you to walk on the lip of the middle falls, which drops 45 feet, and an observation deck that permits a grand view. As you continue your hike to the lower falls, you will pass areas of red shale. The lower falls drops 63 feet, which makes it the second-highest waterfall in West Virginia.

West of High Falls, State Route 39 follows the tortuous course of the North Fork of the Cherry River with outstanding vistas. At the North Bend Campground and Picnic Area, Forest Road 77 to the north will lead to Summit Lake. This 43-acre lake has a campground and is stocked with trout. From the campground are trails into the Cranberry Wilderness.

The northern portion of the Highland Scenic Highway is State Route 150 north of the Cranberry Mountain Visitor Center as it follows the eastern edge of the Cranberry Wilderness. Where the scenic highway crosses the Williams River, Forest Road 86 to the north goes to a secluded campground along Tea Creek. From this junction, the scenic highway curves to the east and soon comes to the Red Lick Picnic Area. Nearby is the Red Lick Overlook.

Between the western edge of the Cranberry Wilderness and the Gauley River are three campgrounds and the Woodbine Picnic Area. Rough Forest Road 272 climbs to Red Oak Knob and the Red Oak Knob Lookout Tower.

State Route 84 branches off State Route 92, providing access to Bird Run Campground. If you continue east on State Route 84, you will come to the top of Allegheny Mountain, which is the dividing line where water to the east flows into the Greenbrier River and water to the west into Knapp Creek and eventually the Ohio River. A narrow back road, Forest Road 55, follows the West Virginia–Virginia state line. This scenic road comes to the site of the Paddy Knob Lookout Tower at an elevation of 4,477 feet.

State Route 92 continues north to the town of Green Bank, home of the Green Bank National Radio Observatory where huge radio telescopes rear high into the air. One of the radio telescopes is 300 feet across with a surface area of 1.8 acres. West of the radio telescopes is the Cass Scenic Railroad, built originally as a logging railroad that ran for 4 miles to the top of Cheat Mountain. Long after the railroad ceased to operate, it was reconstructed and lengthened to 11 miles; now it is used for pleasure rides operated by the state of West Virginia.

South of the Cass Scenic Railroad and reached by State Route 66 is an isolated plateau in the Monongahela National Forest known as Thorny Flat. Because of the high elevation here at 4,839 feet, the vegetation is more reminiscent of that found many miles to the north.

Crooked State Route 1 between Green Bank and the Cass Railroad passes a grand cave system with underground passages of more than 17,000 feet. Two interconnecting caves, Cassell and Cassell Windy, have marvelous formations and are also the home of the very rare minute cave amphipod (*Stygobromus parvus*). You must check with the district forest ranger about access to these caves. The highway roughly follows the course of the Greenbrier River, which winds for 40 miles to the town of Marlinton. This is a good whitewater river that meanders through breathtaking scenery.

At the town of Durbin, where U.S. Highway 250 and State Routes 28 and 92 come together, you have two choices to explore the remainder of the national forest. U.S. Highway 250/State Route 92 to the west will take you to the areas of interest in Cheat Mountain; Forest Road 44 north from Durbin gives access to Shavers Mountain and Middle Mountain and eventually joins State Route 27, which goes to the Otter Creek Wilderness. State Route 28 east from Durbin allows you to explore Spruce Mountain and North Fork Mountain.

Cheat Mountain is a beautiful mountain with knobby peaks, clear streams, and occasional unique forest communities. It is also the home of the rare Cheat Mountain salamander. Shavers Fork River meanders along the eastern edge of the range, separating Cheat Mountain from Shavers Mountain. Just north of the highway is Gaudineer Knob and a stand of virgin red spruce that is the focal point of the Gaudineer Knob Scenic Area. There is a fire tower atop 4,445-foot Gaudineer Knob. The scenic area has 130 acres of virgin spruce, with some of the trees more than 100 feet tall.

After the highway leaves Gaudineer Knob, it crosses Shavers Fork in a picturesque setting and then climbs to the Cheat Summit Historical Site, where Fort Milroy was located during the Civil War. The fort was used to guard this crossing of Cheat Mountain.

Forest Road 44 follows the West Fork of the Greenbrier River, wandering between Shavers Mountain and Middle Mountain, all in the Monongahela National Forest. The forest road ends at the community of Glady where State Route 27 continues north to the south end of Otter Creek Wilderness. If you take a series of forest roads back and forth and up and down south from Glady, you can make your way to the Laurel Fork North Wilderness and the Laurel Fork South Wilderness. These two areas are separated only by State Route 40. Forest Road 14 forms the western border of both wildernesses, and the Laurel Fork River penetrates each wilderness. The river is very narrow in places and heavily forested. Wildlife abounds.

Otter Creek Wilderness lies between Shavers Mountain and McGowan Mountain, with Otter Creek flowing through the heart of the area. A number of trails are in the wilderness. Outside the northern edge of the Otter Wilderness and south of the town of Parsons is the Fernow Experimental Forest where techniques of managing mountain forests are tested.

Between Otter Creek Wilderness and the town of Elkins, on the north side of U.S. Highway 33, is the Stuart Recreation Area nestled in a bend of Shavers Fork River. This area includes a campground, picnic shelter, swimming beach at the river, and a nature trail that consists of several loops, depending on how much you wish to hike. Another trail leads from the campground to the river. The Stuart Memorial Drive is a narrow, 10.7-mile gravel road from the campground that passes through the Shavers Fork Valley. After climbing to 4,000-foot Middle Point Mountain, the road descends into Alpena Gap. From the memorial highway is a spur road to Bickle Knob and access to the Otter Creek Wilderness.

Spruce Mountain and North Fork Mountain lie northeast of Durbin and are reached by side roads off State Route 28. A short distance east of the Greenbrier Ranger Station is the unique Island Campground that is located on an island in the East Fork of the Greenbrier River. If you are up to adventure, a few miles east of Island Campground is Forest Road 112. Follow this road and then the very rough Forest Road 254 for several miles to Blister Swamp, where the balsam fir, a rare northern conifer, covers about 40 acres. Because the fir produces resin that forms swellings on the trunk, the tree is known locally as a *blister pine*. The last few miles are through private property. This is a high-elevation, spring-fed swamp at 3,637 feet, and the water in the swamp forms the headwaters of the East Fork of the Greenbrier River.

Spruce Mountain features Spruce Knob, the highest elevation in West Virginia at 4,861 feet. A low observation tower permits grand 360-degree views on clear days, but often the knob is enshrouded by heavy fog. The vegetation at this high elevation shows a strong relationship to floras found much farther north. Pink lady's-slipper orchids bloom along the trail in May. Climatic conditions on Spruce Knob are harsh, with frequent strong winds. Most of the red spruces on the knob are only about 20 feet tall and are strongly one-sided because of the wind, looking very much like the divi-divi trees of Aruba.

West of Spruce Knob via Forest Road 112 are Spruce Knob Lake as well as the site of the Gatewood Fire Tower. Each area has a campground. The 26-acre lake was formed when Narrow Ridge Run was dammed. At the north end of Spruce Mountain are the 15-foot-high Falls of Seneca Creek. These wide falls may be reached only by a 3-mile hiking trail.

State Route 28 follows the South Fork of the Potomac River that forms a valley between Spruce Mountain and North Fork Mountain. Behind the village of Seneca Rocks are the famous Seneca Rocks themselves. Be sure to stop in the visitor center of the Spruce Knob–Seneca Rocks National Recreation Area to learn about the area and its Indian legend. Seneca Rocks, composed of a gray sandstone with a notch at the very center of the summit, stand nearly 1,000 feet above the North Fork of the Potomac River. Rock climbers attempt to scale the steep front face of the craggy rocks, but a hiking trail up the back side is easier to traverse. From the summit are unexcelled views.

As State Route 28 continues northeast and eventually makes its way out of the Monongahela National Forest, there are interesting areas on either side of the highway.

To the east is Smokehole and some unusual rock formations. Smokehole is an area along the South Fork of the Potomac River where there are numerous caves in the mountainside. When cool air emitted by the caves comes in contact with more moist air above the river, a dense mist forms that looks like smoke. State Route 2 is a scenic drive through Smokehole, and Big Bend Recreation Area has campsites near the river.

South of Smokehole is perpendicular Eagle Rock that stands 300 feet above the South Fork of the Potomac River. Across the river from this imposing sandstone rock is the grave of William Eagle, a soldier in the Revolutionary War. West of Smokehole are Champe Rocks, a series of gray sandstone knobs that tower above the South Fork of the Potomac River. The rocks are also named for a Revolutionary War soldier.

West of State Route 28 at the northern end of the Allegheny Front is the spectacular and mysterious-looking region known as Dolly Sods (pl. 37). This is high-plateau country, with strong, gusty winds and harsh temperatures. As a result, the spruce trees that grow here are dwarfed and gnarly, and the vegetation in places consists of low-growing blueberry and huckleberry bushes. Forest Road 75 passes along the eastern edge of Dolly Sods. The north part of Dolly Sods has been designated as a scenic area, while the south end is now the Dolly Sods Wilderness. Along the forest road are scenic overlooks, a picnic area, and the Red Creek Campground. At the very northern tip of Dolly Sods and the Monongahela National Forest, where Forest Road 75 makes an extremely sharp turn to the south, are Bear Rocks (fig. 20). The views from these whitish, tilted rocks are unsurpassed.

Figure 20. Bear Rocks, Monongahela National Forest (West Virginia).

Southwest of Dolly Sods and the Roaring River Plains is Mount Porte Canyon, which has a knob with an elevation of 4,760 feet. The canyon and knob are named for David Hunter Strother who wrote about the West Virginia mountains long ago; Strother's pen name was Mount Porte. This area may be reached only by a lengthy hiking trail.

Cranberry Glades

Some of West Virginia's Monongahela National Forest is so dense that summer sunlight doesn't reach the forest floor. At higher elevations, where the climate is cool, rainfall abundant, and fog common, the tree cover is provided by red spruces, yellow birches, and balsam firs. Sugar maples, beeches, basswoods, and white pines dominate the mountain ravines. There are even a few virgin stands of red spruce, where trees 300 to 400 years old have been spared from lumbering. With so much unbroken forest, one vast opening, where only the occasional tree rises above a sea of low-growing vegetation, provides a startling inconsistency. This is Cranberry Glades, a bowl-shaped, boggy area at the foot of three mountains: Cranberry, Black, and Kennison.

The basin containing Cranberry Glades is about 3 miles long and up to 1 mile wide and about 3,400 feet above sea level. Penetrating the heart of the glade are Cranberry River and its tributaries, Charles Creek and Yew Creek. Cranberry River, which originates from springs halfway up Cranberry Mountain, never spreads more than 12 feet across. Four treeless bogs, ranging in size from 8 to 60 acres, are scattered through the basin. The rest of Cranberry Glades consists of dense thickets of shrubs or bog forests.

The four bogs have a layer of peat that may reach a depth of 11 feet. Peat (dead sphagnum moss that has sunk beneath a mat of living sedges and heaths) builds up where there is insufficient oxygen to promote the complete decay of vegetable matter into soil. Beneath the peat in Cranberry Glades is usually a layer of algal ooze that may be another 2 feet thick. This is a dark, jellylike material, the decomposed remains of algae that lived tens of thousands of years ago. During spring, the water table may come within an inch of the peaty surface, while in autumn, it drops nearly a foot.

Bogs often form in lakes where poor drainage allows organic detritus to accumulate, principally from sphagnum moss that grows along the water's edge. Early explorers in the glades thought that the wet basin was once a natural mountain lake, probably one that resulted when a landslide created a dam. The evidence, however, casts doubt on their theory. Since the downstream end of the glade is nearly 50 feet lower than the upper end and 500 feet wide, the area could not have been flooded unless a landslide built a dam 50 feet high and 500 feet wide. There are no remnants of any such dam. In addition, the deepest peat is at the upper end of the basin. If lake waters once covered the entire area, the thickest layers of peat would be at the lower end, where the water would have been deepest. Finally, some parts of the glade do not have peat deposits, an unlikely occurrence if the bog had developed in a lake.

H. D. Darlington, a biologist who studied the Cranberry Glades extensively, noted that the peat varied in thickness throughout the glade and was absent under the streams. If the basin had been covered by a lake, peat would be under the streams today. Darlington theorized that sediments formed levees along the streams; whenever the streams then overflowed, water was trapped in low-lying areas outside the stream banks. Thick accumulations of peat developed in these basins. Darlington also proposed that some peat may have been deposited in stream beds abandoned when drainage patterns shifted. Even in recent times, there is evidence that the Cranberry River has been straightening its course, eliminating some of its old meanderings.

Peat is so acidic that most of the plants that flourish in the adjacent mountain forests are unable to grow in the bogs. Plant life in the open bogs is dominated by two species of cranberries, which are trailing evergreens whose rosy summer flowers develop into tart red cranberries by September. (The species producing the larger berries is the same as that harvested commercially.) Both of these bog-inhabiting plants of the heath family are more common in the northern United States and Canada. Another heath that grows in the open bogs is the bog rosemary; its presence marks the southernmost place in the world where it grows naturally. Fungi living on and in the roots of the heaths and other bog-inhabiting plants help them to survive by making available nutrients that, in acidic conditions, are not generally found in usable form. Biologists call this mutually beneficial relationship *mycorrhiza*.

During spring and summer, the open bogs are green, punctuated by occasional flashes of yellow from the showy flowers of the swamp candle and pink from two kinds of wild orchids. In autumn, however, white dominates the vegetation, appearing like a blanket of light snow. Flowers of the white-top aster blend with the smaller milk-white flower clusters of the beaked rush, and the seed heads of cotton grass are entangled in masses of hairs.

Nonflowering plants, such as mosses and lichens, hug the surface of the peat. The mosses are so abundant in places that they form small mounds known as *hummocks*. The mosses and lichens are in constant competition. During extended periods of rain, the fast-growing gray lichens spread over the green mosses, forming an intricate parquet. But in dry weather, the lichens crumble into powder, and the mosses surge over their desiccated tissues.

Surrounding part of the open bogs are almost 400 acres of a nearly impenetrable thicket of slender-branched shrubs. There are 12 different kinds of shrubs, including the swamp rose and the speckled alder, whose stems are covered by conspicuous white-blotched pores that assist the plant in the exchange of gases. In summer, great masses of white woolly aphids cover the alders, giving the thickets a stark and eerie appearance. Beneath the shrubs grow a diversity of wildflowers that provide a year-round display, from the skunk cabbage, which forms its fetid flowers while snow still covers the ground in February, to the aster, turtlehead, and monkshood, which flower in the fall.

Sometimes lying between the wet alder thickets and the rich hardwood forests of the surrounding mountains is a bog forest where red spruce, Canadian hemlock, yellow birch, and black ash grow above a soggy forest floor covered by wood sorrel (pl. 38) and mosses. From the branches of the bog forest trees hang long festoons of the *Usnea* lichen, reminiscent of the totally unrelated Spanish moss.

The encroachment of woody plants from the surrounding shrub and forest communities is a major threat to the open bogs. Shrubby St. John's wort is particularly aggressive, but chokeberry and winterberry are invading at a rapid rate. Most trees cannot establish themselves in the bogs because they can't find firm anchorage, but a few have reached maturity where conditions are favorable. Some West Virginia botanists speculate that in just a few hundred years woody plants will transform the entire basin into a bog forest.

Until a few years ago, Cranberry Glades was several miles from the nearest public road and virtually inaccessible, except to the ardent hiker. Today, however, the Highland Scenic Highway passes near the site, a modern visitor center has been erected nearby, and a short boardwalk has been provided. The boardwalk passes through a bog forest and an alder thicket before cutting across the corners of two of the open bogs. It not only protects the sensitive ecosystem from indiscriminate trampling but also keeps visitors from miring in up to their waists.

Dolly Sods

Walking across the flat terrain of Dolly Sods, a tundralike setting largely within West Virginia's Monongahela National Forest, people are apt to forget they are on a mountaintop until they reach Bear Rocks at the northeast corner: There they can look straight down 2,500 feet to the valley below. Thirteen miles long by 2 to 4 miles wide, Dolly Sods sits atop a ridge of the Allegheny Mountains, which form the eastern section of the 4,000-foot-high Allegheny Plateau. As in the Appalachians to the east, the folds of this ancient mountain terrain run from the northeast to the southwest.

While pine forest covers many of the surrounding mountains and valleys, only a few red spruce trees dot Dolly Sods. They are sometimes called *flagged spruces* because all their branches are on the east side of the trunk and resemble flags blowing in the breeze. Branch growth on the west side is inhibited by the constant drying force of the wind and abrasion of the buds by windblown ice crystals.

Much of Dolly Sods consists of treeless balds, whose shallow, acid soils mainly support shrubby members of the heath family. Blueberries, huckleberries, and cranberries provide ample feasts for wildlife, as well as for hordes of human berry pickers. Some other heaths include the trailing arbutus, which blooms after the snows melt in late April, and mountain laurel, azaleas, and rhododendrons, whose blossoms appear soon thereafter. A smattering of wildflowers grows alongside the heaths. Pink lady's-slipper orchids form their pouchlike flowers in May, and the dainty painted trillium appears by the first of June. Wildflowers continue to bloom throughout the short summer until the deep blue flowers of the wild gentian close out the show. All these plants must survive severe winds and a cold climate. As much as 150 inches of snow may fall in winter, forming drifts that do not completely melt until May. Frost may occur any day of the year, and by September, more snow may be on the way.

Scattered in poorly drained depressions of the high plateau are soggy areas with continuous mounds of sphagnum and haircap mosses. The plant species contained in these sphagnum glades—which include cranberries and tiny insect-consuming sundew plants—are similar to those found in bogs much farther north, but the origins of these habitats are different. Northern bogs usually developed from ponds left during the retreat of the last Ice Age glaciers, while the sphagnum glades on Dolly Sods lie over water-impervious rocks that impede the drainage of mountain streams. The cool, cloudy, and often foggy climate also helps to maintain the high moisture level.

In a few areas of Dolly Sods, where limestone rock lies near the surface, grass grows in dense mats. Early 19th-century settlers, unable to farm the rocky mountaintop successfully, turned to raising grazing animals to take advantage of these isolated grassy patches. One of the earliest settlers in the region was the Dahle ("Dolly") family, and the dense grass on their land claim was referred to as "the sod." Although the original grassland settled by the Dahle family occupied only about one square mile, the name Dolly Sods now applies to the entire tableland.

At the time the early settlers exploited these grass patches, the rest of the mountaintop was nearly covered in forest. Today there are no stumps or other visible traces of this forest, but in 1746, members of the Fairfax Boundary Survey Party reported that as they walked west from Bear Rocks, they passed through extensive stands of red spruce so dense that no sunlight reached the forest floor. Underfoot were thick deposits of peaty soil derived from decomposing spruce needles and mosses. The red spruce forest was a remnant of a tongue of northern coniferous forest carried southward along Appalachian crests. When the forest was in this virgin state, its plants and animals consisted of species now typical of Canada—spruces, dwarf dog-woods, northern warblers and thrushes, and snowshoe hares.

In 1884, a railroad was built from Davis, a few miles northwest of Dolly Sods, to the Eastern Seaboard, opening up the area to lumbering. In a matter of 35 years, the entire mountaintop was denuded of timber, leaving a barren landscape covered with slash. The thick mat of spruce needles that had been accumulating for centuries dried out once the canopy was removed, forming ready tinder. Fires became common. Some were started by sparks from logging trains, others by people attempting to create pasture, still others by lightning. Sometimes ground fires smoldered for weeks or months, feeding on the organic soil until they reached bedrock. Only the sphagnum bogs, which occupied the low areas of the tableland and held great quantities of water, were spared, providing sanctuaries for species like the sundews.

Today, although red spruce is making a slow recovery in sheltered coves, only pioneer mosses and liverworts have been able to etch their home on some of the bare boulder fields. And blueberry pickers continue to set fires to retard the growth of the trees invading the blueberry patches.

Thus Dolly Sods is far from being a pristine area. Nevertheless, naturalists have a reason to flock there: to glimpse the same kinds of plants that also grow naturally in the boreal regions of North America. For example, 1,000 miles northeast and 4,000 feet lower in elevation, near sea level along the eastern coast of Canada, there is relatively flat terrain with scattered boulders. In areas where the climate is extreme, growth of vegetation is sparse, and the resultant landscape often appears similarly barren and bleak.

Gaudineer Knob

Managed in large part by the Monongahela National Forest, the Allegheny Highlands of West Virginia consist of a series of lofty ridges that run northeast to southwest, frequently intersected by rushing mountain streams that have carved deep gorges. When the first European settlers reached this region during the early 18th century, they found it cloaked in some of the densest forests east of the Mississippi. In the gorges and on the steep mountain slopes, mostly below 3,200 feet, was a grand mixture of deciduous trees—sugar maple, American beech, basswood, the wintergreen-scented yellow birch, red maple, sweet birch, and cucumber magnolia—along with white pine and other conifers. On the highest peaks (or knobs, as West Virginians call them) grew nearly pure stands of red spruce.

The spruces grew straight and tightly spaced over a dark and impenetrable growth of white rhododendron. West Virginia botanist Earl Core reports that the first settlers spoke of the dark spruce forests "in dread, as an ill-omened region, filled with bears, panthers, impassable laurel brakes, and dangerous precipices." Remnants of the boreal forest that reached south in Ice Age times, the red spruce stands grew only above 3,200 feet, where the cool climate and abundant rainfall, along with thick daily fogs, emulated conditions far to the north in Canada, where red spruce predominates today.

Core estimated that the red spruce forests in West Virginia originally covered 469,000 acres. From 1770 to 1920, however, lumbering was the chief industry in West Virginia, and nearly every square foot of forest succumbed to the ax and the saw, as detailed in botanist Roy Clarkson's book *Tumult on the Mountains*. Many areas were clear-cut, after which they were usually swept by fires (deliberately set, accidental, or natural) that burned out of control. Among the places so denuded was 4,445-foot Gaudineer Knob, the highest point on Shaver Mountain, which was logged and possibly burned between 1905 and 1915. But about 0.25 mile down from the knob, a 140-acre tract of virgin red spruce was spared cutting—because of a surveyor's error —and today is managed by the Forest Service as the Gaudineer Scenic Area.

Some of the red spruces are more than 100 feet tall and have a diameter of 3 feet when measured four and a half feet above the ground. These rugged monarchs are estimated to be nearly 300 years old; unfortunately, some are beginning to show their age and have dead crowns and upper branches. Others have succumbed to various insect infestations. Many large, fallen logs lie across the forest floor, covered with dense growths of mosses, lichens, and liverworts encouraged by the very moist conditions. Where the canopy has been opened following the fall of a large tree, yellow birch and American beech have become established with a shrub layer of laurel. Handsome ferns with 3-foot fronds grow in openings where sunlight is able to penetrate to the forest floor. The white mountain oxalis is the most common wildflower in the stand.

One notable animal that lives in the spruce forest is the Cheat Mountain salamander, which is on the federal list of threatened species. Discovered in 1935, this salamander is confined to an area of less than 1,000 square miles in West Virginia, living at elevations above 3,120 feet, primarily where red spruce is a prominent part of the vegetation. It spends its days under rocks and logs, coming out at night to forage on the forest floor for mites, small beetles, flies, and other insects. Marshall University herpetologist Thomas Pauley has noted that this salamander survives in places that are cooler and more humid than is optimal for most other salamanders. (However, on a recent trip to Gaudineer Scenic Area when snow was still scattered on the ground, I located a half-frozen four-toed salamander hiding under a birch log.)

Since the uncut spruce forest represents a little bit of Canada in West Virginia, some animals with northern affinities can be found there. The snowshoe hare and the rock vole are two such mammals; among the birds are the chestnut-sided warbler, magnolia warbler, Blackburnian warbler, solitary vireo, Swainson's thrush, northern water thrush, dark-eyed junco, and golden-crowned kinglet. All except the last two are neotropical migrants, overwintering in the rainforests of Mexico, Central and South America, and the Caribbean.

Chemist George A. Hall, who is a noted ornithologist, and other birders have studied the effect on bird populations of vegetation regrowth on Gaudineer Knob, whose original stand of spruce was harvested by 1915. Shortly after the knob was cleared, great numbers of young spruce sprouted from seeds that had blown in from adjacent forests. By 1937, the trees had grown to shoulder height and were densely packed. Beneath them, thickets of mountain laurel made the forest nearly impassable and apparently prevented the regrowth of white rhododendron, lost in the forest cutting. Here and there, yellow birches and mountain ashes found a little space to grow. The magnolia warbler, a ground-inhabiting bird, thrived under these conditions.

By the early 1980s, Hall reported, some of the red spruces had grown 50 feet tall. As their crowns coalesced, the understory of mountain laurel was shaded out and died, leaving the forest floor covered by an accumulation of dead needles, old branchlets, carpets of mosses, and old stumps of white rhododendron. As the understory vegetation was lost, the number of magnolia warblers also declined, down to levels characteristic of the virgin forest below the knob.

Hall has observed a decrease among some bird populations in the virgin spruce stand as well, particularly the neotropical migrants. The solitary vireo and the Blackburnian warbler, as well as the magnolia warbler, have shown a significant decline since 1968. This may result from the rapid clearing of forests in the birds' wintering grounds. Ornithologist John Terborgh has estimated that the loss of 1 acre of tropical forest where there is a concentration of wintering birds is equivalent to removing 5 or 6 acres of breeding habitat in temperate regions. As these bird species have decreased at Gaudineer, the golden-crowned kinglet (not a neotropical migrant) has increased in numbers, perhaps moving into niches vacated by the declining warblers.

Otter Creek

Originating about 3,050 feet above sea level in the Allegheny Mountains of West Virginia, Otter Creek rushes north for 7 miles until, at 1,780 feet, it empties into the Dry Fork of the Cheat River. On the way it passes through a basin about 4 miles wide, flanked by parallel ridges that crest at 3,800 feet: Shavers Mountain on the east and McGowan Mountain on the west. The basin was extensively logged between 1890 and 1914 and sporadically beginning in 1958, but in 1972, the entire 32-square-mile drainage was set aside as Monongahela National Forest's Otter Creek Wilderness.

Because of the prior logging, most of the forested land is second growth, but many beautiful sugar maples, American beeches, yellow birches, and wild black cherries may be found. At higher elevations, particularly on Shavers and McGowan mountains, hemlocks and red spruces intermingle with the hardwood trees. Along Yellow Creek, one of the babbling streams that empties into Otter Creek, there is a stand of red pine. Another tributary of botanical interest is Moore's Run, whose upper reaches flow through a sphagnum bog. The bog is a legacy of the great continental ice sheet that covered the area about 12,000 years ago and scoured out depressions where water accumulates. Some of the wetland plants that have filled these depressions are the same species that grow far to the north. Sundews, gentians, yellow bartonia, white sedge, white beak rush, and skunk currant are just a few of the species recorded in the bog.

One of the animals that lives in the Otter Creek Wilderness is the West Virginia flying squirrel (pl. 39), a subspecies confined to the mountains of West Virginia and adjacent Virginia. A flying squirrel does not actually fly but glides, leaping as much as 150 feet from tree to tree without having to descend to the ground where it is vulnerable to such predators as owls and hawks. The squirrel controls its glide by maneuvering the winglike flaps of skin that extend from its wrists to its ankles and using its broad, flat tail as a rudder. Wildlife photographer Tom Ulrich, who has patiently observed these secretive and nocturnal animals, notes that before taking off, the squirrel checks the distance and elevation of its target with quick head movements in a triangular motion. Once in the air, the descending squirrel rapidly gains speed, but as it approaches the landing tree, it inclines its body upward, resulting in a relatively gentle, four-point landing against the trunk.

Northern flying squirrels, which range across Canada, with extensions into the Appalachian Mountains and the Sierra Nevadas, are one of two flying squirrel species found in North America. They generally live in coniferous forests or in forests with a mixture of conifers and hardwoods, and their diet consists mostly of lichens and fungi. They are also reported to drink tree sap. The second species, the slightly smaller and lighter southern flying squirrel, lives primarily in deciduous hardwood forests from southern Maine to Florida and from the eastern Dakotas to southeastern Texas. Unlike its northern cousin, it gets most of its nutrition from seeds and nuts.

Because there are some differences among the northern flying squirrels inhabiting the various regions of North America, zoologists recognize 25 subspecies, including that of West Virginia. Most likely, this diversity arose as a result of the last Ice Age, when plant and animal species in the far north were displaced south by great glaciers. As the glaciers retreated, some species, such as the red spruce and the northern flying squirrel, were left behind on the tops of the highest mountains, which became coniferous islands surrounded by lower-elevation hardwood forests. Because of their relative isolation, the small populations of northern flying squirrels left in different regions became distinct.

The West Virginia northern flying squirrel is now so rare that it has been declared an endangered animal by the U.S. Fish and Wildlife Service and is subject to all the protection provided by the Endangered Species Act. Wildlife biologist Craig Stihler and his West Virginia colleagues have been studying the habitat requirements and lifestyle of the subspecies in order to prepare a management plan that will prevent its extinction. Despite intensive efforts to capture, examine, tag, and release as many animals as possible, they have located fewer than 100 specimens.

Most of the squirrels live above 3,300 feet on cool, moist, north-facing slopes in a transition zone where spruce is intermixed with several kinds of deciduous hardwoods. The squirrels require a forest zone with mature, widely spaced trees whose dense shade restricts the understory vegetation (the impenetrable tangles of rhododendron that grow beneath younger forest trees interfere with the ability of the squirrels to glide freely). The natural cavities in older, decaying hardwoods also provide the nesting sites the squirrels need, at least in the cooler seasons. Stihler has observed as many as seven adults and juveniles sharing the same nest.

One of the complications for those concerned with preserving the West Virginia northern flying squirrel is that the southern species, which inhabits hardwood forest, sometimes lives in the same area. Even though slightly smaller, the more aggressive southern flying squirrels tend to drive the northern flying squirrels higher up the mountain, restricting them to the limited tracts of conifer-dominated forest. In addition, researchers have found that the southern flying squirrels harbor parasitic roundworms, which are probably transmitted from animal to animal by means of an intermediate host. The southern species seems to tolerate the parasite well, but it may prove harmful if transferred to the northern flying squirrels.

NATIONAL FORESTS IN WISCONSIN

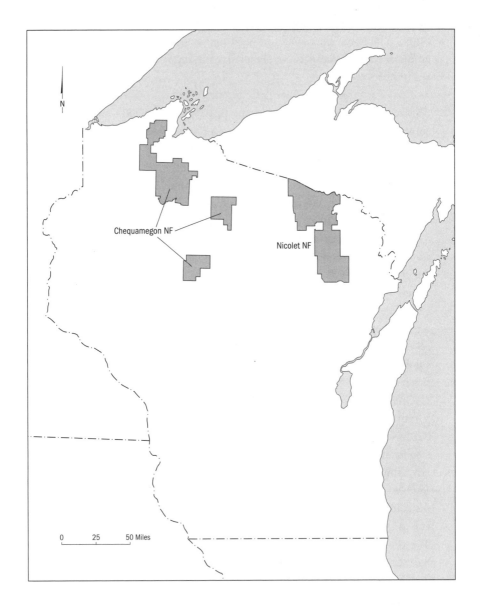

Wisconsin has two national forests, but in 1998, they were both managed as one administrative unit, with each maintaining its own headquarters. The national forests of Wisconsin are in the Eastern Region of the United States Forest Service. The Eastern Region Office is located at 626 W. Wisconsin Avenue, Milwaukee, WI 53202.

Chequamegon National Forest

SIZE AND LOCATION: 861,150 acres in north-central Wisconsin, in the north woods country. Major access routes are U.S. Highways 2 and 63 and State Routes 13, 64, 70, 77, and 182. District Ranger Stations: Glidden, Hayward, Medford, and Park Falls. Forest Supervisor's Office: 1170 Fourth Avenue, Park Falls, WI 54552, www.fs.fed.us/r9/cnnf.

SPECIAL FEATURES: Winter sports areas; boat ramps; mountain biking.

SPECIAL ATTRACTIONS: Moquah Barrens; Great Divide National Scenic Byway; North Country National Recreation Trail; Ice Age National Scenic Trail; Morgan Falls/St. Peter's Dome.

WILDERNESS AREAS: Porcupine Lake (4,292 acres); Rainbow Lake (6,583 acres).

The Chequamegon is the more westerly and northerly of the two national forests in Wisconsin. As a result, a variety of vegetational communities may be found, including several with truly northern affinities. In some places where there are exposed uplands and cold, wet valleys, there is typical northern coniferous forests where balsam fir, black spruce, and white spruce occur. Nutrient-rich and well-drained soil often has a northern mixed hardwood forest. Species encountered here include hemlock, sugar maple, yellow birch, basswood, aspen, white pine, and white ash. Forested wetlands occur in basins or depressions where the soils are saturated and contain little peat. Trees in this forest often consist of black ash, red maple, and many species of willow, with smaller trees and shrubs of tag alder and winterberry present. Marsh marigold is usually a prominent early spring wildflower.

Permanently wet areas often develop on shallow soil that leads to poor drainage. Where there is shallow standing water for most of the year, marshes develop. In poorly drained soil with sphagnum usually present, bogs form, and where the bedrock is mostly composed of calcium and magnesium, fens are found. Old bogs may fill in with larches and black spruces. Marshes usually have wild blue iris, swamp milkweed, spotted joepyeweed, wild calla, and several species of beggarticks (*Bidens*).

Bogs contain a highly interesting flora that usually consists of bog rosemary, Labrador tea, pitcher plant, sundew, bog candles, bog laurel, and often two species of wild cranberry. Fens may often be distinguished by the occurrence of a native thistle, bog lobelia, and shrubby cinquefoil.

One area in the Chequamegon that is unique is the Moquah Barrens. This dry sandy area of scrubby pines is the northernmost of 10 such areas in northwestern Wisconsin. Most of the barrens, which are located northeast of Iron River, are maintained by the Forest Service by prescribed fires and timber harvests. If it were not for these management techniques, the pine barrens would soon lose their unique identity.

This type of barrens at one time covered a much larger area of northern Wisconsin. Among the plants that find the dry, sandy habitat of the barrens ideal beneath the jack pines are bearberry, cowwheat (a small, yellow-flowered member of the snapdragon family), sand cherry, sweet fern, several kinds of grapeferns, and jointweed, whose relatives include the wetland plants known as smartweeds. Blueberries are common on the barrens, attracting many berry pickers.

The 861,150 acres of the Chequamegon National Forest contain more than 400 lakes that fill depressions gouged out by previous glaciers. The glaciers, which covered most of the area as recently as 12,000 years ago, have left several other significant landforms. Many of these can be observed along the remarkable Ice Age National Scenic Trail. This 100-mile-long trail winds through public and private lands, with some of it in the Chequamegon National Forest. The trail follows the end moraines from the most recent glaciation about 12,000 years ago. The glaciers had been advancing and retreating for 100,000 years. Several lobes of ice extended south into Wisconsin, carrying with it sand, pebbles, and boulders. Where the ice finally receded, it left behind rocky, sandy, and gravelly ridges known as *moraines*. Large blocks of rocks, called *erratics*, were also left behind here and there. As you hike the trail, you will traverse eskers and kettle holes.

The Ice Age Trail crosses the lowest district of the Chequamegon, entering north of the village of Lublin. The trail passes through a mature hemlock forest south of the Chequamegon Water Flowage and then circles through an area of high ridges and bog eskers south of Perkinstown. The Chequamegon Water Flowage has a wildlife observation area at the south end of the lake as well as areas for camping, picnicking, and boating. After circling east of Richter Lake, the Ice Age Trail goes through a red pine plantation and over White Birch Ridge and Yellow Birch Ridge before reaching Jenny Lake, then winding north and then east past Lost Lake Esker and a large bog immediately south of the trail. It then goes across the top of the Ice Age Non-motorized Area and through a mature white pine stand to the Mondeaux Flowage. Near Perkinstown is a popular cross-country ski area, and there is a campground and boat launch at Kathryn Lake. Several points of interest are near the Mondeaux Flowage, including the historic dam built by the Civilian Conservation Corps in the 1930s, several campgrounds, picnic areas, and boat ramps. The Mondeaux Lake hiking trail follows the west side of the lake for about 2.5 miles, across the Mondeaux Esker.

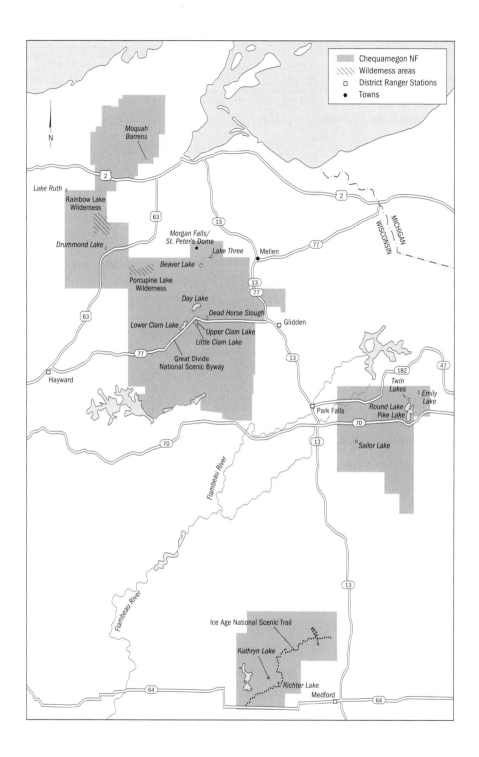

The largest section of the Chequamegon National Forest is crossed by the Great Divide Scenic Byway (State Route 77) that bisects the national forest from east to west from Glidden to Hayward. The byway roughly follows the divide between waters flowing north to Lake Superior and south to the Mississippi River. On both sides of the scenic byway are countless marshes, lakes, and other wetlands where waterfowl and aquatic animal life abound. Fishing is good for muskellunge, trout, walleye, and northern pike.

Several recreation sites are clustered near the center of the scenic byway and also along the western edge. The central area includes the Twin Lakes, Day Lake, Dead Horse Slough, Upper Clam Lake, Little Clam Lake, and Lower Clam Lake. Several hiking and ski trails are in this area, including the popular West Torch Ski Trail. Activities north of the western edge are centered around another group of lakes, although several of the lakes are not on national forest land. The Mukwonago Ski Trail is here.

About 15 miles north of the Mukwonago Ski Trail is the Porcupine Lake Wilderness. The wilderness provides a good cross section of most of the forest types in the Chequamegon. In moist but not wet areas are woods dominated by sugar and red maples and yellow birch, all providing vivid autumnal coloration. Drier regions support a forest of red oak, quaking aspen, white pine, and hemlock, whereas the wettest areas contain white cedar, larch, and white spruce. The western part of the wilderness is hilly, while the eastern side has much of the wet forest habitats. A few lakes are located in the wilderness, including Porcupine Lake toward the western side. The North Country National Recreation Trail passes through the wilderness and around the upper end of Porcupine Lake.

The North Country National Recreation Trail actually begins at the northwestern corner of the Chequamegon National Forest at Lake Ruth nearly 20 miles northwest of the Porcupine Lake Wilderness. After leaving Lake Ruth, the trail cuts diagonally across another wilderness area, the Rainbow Lake Wilderness. After leaving the wilderness, the trail in less than 5 miles comes to an interpretive trail at Drummond Woods. This interesting trail passes through a very pretty forest. A short distance to the west of the North Country National Recreation Trail is Drummond Lake and the northwestern end of the Drummond Cluster of the Camba Mountain Bike Trail. The scenic trail then circles north of Two Lakes and their campground before reaching the Porcupine Lake Wilderness. After entering the wilderness on the west side, the trail continues eastward, eventually becoming more rocky and rugged when it reaches the Penokee Hills and exiting the national forest just before the town of Mellen.

The North Country National Recreation Trail passes through campgrounds at Beaver Lake and Lake Three. From Lake Three there is a forest road to two of the more unusual features in the Chequamegon National Forest, St. Peter's Dome and Morgan Falls. A 1.5-mile trail to the top of St. Peter's Dome, which is a high granite knob, results in a 360-degree overview of the region. Morgan Falls has a spectacular scenic drop of sparkling water.

Fewer than 15 miles south of Lake Superior, the Moquah Barrens lies in the most northern section of the Chequamegon National Forest. The Old Baldy Overlook northwest of the barrens has a spectacular view, as does Mount Valhalla to the northeast. Mount Valhalla also has a winter sports area and a hiking trail. Two miles south of Mount Valhalla, at the Birch Grove Campground on the shore of Twin Lakes, is the 0.75-mile West Twin Lake Hiking Trail.

A disjunct section of the Chequamegon National Forest lies east of Park Falls. State Route 70 east from Fifield crosses that part of the national forest and provides access to all of the features. South of this highway are the significant Schmuland Wetland Area and the Riley Lake Wildlife Management Area. There is a nice campground and boat launch area at Sailor Lake at the northern end of the Schmuland Wetland Area. Several lakes are north of State Route 70, with campgrounds at Emily Lake, Twin Lake, and Smith Rapids. Smith Rapids has a covered bridge over the South Fork of the Flambeau River, which is a great canoe stream. Although built in 1991, the covered bridge is worth seeing because of its diamond-shaped truss pattern known as Town Lattice, named for Itel Town, the designer of this type of truss.

Near the eastern edge of this section of the national forest are Round Lake and Pike Lake. Round Lake is significant because of its historic logging dam on its western side. A 0.25-mile east of the north shore of Round Lake is the Tucker Lake Hemlocks Natural Area, an old-growth hemlock and yellow birch stand. Along with these trees, other canopy species include basswood, red maple, white ash, balsam fir, and white pine. A shrub layer of mountain maple, beaked hazelnut, fly honeysuckle, and leatherwood is also present. On the forest floor is an abundance of intermediate wood fern and attractive wildflowers such as starflower, Canada mayflower, rosy twisted stalk, and small enchanter's nightshade. Southeast of Pike Lake is the impressive Memorial Grove and Caro Forest Trail. The Memorial Grove is a 64-acre tract of virgin timber that its owner, Anne Caro, gave to the United States Forest Service to protect it from private timber interests. The Caro Forest Trail has a short loop and an optional difficult, longer side loop. The short loop passes a glacial kettle hole and crosses two eskers. There are large, handsome basswood trees along this short loop, as well as beautiful sugar maples. The longer loop winds through yellow birches and hemlocks. A picnic area is near the trailhead.

Nicolet National Forest

SIZE AND LOCATION: 664,169 acres in northeastern Wisconsin. Major access routes are U.S. Highway 8 and State Routes 32, 52, 70, and 139. District Ranger Stations: Eagle River, Florence, Lakewood, and Laona. Forest Supervisor's Office: 68 S. Stevens Street, Rhinelander, WI 54501, www.fs.fed.us/r9/cnnf.

SPECIAL FEATURES: Boat ramps; campgrounds on the water.

SPECIAL ATTRACTIONS: Cathedral Pines; Lakewood Auto Tour; Assessor's Interpretive Trail; Franklin Nature Trail.

WILDERNESS AREAS: Blackjack Springs (5,900 acres); Headwaters (20,000 acres); Whisker Lake (7,300 acres).

The Nicolet is the easternmost of Wisconsin's two national forests, and it is far enough north to have fine stands of mixed hardwood forest species. Several areas have mature, old-growth forests where some of the most majestic white pines, hemlocks, basswoods, sugar maples, and specimens of other species may be admired. Those regions with trees that seem to reach to the sky are referred to as *cathedral forests*, and one of the prettiest and easiest to reach is the Cathedral Pines in the southern part of the Nicolet National Forest, about 2 miles southwest of State Route 32 and a few miles west of Lakewood.

Southeast of Cathedral Pines on a side road east of State Route 32 is an observation point at Mountain Lookout Tower offering fine views. North of Cathedral Pines and on a short spur road to the east of State Route 32 is another good observation point on Quartz Hill. Fanny Lake is a pleasant lake with a popular combined cross-country ski trail and hiking trail, and small walk-in campsites. One of the best trout streams in the national forest is Jones Creek south of Jones Springs, while Fanny Lake is great for panfish and bass.

One of the best ways to get a feel for the southern part of the Nicolet National Forest is to drive the 65-mile Lakewood Auto Tour. A brochure describing 17 tour stops along the Lakewood Auto Tour is available at the Lakewood Ranger Station.

The northern district of the Nicolet National Forest lies east of the communities of Eagle River and Three Lakes and extends to the Michigan state line, where it adjoins the Ottawa National Forest. The western side of this district contains several important recreation and natural areas. Lac Vieux Desert, a lake partly shared by the state of Michigan, is at the northwest corner of the national forest. There is a campground, picnic area, and boat launch area at the western edge of the lake. Ten miles south is Blackjack Springs Wilderness, a densely forested region centered around four crystal-clear springs that form the headwaters of Blackjack Creek. The hilly terrain in the wilderness is a direct result of the glacial history of the area.

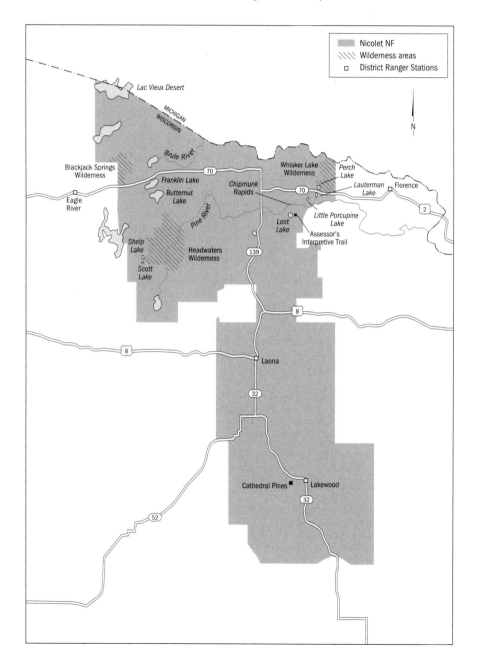

Southeast of the wilderness is Anvil Lake, which has a campground, boat ramp, and a number of hiking trails into the surrounding forest. At the southern end of the lake is the Old Military Trail, now designated as the Heritage Drive National Scenic Byway. This road heads due south for 11 miles, terminating at Virgin Lake at the edge of the national forest. After the scenic byway crosses Ninemile Creek, a branch of it to the northeast goes to the Franklin Lake–Butternut Lake area. The road actually goes between the two lakes. This is one of the finest parts of the national forest. The Franklin Nature Trail is a great way to become familiar with the natural features of the Nicolet. The trail passes through the Avenue of Giants where some of the largest hemlocks, pines, basswoods, and oaks are more than 400 years old. Another part of the trail goes through the Hemlock Cathedral where some of the tallest hemlocks in the forest live. There is a very short side trail, about 1 mile long, to Butternut Lake. A boardwalk allows you to hike across a part of a good tamarack swamp where, in season, you may see lady's slipper orchids, pitcher plants, and other attractive bog species. Poison sumac is in the bog as well. At one point the trail passes beneath a natural arch formed by evergreen trees. Where the scenic byway makes a sharp turn to the east, a forest road heads due south. In about 1 mile, this forest road comes to the Sam Campbell Road, named for a local writer, philosopher, and lover of nature. The 1.75-mile self-guiding Sam Campbell Memorial Trail begins here and passes through a serene forest of pines, balsam firs, maples, and basswoods. The trail swings down to Wegimind Point on the north bank of Four Mile Lake and then to tiny Vanishing Lake, both areas prominent in Campbell's writings. Conveniently placed benches are along the trail.

Less than 5 miles to the east of Sam Campbell Memorial Trail is the Headwaters Wilderness, the largest in the Nicolet National Forest. Two forest roads cross the area, dividing it into three units. The wilderness gets its name because the headwaters of the Pine River, a National Wild and Scenic River candidate, are in the southeastern unit of the wilderness. Although all three units have some ridges that support hardwood forest species, much of the area contains bogs and forested wetlands. Quiet and peaceful Shelp Lake is at the southwestern corner of the wilderness, as is the Giant Pine Trail. The trail may be reached by one of the roads that bisects the wilderness. This is a fabulous trail to see what the wilderness is about. From the parking area along the forest road, you may take the 1.5-mile Hemlock Ridge Loop of the Giant Pine Trail that completely encircles a fine example of a spruce swamp. The back end of the trail follows a ridge where hemlock trees abound. A forest of magnificent and majestic giant white pines is at the northeastern corner of the loop before the trail climbs onto the beautiful ridge. The main loop trail is a ski-only trail that branches north at the upper end of the Giant Pine forest and makes a 6-mile loop, eventually encircling an even larger spruce swamp. Just outside the southern boundary of the wilderness, at the northern edge of Scott Lake, is another hiking trail through a forest of giant hemlocks. The northern part of the wilderness contains Kimball Creek, which eventually empties into the Pine River. There are informal unmarked hiking trails along the creek.

The northeast corner of the Nicolet National Forest lies west of the town of Florence and contains other points of interest including the Whisker Lake Wilderness. Six nice trout-filled lakes are located within the wilderness, including Whisker Lake, bordered on some of its shorelines by stands of large white pines that the locals refer to as "chin whiskers" because of their needles. The Brule River forms the northern boundary of the wilderness, and it also serves as the boundary between Wisconsin and Michigan. The Brule River is a great one for canoeing.

Fifty-one-acre Perch Lake, just outside the western edge of Whisker Lake Wilderness, is a place for isolation and seclusion from the outside world. The five campsites around the lake can be reached only by trail from the parking area. The forest around Perch Lake contains hemlock, white pine, and spruces, along with deciduous trees such as white birch, yellow birch, sugar maple, red maple, and basswood. Listen and watch for loons (pl. 40) on the lake. Barely 1 mile south of Perch Lake is Lauterman Lake where you may hike the Lauterman National Recreation Trail. This trail has loops around Lauterman Lake, Little Porcupine Lake, and Lost Lake. The trail also goes past Chipmunk Rapids on the Pine River. Lost Lake features an interpretive nature trail.

ART CREDITS

Special thanks to Elton Sonny Cudabac at the USDA Forest Service, Southern Region, for his assistance in providing photos for this book.

Plates

Kevin Adams, 1, 5, 22, 23

Bill and Laura Hodge, 31, 32

Joanne Miller, 26

Robert Mohlenbrock, 2, 4, 6, 7, 9, 10, 11, 12, 13, 14, 15, 17, 18, 24, 30, 34, 36

Jerry and Marcy Monkman/Ecophotography.com, 19, 20, 21, 33

Bill Schmoker, 16, 28, 40

Craig W. Stihler, West Virginia Department of Natural Resources, 37, 38, 39

USDA Forest Service, Southern Region, 3, 25, 27, 29

John White, 8, 35

Figures

Robert Mohlenbrock, title page, 1, 3, 4, 5, 6, 7, 8, 9, 12, 13, 16, 17, 19, 20

Jerry and Marcy Monkman/Ecophotography.com, 18

Bill Schmoker, 11

USDA Forest Service, Southern Region, 2, 10, 15

John White, 14

INDEX OF PLANT NAMES

bush clover, Virginia *(Lespedeza virginica),* 84

buttercup,
  creeping *(Ranunculus repens),* 130
  hooked *(Ranunculus recurvatus),* 172
butterfly weed *(Asclepias tuberosa),* 87
butternut *(Juglans cinerea),* 102
butterweed *(Senecio glabellus),* 169
butterwort, blue *(Pinguicula caerulea),* 4, 27, 31, 134, 221
buttonbush *(Cephalanthus occidentalis),* 24, 36, 106, 118, 217, 321

cactus, prickly pear *(Opuntia humifusa),* 81, 90, 94, 244
camellia,
  mountain *(Stewartia ovata),* 59
  silky *(Steuartia ovata),* 185, 284
  wild *(Stewartia malacodendron),* 14
  wild *(Stewartia ovata),* 14
camphor weed *(Pluchea camphorata),* 170
campion,
  moss *(Silene acaule),* 202
  starry *(Silene stellata),* 59, 114
cane
  giant *(Arundinaria gigantea),* 25, 169, 174, 182, 216, 294
  maiden *(Panicum hemitomon),* 40, 331
cardinal flower *(Lobelia cardinalis),* 118
cassiope, moss *(Cassiope hypnoides),* 200, 202
catchfly
  round-leaved *(Silene rotundifolia),* 118
cat-tail,
  common *(Typha latifolia),* 86, 143, 265
  narrow-leaved *(Typha angustifolia),* 212
cecropia *(Cecropia peltata),* 271
cedar,
  Atlantic white *(Chamaecyparis thyoides),* 24, 40, 48, 50, 216
  eastern red *(Juniperus virginiana),* 92
  red *(Juniperus virginiana),* 78, 83, 87, 94, 102, 244
  southern red *(Juniperus silicicola),* 217
  white *(Thuja occidentalis),* 51, 121, 124, 131, 139, 149, 156, 199, 366
cherry,
  pin *(Prunus pensylvanica),* 141, 207, 256
  wild black *(Prunus serotina),* 23, 36, 42, 65, 100, 149, 174, 176, 178, 182, 217
chestnut *(Castanea dentata),* 231, 239
  American *(Castanea dentata),* 105, 330
chickweed
  giant *(Ceratium vulgatum),* 114
  star *(Cerastium arvense),* 240
chokeberry
  black *(Aronia melanocarpa),* 159, 79, 149, 208

red *(Aronia arbutifolia),* 25, 131, 318
red *(Aronia prunifolia),* 25, 131, 318
cicely, sweet *(Osmorhiza claytonii),* 67, 257
cinquefoil
  prairie *(Potentilla hippeana),* 151
  shrubby *(Potentilla fruticosa),* 132, 363
  three-toothed *(Potentilla tridentata),* 62, 64, 202
clearweed *(Pilea pumila),* 149, 170
clematis,
  Appalachian *(Clematis verticillata),* 326
  blue-flowered *(Clematis pitcheri),* 169
  hairy *(Clematis albicoma),* 319
  shale barren *(Clematis albicoma),* 326
  white-tailed *(Clematis albicoma),* 327
clintonia *(Clintonia borealis),* 333
clover
  buffalo *(Trifolium reflexum),* 327, 340
  Kates Mountain shale barren *(Trifolium virginianum),* 320, 327
  sweet *(Melilotus officinalis),* 106
  Virginia bush *(Lespedeza virginica),* 84
clubmoss
  fan-leaved *(Lycopodium complanatum),* 79
  obscure *(Lycopodium obscurum),* 61
  rock *(Selaginella rupestris),* 95
coffee tree, Kentucky *(Gymnocladus dioica),* 100
cohosh
  black *(Cimiciga racemosa),* 59, 61, 77, 343
  blue *(Caulophyllum thalictroides),* 14, 17, 67, 91, 207, 256
coinwort *(Centella erecta),* 176
colicroot *(Aletris aurea),* 24
  yellow *(Aletris lutea),* 25, 175
columbine, wild *(Aquilegia canadensis),* 303
coneflower
  purple *(Echinacea purpurea),* 80, 92, 104, 163
  yellow *(Ratibida pinnata),* 93
coontail *(Ceratophyllum demersum),* 49
coontie *(Zamia floridana),* 44, 45
cordgrass, salt marsh *(Spartina alterniflora),* 216, 217
coreopsis, large *(Coreopsis grandiflora),* 58
corydalis
  golden *(Corydalis aurea),* 60
  spring *(Corydalis flavula),* 170
cottonwood, swamp *(Populus heterophylla),* 2, 25, 74, 85, 284
cowwheat *(Melampyrum lineare),* 148, 364
cranberry,
  large *(Vaccinium macrocarpon),* 139, 208, 331, 363
  mountain *(Vaccinium vitis-idaea),* 202, 203

intermediate *(Dryopteris intermedia)*, 60, 64, 207, 367

interrupted *(Osmunda claytoniana)*, 106, 149

Japanese climbing *(Lygodium japonicum)*, 174

lip *(Cheilanthes lanosa)*, 95

Louisiana shield *(Dryopteris ludoviciana)*, 174

maidenhair *(Adiantum pedatum)*, 67, 114, 184, 207

marginal shield *(Dryopteris marginalis)*, 91, 105

marsh *(Thelypteris palustris)*, 44, 106, 149

mosquito *(Azolla mexicana)*, 88, 286

netted chain *(Woodwardia areolata)*, 22, 34, 44

oak *(Gymnocarpium dryopteris)*, 149, 156

Peters' filmy *(Trichomanes petersii)*, 17

purple cliffbrake *(Pellaea atropurpurea)*, 75

rattlesnake *(Botrychium virginianum)*, 114, 149, 219

royal *(Osmunda regalis)*, 21, 34, 40, 44, 79, 106, 131, 182

sensitive *(Onoclea sensibilis)*, 34, 106, 149, 219

southern lady *(Athyrium asplenioides)*, 34

southern shield *(Dryopteris ludoviciana)*, 34

spinulose wood *(Dryopteris carthusiana)*, 114, 149, 266, 333, 343

sweet *(Comptonia peregrina)*, 141, 149, 364

Virginia chain *(Woodwardia virginica)*, 149, 331

Walking *(Asplenium rhizophyllum)*, 67

fescue

meadow *(Festuca pratensis)*, 106

six-weeks *(Vulpia octoflora)*, 82

fetterbush *(Lyonia lucida)*, 25, 53, 176, 215

fetterbush *(Lyonia mariana)*, 25, 53, 176, 215, 329

fir

balsam *(Abies balsamea)*, 121, 127, 139, 143, 156, 194, 203, 312, 351, 363

Fraser *(Abies fraseri)*, 234, 240, 343

firepink *(Silene virginica)*, 59, 105, 114, 234

round-leaved *(Silene rotundifolia)*, 114

fleabane

daisy *(Erigeron philadelphicus)*, 170

hyssop-leaved *(Erigeron hyssopifolium)*, 134

flower of an hour *(Talinum parviflorum)*, 81, 95,

fly poison *(Amianthium muscaetoxicum)*, 59, 328

flytrap, Venus' *(Dionaea muscipula)*, 220, 285

foamflower *(Tiarella cordifolia)*, 16, 114, 207, 240, 256, 343

forget-me-not *(Myosotis macrosperma)*, 170, 295

white *(Myosotis verna)*, 172

frog-fruit, nodding *(Phyla nodiflora)*, 44

galax, mountain *(Galax aphylla)*, 118

gallberry, *(Ilex glabra)*, 24, 36, 48, 54

large *(Ilex coriacea)*, 23, 25

garbera(Garbera heterophylla), 38

gayfeather, Heller's *(Liatris helleri)*, 237

gentian

fringed *(Gentiana crinita)*, 62

hairy *(Gentiana villosa)*, 163

yellow *(Gentiana alba)*, 104, 251

geranium, wild *(Geranium maculatum)*, 12, 16, 80, 99, 114, 117, 182, 210, 333

ginger,

Shuttleworth's wild *(Asarum shuttleworthii)*, 60

wild *(Asarum canadense)*, 12, 99, 114, 117, 182, 219, 234, 256, 343

ginseng *(Panax quinquefolia)*, 67

small *(Panax trifolia)*, 116

glasswort *(Salicornia virginiana)*, 154, 217, 289

goat's-rue *(Tephrosia virginiana)*, 24, 82, 84, 91

golden Alexanders *(Zizia aurea)*, 251

golden-glow *(Rudbeckia laciniata)*, 257, 343

goldenrod

Blue Ridge *(Solidago roanensis)*, 236

Boott's *(Solidago boottii)*, 87

Drummond's *(Solidago drummondii)*, 75

grassleaf *(Euthamia remota)*, 104, 130

Houghton's *(Solidago houthtonii)*, 127

large-leaved (Solidago macrophylla), 208

late *(Solidago gigantea)*, 170

roughleaf *(Solidago rugosa)*, 174

seaside *(Solidago sempervirens)*, 216, 289

shale barren *(Solidago harrisii)*, 324, 326

stiff *(Solidago rigida)*, 251

swamp *(Solidago patula)*, 106, 172

threadleaf *(Euthamia tenuifolia)*, 176

whitehaired *(Solidago albopilosa)*, 112, 116

zigzag *(Solidago flexicaulis)*, 114, 116

goldenseal *(Hydrastis canadensis)*, 59

goldthread *(Coptis groenlandica)*, 134, 139, 156, 208

gooseberry, wild *(Ribes cynosbati)*, 343

gourd, Boykin's *(Cayaponia boykinii)*, 169

grama, sideoats *(Bouteloua curtipendula)*, 87

grape

fox *(Vitis vulpina)*, 174

muscadine *(Vitis rotundifolia)*, 216
winter *(Vitis cinerea)*, 169
grass
    Conecuh yellow-eyed *(Xyris drummondii)*, 3
    Elliott's yellow-eyed *(Xyris elliottii)*, 40
    Indian *(Sorghastrum nutans)*, 84, 87, 92, 104, 151, 163, 251
    June *(Koeleria macrantha)*, 151
    maidencane *(Panicum hemitomon)*, 40, 331
    marram *(Ammophila breviligulata)*, 153
    melic (Melica nitens), 217
    mountain oat *(Danthonia compressa)*, 241
    poverty oat *(Danthonia spicata)*, 105
    salt *(Distichlis spicata)*, 217, 289
    Small's yellow-eyed *(Xyris smalliana)*, 40
    switch *(Panicum virgatum)*, 163, 216
    tickle *(Agrostis hyemalis)*, 241
    Tuckerman's panic *(Panicum tuckermanii)*, 129
    wire *(Aristida beyrichiana)*, 19, 21, 24, 29, 54
    yellow star *(Hypoxis hirsuta)*, 34, 333
grass-of-Parnassus
    ginger-leaved *(Parnassus asarifolia)*, 60, 65
    large-leaved *(Parnassus grandifolius)*, 40, 49, 50
greenbrier
    bristly *(Smilax tamnoides)*, 36, 174
    round-leaved *(Smilax rotundifolia)*, 216
    Small's *(Smilax smallii)*, 174
    Walter's *(Smilax walteri)*, 174
greeneyes *(Berlandiera subacaulis)*, 24
Guarea *(Guarea glabra)*, 272
gum
    black *(Nyssa sylvatica)*, 24, 26, 30, 49, 54, 78, 100, 114, 164, 217, 292
    swamp *(Nyssa biflora)*, 26, 174
    sweet *(Liquidambar styraciflua)*, 11, 25, 34, 48, 54, 67, 91, 102, 114, 164
    tupelo *(Nyssa aquatica)*, 2, 36, 42, 80, 174, 284

hackberry *(Celtis occidentalis)*, 102, 178, 286, 293
harbinger of spring *(Erigenia bulbosa)*, 117
hardhack *(Spiraea tomentosa)*, 129
harebell, American *(Campanula rotundifolia)*, 80
hatpin(Eriocaulon septangulare), 129
hatpins, lady's *(Syngonanthus flavidula)*, 31
hawksbeard *(Hypochoeris radicata)*, 342
hawkweed *(Hieracium gronovii)*, 342
haw, possum *(Ilex decidua)*, 172

hawthorn
    green *(Crataegus viridis)*, 169, 172
    parsley *(Crataegus marshallii)*, 74
Hazelnut, beaked *(Corylus cornuta)*, 199, 367
heath
    climbing *(Pieris phylleriaefolia)*, 49
    mountain *(Rhodora canadensis)*, 200, 319
heather, mountain golden *(Hudsonia montana)*, 237
hellebore, green *(Veratrum viride)*, 59
hemlock,
    bulblet *(Cicuta bulbifera)*, 89
    Canadian *(Tsuga canadensis)*, 14, 58, 257, 114, 120, 124, 149, 157, 194, 226, 247
    Carolina *(Tsuga carolinensis)*, 234
hempvine, climbing *(Mikania scandens)*, 169, 183
hepatica *(Hepatica acutiloba)*, 16, 59, 67, 114, 117, 219
Hercules'-club *(Aralia spinosa)*, 216, 295
hickory,
    bitternut *(Carya cordiformis)*, 114, 178, 256, 284, 289
    kingnut *(Carya laciniosa)*, 74
    mockernut *(Carya tomentosa)*, 8, 15, 24, 36, 59, 78, 102, 114, 118, 178
    pale *(Carya pallida)*, 58, 292
    pignut *(Carya glabra)*, 8, 15, 24, 36, 78, 91, 118
    shagbark *(Carya ovata)*, 78, 81, 91, 102, 115, 212, 219, 244, 293
    shellbark *(Carya laciniosa)*, 293
    water *(Carya aquatica)*, 25, 36, 80, 85, 169, 284
hobble
    dog *(Leucothoe editorum)*, 34
    witch *(Viburnum alnifolium)*, 59, 311
holly
    American *(Ilex opaca)*, 15, 33, 36, 48, 76, 173, 176, 182, 217, 284
    Carolina *(Ilex ambigua)*, 38
    dahoon *(Ilex cassine)*, 36, 53, 174, 176
    deciduous *(Ilex decidua)*, 34, 163, 169, 172, 176, 321
    mountain *(Nemopanthus mucronatus)*, 14, 59, 131, 159
    myrtle-leaved *(Ilex myrtifolia)*, 23, 25, 27, 53, 176
    scrub *(Ilex cumulicola)*, 38
    yaupon *(Ilex vomitoria)*, 53, 173, 176, 216
honeysuckle
    bear *(Lonicera dioica)*, 59
    bush *(Diervilla lonicera)*, 14, 59, 148, 256
    fly *(Lonicera villosa)*, 367
hop hornbeam *(Ostrya virginiana)*, 15, 102, 184, 219, 295
horehound, water *(Lycopus virginicus)*, 130, 149, 286

rushfoil *(Crotonopsis elliptica)*, 82, 95
rye, Canada wild *(Elymus canadensis)*, 84

sage, lyreleaf *(Salvia lyrata)*, 174
sandwort, mountain *(Arenaria montana)*,
   199, 202, 207
sassafras *(Sassafras albidum)*, 16, 23, 102,
   114, 148, 174, 217, 248, 265, 330
saxifrage
   brook *(Saxifraga arguta)*, 14
   rock *(Saxifraga michauxii)*, 58, 343
   Virginia *(Saxifraga virginiensis)*, 328
schisandra, climbing *(Schisandra glabra)*, 59
Sebastian-bush *(Sebastiana fruticosa)*, 176
Sedge *(Carex exilis)*, 132
   Biltmore *(Carex biltmoreana)*, 60, 234
   blue *(Carex glaucescens)*, 176
   Chapman's *(Carex chapmanii)*, 40, 49
   cypress-knee *(Carex decomposita)*, 182
   Emmons' *(Carex emmonsii)*, 114
   Fernald's hay *(Carex aenea)*, 236
   Fraser's broad-leaved *(Cymophyllus
      fraseri)*, 61
   hop *(Carex lupulina)*, 149
   Howe's *(Carex howei)*, 174
   inflated *(Carex vesicaria)*, 321
   Mead's *(Carex meadii)*, 93
   Pennsylvania *(Carex pensylvanica)*, 328
   three-seeded *(Carex trisperma)*, 139, 322
   three-way *(Dulichium arundinaceum)*,
      149, 321, 331
   Tuckerman's *(Carex tuckermanii)*, 149
   two-seeded *(Carex disperma)*, 139
   variable *(Carex polymorpha)*, 319
   white *(Dichromena colorata)*, 360
   white nut *(Scleria triglomerata)*, 77
sedum, blue-leaved *(Sedum glaucophyllum)*,
   328
serviceberry *(Amelanchier arborea)*, 114,
   248
   smooth *(Amelanchier laevis)*, 240
shadbush *(Amelanchier arborea)*, 91, 141,
   292
shooting star, French's *(Dodecatheon
   frenchii)*, 78
shoregrass *(Littorella americana)*, 130
silverbell *(Halesia carolina)*, 16, 33, 65, 68,
   229
silverbell *(Halesia diptera)*, 16, 33, 65, 68
silverling, White Mountain *(Paronychia
   argyrocoma)*, 199, 207
skullcap
   blue-flowered *(Scutellaria incana)*, 31,
      114, 257
   mad dog *(Scutellaria lateriflora)*, 139
   Small flower *(Scutellaria parvula)*, 149,
skunk cabbage *(Symplocarpus foetidus)*, 331,
   355

smartweed, water *(Polygonum amphibium)*,
   149, 174, 208, 364
snailseed *(Cocculus carolinus)*, 163, 169
snakeroot
   black *(Sanicula odorata)*, 170
   Lucy Braun's white *(Eupatorium lucy-
      brauniae)*, 112
   Virginia *(Aristolochia serpentaria)*, 295
   white *(Eupatorium rugosum)*, 212
sneezeweed, Virginia *(Helenium
   virginianum)*, 331
snowbell, American *(Styrax americana)*,
   172
snowbell bush
   bigleaf *(Styrax grandifolia)*, 76
   large-flowered *(Styrax grandifolia)*, 16
   shrubby *(Styrax grandifolia)*, 33
snowberry, creeping *(Gaultheria hispidula)*,
   157, 208
Solomon's-seal *(Polygonatum commutatum)*,
   99, 114, 117, 210, 234, 256, 333
   downy *(Smilacina pubescens)*, 207
   false *(Smilacina racemosa)*, 99, 114, 117,
      210, 219, 256
   three-leaved *(Smilacina trifoliata)*, 159
sorrel
   mountain *(Oxalis Montana)*, 207, 240
   sheep *(Rumex acetosella)*, 241
   white wood *(Oxalis montana)*, 118
   wood *(Oxalis violacea)*, 174, 355
sourwood *(Oxydendrum arboretum)*, 16, 58,
   114, 118, 176, 330, 339
spatterdock *(Nuphar luteum)*, 208
spear scale *(Atriplex patula)*, 80
spicebush *(Lindera benzoin)*, 16, 59, 91, 117,
   169, 219
spiderwort *(Tradescantia ohiensis)*, 58
spikemoss,
   twisted *(Selaginella tortifolia)*, 58
spikenard *(Aralia racemosa)*, 14
spikerush, Robbins' *(Eleocharis robbinsii)*,
   176, 212, 330
spleenwort,
   ebony *(Asplenium platyneuron)*, 114
   maidenhair *(Asplenium trichomanes)*, 91,
      118
   mountain *(Asplenium montanum)*, 14,
      118
   silvery *(Diplazium thelypterioides)*, 14,
      343
   Trudell's *(Asplenium trudellii)*, 14
spruce
   black *(Picea mariana)*, 123, 127, 131, 134,
      143, 149, 156, 201, 363
   red *(Picea rubens)*, 194, 199, 203, 240,
      312, 322, 331, 339, 343, 350
   white *(Picea glauca)*, 158, 199, 208, 363,
      366

whitlowwort, silvery *(Paronychia argyrocoma)*, 64
widowscross *(Sedum pulchellum)*, 91, 95

willow
  bearberry *(Salix uva-ursi)*, 202
  black *(Salix nigra)*, 25, 118, 212, 292, 295
  Carolina *(Salix caroliniana)*, 285
  dwarf pussy *(Salix humilis)*, 59
  Florida *(Salix floridana)*, 40, 49
  shining *(Salix lucida)*, 208
  silky *(Salix sericea)*, 59
  tealeaf *(Salix herbacea)*, 203
  Virginia *(Itea virginica)*, 34, 106
wintergreen *(Chimaphila maculata)*, 60, 101, 132, 134, 141, 148, 152, 219, 357

yam, Florida wild *(Dioscorea floridana)*, 44
yaupon *(Ilex vomitoria)*, 53, 173, 176, 216
yellowwood *(Cladrastis kentukea)*, 60, 62, 64, 76

GENERAL INDEX

ABOUT THE AUTHOR

Robert H. Mohlenbrock is Distinguished Professor Emeritus at Southern Illinois University, Carbondale, where he taught botany for 34 years. He is also Senior Scientist for Biotic Consultants, Inc. He is the author of 50 books, most of them field guides, and is a contributing editor to *Natural History* magazine, which has just published the 201st article in his *This Land* series. He served for 16 years as Chairman of the North American Plant Specialists Group of the Species Survival Commission of the International Union for the Conservation of Nature. Since his retirement, he has taught over 180 week-long wetland plant identification classes in 29 states.

Series Design:	Barbara Jellow
Design Development:	Jane Tenenbaum
Cartographer:	Bill Nelson
Compositor:	Michael Bass Associates
Text:	10/13.5 Minion
Display:	Franklin Gothic Book and Demi
Printer and Binder:	Friesens